Smart Innovation, Systems and Technologies

Volume 86

Series editors

Robert James Howlett, Bournemouth University and KES International,
Shoreham-by-sea, UK
e-mail: rjhowlett@kesinternational.org

Lakhmi C. Jain, University of Canberra, Canberra, Australia;
Bournemouth University, UK;
KES International, UK
e-mails: jainlc2002@yahoo.co.uk; Lakhmi.Jain@canberra.edu.au

About this Series

The Smart Innovation, Systems and Technologies book series encompasses the topics of knowledge, intelligence, innovation and sustainability. The aim of the series is to make available a platform for the publication of books on all aspects of single and multi-disciplinary research on these themes in order to make the latest results available in a readily-accessible form. Volumes on interdisciplinary research combining two or more of these areas is particularly sought.

The series covers systems and paradigms that employ knowledge and intelligence in a broad sense. Its scope is systems having embedded knowledge and intelligence, which may be applied to the solution of world problems in industry, the environment and the community. It also focusses on the knowledge-transfer methodologies and innovation strategies employed to make this happen effectively. The combination of intelligent systems tools and a broad range of applications introduces a need for a synergy of disciplines from science, technology, business and the humanities. The series will include conference proceedings, edited collections, monographs, handbooks, reference books, and other relevant types of book in areas of science and technology where smart systems and technologies can offer innovative solutions.

High quality content is an essential feature for all book proposals accepted for the series. It is expected that editors of all accepted volumes will ensure that contributions are subjected to an appropriate level of reviewing process and adhere to KES quality principles.

More information about this series at http://www.springer.com/series/8767

Jeng-Shyang Pan · Tsu-Yang Wu
Yong Zhao · Lakhmi C. Jain
Editors

Advances in Smart Vehicular Technology, Transportation, Communication and Applications

Proceedings of the First International Conference on Smart Vehicular Technology, Transportation, Communication and Applications, November 6–8, 2017 Kaohsiung, Taiwan

Springer

Editors
Jeng-Shyang Pan
Fujian University of Technology
Fuzhou, Fujian
China

Tsu-Yang Wu
Fujian University of Technology
Fuzhou, Fujian
China

Yong Zhao
Superconductivity R&D Center
Southwest Jiaotong University
Chengdu, Sichuan
China

Lakhmi C. Jain
University of Canberra
Canberra, ACT
Australia

ISSN 2190-3018 ISSN 2190-3026 (electronic)
Smart Innovation, Systems and Technologies
ISBN 978-3-319-70729-7 ISBN 978-3-319-70730-3 (eBook)
https://doi.org/10.1007/978-3-319-70730-3

Library of Congress Control Number: 2017957863

Printed on acid-free paper

This Springer imprint is published by Springer Nature
The registered company is Springer International Publishing AG
The registered company address is: Gewerbestrasse 11, 6330 Cham, Switzerland

Preface

This volume composes the proceedings of First International Conference on Smart Vehicular Technology, Transportation, Communication and Applications (VTCA 2017), which is hosted by Fujian University of Technology and is held in Kaohsiung, Taiwan, on November 6–8, 2017. VTCA 2017 is technically co-sponsored by Springer, Fujian University of Technology, Fujian Provincial Key Laboratory of Digital Equipment, Fujian Provincial Key Lab of Big Data Mining and Applications, and National Kaohsiung University of Applied Sciences in Taiwan. It aims to bring together researchers, engineers, and policymakers to discuss the related techniques, to exchange research ideas, and to make friends.

Forty-seven excellent papers were accepted for the final proceeding. We would like to thank the authors for their tremendous contributions. We would also express our sincere appreciation to the reviewers, Program Committee members, and the Local Committee members for making this conference successful. Finally, we would like to express special thanks for the financial support from Fujian University of Technology, China, in making VTCA 2017 possible and also appreciate the great help from National Kaohsiung University of Applied Science for the local organization.

September 2017

Jeng-Shyang Pan
Tsu-Yang Wu
Yong Zhao
Lakhmi C. Jain

Preface

Organization

Organizing Committee

Honorary Chairs

Xinhua Jiang	Fujian University of Technology, China
Lakhmi C. Jain	University of Canberra, Australia
Xin Tong	Fujian University of Technology, China
Russell Hsing	The OpenFog Consortium, USA

Advisory Committee Chairs

Yong Zhao	Southwest Jiaotong University, China
Tang-Hsien Chang	National Taiwan University, Taiwan
Bin-Yih Liao	National Kaohsiung University of Applied Sciences, Taiwan
Zhangxi Lin	Texas Tech University, USA

Conference Chairs

Jeng-Shyang Pan	Fujian University of Technology, China
Shaoquan Ni	Southwest Jiaotong University, China

Program Chairs

Shi-Huang Chen	Shu-Te University, Taiwan
Fumin Zou	Fujian University of Technology, China
Yihang Zhang	Chinese Academy of Sciences, China

Publication Chairs

Tsu-Yang Wu	Fujian University of Technology, China
Zhihong Xu	Fujian University of Technology, China
Trong-The Nguyen	Haiphong Private University, Vietnam

Local Organization Chairs

Mong-Fong Horng	National Kaohsiung University of Applied Sciences, Taiwan
Ming-Der Jean	Fujian University of Technology, Taiwan

Finance Chair

Jui-Fang Chang	National Kaohsiung University of Applied Sciences, Taiwan

Program Committees

Chien-Ming Chen	Harbin Institute of Technology - Shenzhen, China
Shu-Chuan Chu	Flinders University, Australia
Falko Dressler	University of Paderborn, Germany
Javier Gozalvez	University Miguel Hernandez, Spain
Shih-Hsu Huang	Chung Yuan Christian University, Taiwan
Hsi-Shan Hwang	Fujian University of Technology, China
Yusheng Ji	National Institute of Informatics, Japan
Alireza Khaligh	University of Maryland, USA
Cheng-Chi Lee	Fu Jen Catholic University, Taiwan
Chun-Ta Li	Tainan University of Technology, Taiwan
Lyuchao Liao	Fujian University of Technology, China
Jerry Chun-Wei Lin	Harbin Institute of Technology - Shenzhen, China
Jianhua Liu	Fujian University of Technology, China

Isamu Matsunami	The University of Kitakyushu, Japan
Jelena Misic	Ryerson University, Canada
Kaoru Ota	Muroran Institute of Technology, Japan
Tien-Wen Sung	Fujian University of Technology, China
Jing-Jou Tang	Southern Taiwan University of Science and Technology, Taiwan
Pei-Wei Tsai	Swinburne University of Technology, Australia
Yen-Ming Tseng	Fujian University of Technology, China
Yuh-Ming Tseng	National Changhua University of Education, Taiwan
Christian Wietfeld	Dortmund University of Technology, Germany
Xiaolei Yan	Fujian University of Technology, China
Jingang Yi	Rutgers University, USA
Qingyong Zhang	Fujian University of Technology, China

Contents

Smart Vehicular Technology

Intelligent Railway Technology

Innovative Electrical Technology

Innovative Intelligent Computation Technology

Applied Mathematics and Its Applications

Smart Wireless Access Technology

Smart Vehicular Technology

Easy *i*-Move: Structured Image Recognition Solution for Automatic Guided Vehicles

Yen-Hui Kuo[1,2(✉)], Shu-Hui Lin[1,2(✉)], and Eric Hsiao-Kuang Wu[2(✉)]

[1] Institute for Information Industry, National Central University, Taoyuan, Jhongli, Taiwan
{yenhui,vickylin}@iii.org.tw
[2] Department of Computer Science and Information Engineering, National Central University, Taoyuan, Jhongli, Taiwan
hsiao@csie.ncu.edu.tw

Abstract. There are many kinds of image processing application in the industry, to improve better performance of producing, Automatic Guide vehicles (AGV) solution have been driven in service, for instance, the expensive solution such as light detection and ranging (LIDAR), magnetic tape guided and RFID are used. In the industry which not only search for better delivery performance but also search for economic efficiency. Here we present an economic efficiency solution, Easy *i*-Move for AGVs guiding system to fulfill customers' necessity. In this study, we focus on color tape image recognition technology based on HSI color model. The implementation of the system architecture and the technology of image recognition in the proposed system are described and validated in this paper.

Keywords: Image recognition · Automatic Guided Vehicle · Color model

1 Introduction

There are many kinds of d Guide vehicles guiding solution, for instance, the light detection and ranging (LIDAR), a remote sensing method used to examine the surface of the earth, and magnetic tape guided are used. In the industry section which not only search for better delivery performance but also search for economic efficiency. Usually, manufacture working area reorganized from time to time. Working area reorganized which caused a very much housekeeping job to do subsequently. When reorganization occurs, the course needs to change. Therefore, we are trying to apply the color tape image recognition AGVs guiding strategy to handle such situations. There are some problems exist when we apply image recognition solution for AGVs guidance. We have to solve (1) Image processing time must short enough for guiding car:

The most important things in the Automatic Guide Vehicles guiding system is guiding response time. Car is running in real time, about 50 cm per second, image recognition system must take a quick response to guiding control module, which generates control instruction to AGV car. (2) Lighting noise proof: In the manufacturing there are a lot of lighting noise, such as the light from fluorescent, LED light

J.-S. Pan et al. (eds.), *Advances in Smart Vehicular Technology, Transportation, Communication and Applications*, Smart Innovation, Systems and Technologies 86,
https://doi.org/10.1007/978-3-319-70730-3_1

from instruments, sun light from outside etc., the image recognition algorithm not only needs to quick response but also need to analyze the picture accurately.

Because of the previously stated concerns, this paper will propose an AGV system using image recognition technology. We implement an AGVs system to verify this solution.

The rest of this paper is organized as follows. In Sect. 2, the RGB and HSI color model is introduced. In Sect. 3, the challenges of color tape image recognition solution for AGV system will be discussed. In Sect. 4, the Image recognition strategy for AGV is exploited. In Sect. 5 deployment of the new intelligent AGV system are presented. We conclude the paper in Sect. 6.

2 RGB and HSI Color Model

2.1 RGB Color Model

The RGB color model is an additive color model in which red, green and blue light are added together. The name of the model comes from the initials of the three primary colors, red, green and blue. Initially, we used the RGB color model to implement image recognition system. Our development platform is windows visual studio using Emgucv library [1, 2], in c# language, we get RGB color attributes from a picture of each pixel though GetPixel(x, y) method, which return R, G, B value for red, green, and blue respectively, which value vary from 0–255. This method seems very straight forward to tell a specific color tape apart from a picture which we got from camera, but The AGV System running in manufacturing which environment is complicated, there are a lot of lighting noise as mention before, I cause the picture catch from the camera has color distortion. The ground color varies from grey to blue, as shown in Fig. 1.

normal color distorted color

Fig. 1. Color distorted by light (Color figure online)

Initially, We used the blue color tape as guiding line, that color distortion like this causes a lot of trouble on image recognition, we can't tell part the guiding line from ground simply use R, G and B value from RGB color model, so we try HSI color model solution [3].

2.2 HSI Color Model

The HSI color model represents every color with three components: hue (H), saturation (S), intensity (I). When humans view a color object it is described by its hue, saturation and brightness. Hue is a color attribute that describes a pure color (pure green, blue or red). Saturation gives a measure of the degree to which a pure color is diluted by white light. Intensity is key factor in describing color sensation. We convert RGB color space to HSI color space by using the following formulas [4–6].

$$\text{numerator} = [(\mathrm{R}-\mathrm{G})+(\mathrm{R}-\mathrm{B})] \div 2$$

$$\text{denominator} = \sqrt{(\mathrm{R}-\mathrm{G})^2+(\mathrm{R}-\mathrm{B})(\mathrm{G}-\mathrm{B})}$$

$$\begin{cases} \mathit{if}(\text{denominator}) \neq 0 \\ \quad \text{Theta} = \text{Acos}(\text{numerator} \div \text{denominator}) \times 180 \div 3.14 \\ \quad \mathit{if}(B < G) \\ \quad\quad \mathit{Hue} = \text{Round}(\text{Theta}, 2) \\ \quad \mathit{else} \\ \quad\quad \mathit{Hue} = \text{Round}((1-\text{Theta}), 2) \\ \quad \mathit{else} \\ \quad\quad\quad \mathit{Hue} = 0 \end{cases}$$

$$\begin{cases} \text{if}(\mathrm{R}+\mathrm{G}+\mathrm{B}) \neq 0 \\ \quad S = 1 - [3 \times \min(\mathrm{R,G,B})] \div (\mathrm{R}+\mathrm{G}+\mathrm{B}) \\ \text{else} \\ \quad S = -1.0 \end{cases}$$

$$\mathrm{I} = (\mathrm{R}+\mathrm{G}+\mathrm{B}) \div 3$$

In c# language, we get RGB color attributes from a picture of each pixel though GetPixel(x, y) method, which return R, G and B. We feed R, G, B value into above formula then can get H, S, I value easily. The value of Hue varies from −180–180, Saturation varies from 0–1, Intensity varies from 0–255, that gives much space to justify pixel color. We will exploit more detail in the fourth section. The Challenges of color tape image recognition solution for AGV system.

3 System Operation Components

AGV car consists of a front end camera, image recognition module and driving control unit. The front end camera captures the picture of course while running, image recognition module analyzes the colored tape position and based on the color tape position generates a direction calibration value or gives a reach the anchor point instruction to drive the AGV car.

3.1 Challenges

The image recognition module plays the major game in driving AGV car, image recognition module must give an accurate instruction for AGV car. There are implementation challenges involving potential lighting noise, image recognition errors, image analyze processing speed, which will cause AGV driving malfunction.

With the HSI color model technology, we get the maximum image recognition accuracy, proposed an image down size solution to speed up system performance, every driving control cycle keep within 50 ms, to meet the car stable speed on 50 cm per second of our system. Our proposed scheme not only maintains the driving security, but it also maximizes image recognition accuracy and stable driving speed.

4 The Image Recognition Strategy for AGV System

The Image recognition strategy for AGV System is based on color recognition of each pixel [7]. The Image recognition process is shown as Fig. 2.

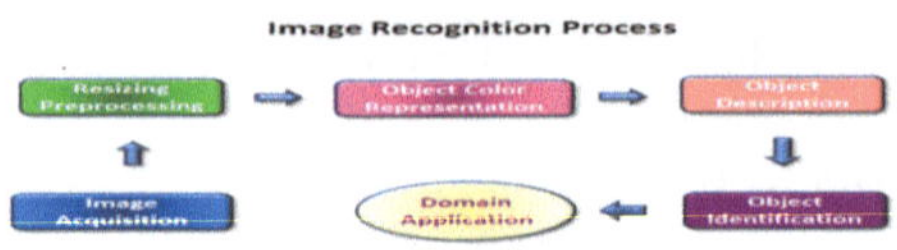

Fig. 2. Image recognition process

We get an image from camera, down resizing, read pixel color, telling object color apart from others, identify the right object, finally, output control instructions. We get the a picture from the camera consists of 640 × 480 pixels, if we used the raw picture to do Image analyze processing, it takes about 195.07 ms per picture, it takes too much time in processing the image analysis, it can't meet the AGV running speed requirement, it must less than 50 ms per analysis operation, we down size the picture to 20 × 20 pixels, it takes about 0.51 ms per picture, in our system the average operation time from take a picture to Image analyze and give an instruction to AGV car, it takes 2.08 ms from our 2002 sample pictures. The size of picture affected the Image analyze processing time shown in Table 1.

Table 1. Image analyze processing time

Platform	Image size	Analysis time IV (millisecond)	Meet requirement (50 ms<)
Intel Core i7-3540 M CPU: 3 GHz RAM: 8 GB OS: windows7	640 × 480	=195.07	✗
	100 × 100	=6.35	✓
	20 × 20	=0.51	✓

Our system must function normally despite the fact that the manufacturing, warehousing lighting noise. Furthermore, we want to maximize the system's accuracy as much as possible. Order to meet this requirement, we propose a HSI color model technology based on tape color. The tape color could be affected by a wide range of factors, such as fluorescent or LED color, the light from machine or equipment, and any lighting source. Accordingly, an accurate and reliable algorithm for Image analysis, the tape color must take into account the effects of these factors.

We exploit the detail of the image recognition algorithm, which we implement in our system. In the experimental manufacture, there is yellow color tape already, so we use yellow color for image recognition. After we captured a picture from end camera, we down size the picture from 640 × 480 pixel to 20 × 20 pixel, the picture shown as Fig. 3.

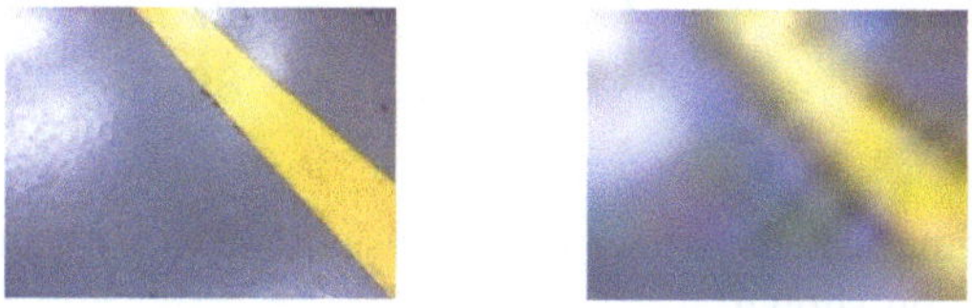

Original picture down size picture

Fig. 3. Picture down size before and after (Color figure online)

After the picture downsizing, it become blurred like the above right side picture, although the image becomes blurred, it is still good for recognition, and the most important is that the Image analyze processing time is short enough, so in the experiment 20 × 20 pixel is a good size for recognition. Based on Sect. 2.2 HSI color model formula, we caught each H, S, and I value of pixel[x, y] and saved in Hue[x, y], Sat[x, y], and Inten[x, y] array respectively. Here I draw 20 × 20 circle to represent the 20 × 20 pixel, as shown in Fig. 4.

Fig. 4. 20 × 20 pixel picture

In real environment only H, S, and I value is not enough to tell similar pixel apart. Therefore, we propose additional HSI formula:

$$\mathrm{HSI} = \mathrm{H} \times \mathrm{S} \times \mathrm{I}$$

We further multiple the H, S, and I value generate a much more range and unique value to ensure color representation. With HSI model produces a representation of HSI, H, S, and I value. The Hue values of yellow color are positive around 50–60. In this example, we are lucky, we can tell the yellow color apart from Hue attribute as f(HSI,H) function shown.

$$
f(\mathrm{HSI,H})\begin{cases} \text{if } HSI > HSI_{min} \\ \quad \text{then key factor is} \\ \quad H_{min} < H < H_{max} \\ \qquad where \\ \quad HSI_{min} = 1400 \\ \quad H_{min} = 30 \\ \quad H_{max} = 80 \end{cases}
$$

In some other complicated case we have to use more attributes to tell them apart, as g (HSI, H,S,I) function shown.

$$
g\,(\mathrm{HSI,H,S,I})\begin{cases} \text{if } \; 0 < HSI \leqq \mathrm{HSI}_{min} \\ then\ key\ factors\ are \\ \quad H_{min} < H < H_{max} \\ \quad S > S_{min} \\ \quad I > I_{min} \\ \quad where\ \ \mathrm{HSI}_{\mathrm{min}} = 1400 \\ A{:}\,H_{min} = 30,\ H_{max} = 50, S_{min} = 0.15, I_{min} = 100 \\ B{:}\,H_{min} = 50,\ H_{max} = 65, S_{min} = 0.05, I_{min} = 100 \end{cases}
$$

Based on f(HSI,H) and g(HSI,H,S,I) function to represent the pixel color, but that can't descript the pixel is right in the line or caused by environment lighting noise, so our recognition algorithm has to add on line characteristics to verify line pixel. Both the side by side pixels of a line are not apart from 2 or more pixels, they must close together, nevertheless left or right, up and down. Apply this line characteristics we can descript the line image object, which

Cnt:	Total chosen color count of row
Line pixel:	(LineRightPos- LineLeftPos) ÷ Cnt < 1
Noise:	(LineRightPos-LineLeftPos) ÷ Cnt ≧ 1

On behalf of the above rules, we get rid of some light noise from environment lighting or instruments light successfully.

Before we apply this solution in the system, we capture the whole course image at the stable speed say 50 cm per second, there are more than 10,000 pictures, and evaluated every misrecognized image carefully, then modify the parameters of f(HSI,H) and g(HSI,H,S,I) functions to make function more accurately.

We get the image structure of Fig. 4 With f(HSI,H) and g(HSI,H,S,I) functions as shown in Table 2.

Table 2. Image structure

Y	XI	X2	Line width
0	7	9	3
1	8	10	3
2	8	11	4
3	9	12	4
4	9	12	4
5	10	13	4
6	11	15	5
7	11	15	5
8	12	16	5
9	13	17	5
10	13	18	6
11	14	19	6
12	15	19	5
13	15	19	5
14	16	19	4
15	17	19	3
16	17	19	3
17	18	19	2
18	19	19	1
19	−1	−1	0

y: row number
x1: tape left side position
x2: tape right side position
line width: tape width

From Table 2 we mapped with the yellow tape image as Fig. 5. We can get a perfect pixel structure of yellow color matched with image picture.

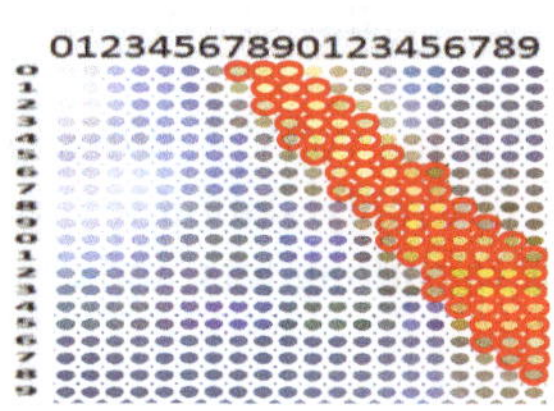

Fig. 5. Image structure data mapping (Color figure online)

5 Application of Image Recognition

At this point, we transfer the image picture to a structure of data table. How can we use these data for the application of AGV system? From the yellow color position in the 20×20 structure, we can conduct the car to follow. Before we use the yellow line position data, we have to make sure the image structure data is yellow line structure not from light noise or something else. In my algorithm, after we got the line image structure, we check the line length and line width first, if the line width great than 0 pixel and line length is great than 9 pixels, then we can sure it is a conduct line image, which we can use to conduct the car. There is a rule of the guiding line as below function.

Guiding line rules:

- Anchor line: line width > 16 and line length < 9; car is approaching the workstation
- Guiding line type1: line width > 0 and line length > 9; guiding line is located at the middle of picture
- Guiding line type2: line width > 0 and 3 < line length $\leq$ 9; guiding line is located at the corner of picture
- Guiding line type3: line width > 0 and 0 < line length $\leq$ 3; guiding line is located at the corner of picture.

From the above rules we identify the line type then we check the column position of the guiding line at the top, based on the position of guiding line to give the direction instruction. The guiding line can be set at the middle of lane or at the side of lane, the rule of at the middle of lane is:

(1) Forward: 7 ≦ guiding line top position ≦ 13
(2) Left: 0 < guiding line top position < 7
(3) Right: 13 < guiding line top position

Next, we carried out experiments in a real environment to verify the usability of the AGV system. Here, we implemented our system on a manufacture, which produces geared motors in Taiwan. We use an AGV car with a front end camera, so this AGV is implemented forwarding only, the driving instructions are send from recognition module to AGV control unit through USB port, the control unit is an Arduino module [8–10]. In the manufacture there are 15 Assembly stations in the manufacture, with a single route length of 180 m, about 6 min per circle. To investigate the performance of the proposed AGV system, we use the real manufacture lay out road map in the experiments.

Initially, AGV car is parked at the AGV parking lot, user asks AGV car to deliver products or materials from any assembly station to any station. After the car finish the delivery job, which will go back to parking lots, and check is there any job waiting for it.

6 Conclusions

In this experiment we take care of the Image processing time to meet the system requirement which around 0.5 ms on our platform. We also get rid of the lighting noise from outside to make the AGV system work in manufactory. There are many kinds of image recognition approach and application, some of further research applies machine learning or deep learning with image recognition technology to enhance image recognition system to specific domain application. Furthermore, image processing can be applied to many domain usages such as: satellite imagery, medical image, photo reconstruction, Automatic identification, defect detection, face discrimination, virtual reality, there are different domain knowledge inside. Here we have demonstrated the feasibility and usability of the proposed AGV system. After further investigation of image recognition solution performance, it will be possible to redesign the system architecture to more secure traffic control system. Many more new applications will then be developed in the near future.

References

1. Emgu CV Essentials. Packt Publishing, 14 November 2013
2. Emgu.CV.example CameraCapture. http://me1237guy.pixnet.net/blog/post/61361335
3. Converting from RGB to HSV. http://coecsl.ece.illinois.edu/ge423/spring05/group8/finalproject/hsv_writeup.pdf
4. Video color conversion. Introduction to Automotive Electronic Systems pdf. http://140.117.156.238/course/IAE/2013/
5. Gonzalez, R.C., Woods, R.E.: Digital Image Processing, 3rd edn. Prentice Hall, Upper Saddle River (2008)
6. Setting up Emgu CV Project with Visual Studio. https://notebookbft.wordpress.com/2015/03/08/setting-up-emgu-cv-project-with-visual-studio/
7. The Principle of Image Processing Technology and Application. http://web.ncyu.edu.tw/~lanjc/lesson/C9/class/11.pdf
8. Arduino Mini USB Adapter. http://elesson.tc.edu.tw/md221/pluginfile.php/4151/mod_resource/content/1/arduino.pdf
9. Online electricity knowledge. http://bbs.audiohall.net/viewtopic.php?t=1337&sid=999
10. http://elesson.tc.edu.tw/md221/pluginfile.php/4151/mod_resource/content/1/arduino.pdf

Evaluating Multi-dimensional Abilities of Bus Drivers

Ting-An Kuo[1], Chiuhsiang Joe Lin[1], and Po-Hsiang Liu[2(✉)]

[1] Department of Industrial Management, National Taiwan University of Science and Technology, No. 43, Keelung Road, Section 4, Da'an District, Taipei City 10607, Taiwan
a76526@yahoo.com.tw, cjoelin@mail.ntust.edu.tw

[2] Department of Industrial Engineering and Management, St. John's University, No. 499, Tam King Road, Section 4, Tamsui District, New Taipei City 25135, Taiwan
bsliu@mail.sju.edu.tw

Abstract. As professional drivers typically were driving more time on the road than general drivers, the relationship between age and multi-dimensional driving abilities of bus drivers should be of concern to transportation officials tasked with improving workplace safety. This study examined the multi-dimensional abilities of bus drivers through their responses to a self-assessment questionnaire and on tests of hand-eye coordination, balance ability and hand grip strength. Among sixteen participants recruited from an urban bus company, gender and age significantly correlated with self-rating evaluation, hand-eye coordination tests and grip strength capabilities. Recommended of this study were conducted the self-rating evaluation, hand-eye coordination, balance ability and hand grip strength in the process of licensing exams test to assess driving qualification, renew driving license particularly. Bus drivers should also regularly take these ability tests to assess their driving quality and ability. In this way, an employer can accurately assess drivers' multi-dimensional abilities and verify whether they meet the demands of their jobs.

Keywords: Psychological assessment · Occupational drivers · Driving abilities · Safety management · Aging

1 Introduction

Driving is the primary mode of transportation in many countries and the driving ability is intimately associated with health-related quality of life [1]. Further, driving is one of the most complex tasks that humans perform on a regular basis, placing significant demands on sensory, perceptual, cognitive, and motor capabilities. However, age-related declines in these capabilities negatively affect driving performance [2]. Several eye diseases that including glaucoma have been associated with increased risk for motor vehicle accidents [3]. With an aging population, it is estimated that over 58 million people will have open angle glaucoma by the year 2020. Although it is clear that vision is essential for driving, it is not clear what vision skills and tests are actually more closely related to the ability of driving safely [4]. Gentzler and Smither [5] have

J.-S. Pan et al. (eds.), *Advances in Smart Vehicular Technology, Transportation, Communication and Applications*, Smart Innovation, Systems and Technologies 86,
https://doi.org/10.1007/978-3-319-70730-3_2

developed test batteries to identify at-risk older drivers. Wood et al. [6] reported that age-related changes in sensory abilities, such as visual impairment, can potentially influence the ability to perform visual tasks and simultaneous dual tasks not directly related to vision. Indices of unsafe driving performance, including increased crash risk and impaired on-road driving performance, have been reported in older drivers with cataracts and in those drivers with reductions in specific visual functions including visual fields, motion sensitivity, contrast sensitivity and visual attention [7].

In Taiwan, in addition to measurements of cognitive abilities and motor skills, a physical examination is required when drivers apply for an occupational license, and it includes an evaluation of visual abilities (visual acuity, visual field, and night vision), heart functioning and blood pressure. Drivers younger than 60 years are required to retake the physical examination every three years to maintain license validity while those 60 years and older need to be re-examined every year. However, cognitive abilities and motor skills of professional drivers are not re-examined as part of the license renewal process in either age group.

Gentzler and Smither [5] have used various criteria to assess driving performance, including (1) the selective and divided attention abilities, processing speed, memory, accuracy, reaction time as cognitive functions; (2) the range of motion, strength in the extremities such as the foot, mobility as physical functions and (3) static visual acuity, contrast sensitivity, visual field, visuospatial ability, dynamic acuity, useful field of view as visual function. Professional drivers were usually driving more time on the road than general drivers. The purpose of the present study was to examine the multi-dimensional abilities of professional bus drivers using self-assessment tests, hand-eye coordination tests and grip strength and balance ability test. In addition, the study wanted to establish appropriate assessment mechanisms to promote safer driving and, thus, safer roadways.

2 Methods

2.1 Participants

This study recruited sixteen bus drivers that f including 10 males and 6 females from an urban bus company in Taipei city. The mean age of the male drivers was 46 years old, the mean height was 170.5 $\pm$ 6.5 cm, and the mean weight was 73.9 $\pm$ 8.8 kg. The mean age of the female drivers was 43 years old, the mean height was 159.5 $\pm$ 2.1 cm, and the mean weight was 63.9 $\pm$ 10.4 kg.

2.2 Self-rating Questionnaires

The self-rating form is designed to help driver examine driving ability to keep driving safety from American Automobile Association Foundation for Traffic Safety (version 2010). The self-rating tool provides facts and suggestions for safe driving. Although the questionnaire was designed for people over the age of 65, it is important to recognize all driver limitations and to be aware of everything a driver can do to be safe on the road.

Items 1–13 were scored as ‘always or almost always,’ ‘sometimes,’ or ‘never or almost never’ and questions 14 and 15 were scored as ‘none,’ ‘one or two,’ or ‘three or more’. Subjects checked the box next to each appropriate response. The questionnaires were scored by (1) writing the check mark totals recorded in the squares on the question form in the square to the right; (2) writing the check mark totals recorded in the triangle on the question form in the square to the right; (3) multiplying the number in the square by 5; (4) multiplying the number in the triangle by 3; and (5) adding the results of steps 3 and 4 together. Lower scores (0–15) indicate safer drivers and scores ranging from 16 to 34 indicate that the driver is engaging in some practices that need improvement to ensure safety. Scores higher than 35 mean the driver is engaging in too many unsafe driving practices and should stop driving.

2.3 Hand-Eye Coordination Test

The hand-eye coordination test comprised the dynamic stability test and the electrical-mirror tracer test. During the dynamic stability test, a stabilimeter (TKK 1211, Takei Scientific Instruments Co., Ltd., Niigata city, Japan) attached to the arm of the participant is used to monitor stability as the subject traces a groove on a panel with a stylus in a specified period of time (Fig. 1 left). When the stylus makes contact with the edge of the groove, a buzzer is sounded. The paths of the tracer include linear paths, curved paths, and square paths.

Fig. 1. Dynamic stability test and auto-scoring mirror tracing test

The mirror tracing apparatus is a standard psychology laboratory instrument that has been used in a variety of experimental studies (Model 58042A, Lafayette Instrument Co., Indiana, USA). The subject’s task is to trace, by looking in a mirror, a pattern which is on the board. The subject has to trace between the lines with an electric pencil without touching the sides of the star. Errors are tallied by counting the number of times the subject touches the lines. In addition, total time required to trace the complete pattern is recorded (Fig. 1 right).

2.4 Grip Strength and Balance Ability Test

Maximum grip strength across a 50 mm span was measured using a digital grip-strength dynamometer (TKK 5401, Takei Scientific Instruments Co., Ltd., Niigata city, Japan). Measurements ranged from 5.0–100 kg and the minimum measurement unit was 0.1 kg (accuracy, ±2.0 kg). Each measurement continued for approximately 3 s, or until the participant could no longer resist while standing with forearm straight down. The participants rested at least 60 s between trials, and the highest value of three strength measurements was considered to be the subject's maximum strength.

The balance ability test was conducted with both eyes closed while standing on one foot (left foot and right foot were tested separately). The time it took to touch the floor with the other foot was recorded. A standing time greater than one minute was considered to indicate good balance ability.

2.5 Experimental Procedures

The study was approved by the Research Ethics Committee of the researcher's institution. Subjects participated in the experiments after providing informed consent with respect to the investigative procedures. First, subjects need to rate their driving ability by self-rating questionnaires. Further, hand-eye coordination test involved the dynamic stability test and auto-scoring mirror tracing test. Finally, grip strength and balance ability test were conducted. A minimum rest period of 5 min or more, if required, was provided between trials. During the rest periods, participants were asked to stay seated, relaxed and to remain silent.

2.6 Data Analysis

A randomized complete design with between-subject factors (age and gender) was used in this study. In addition, age could be divided into three groups, i.e. younger groups (<30 years old) middle groups (31–54 years old) and older groups (more than 55 years old). Further, multivariate analysis of variance (MANOVA) was utilized to identify significant differences between conditions for AAA self-rating questionnaire, hand-eye coordination test and grip strength and balance ability test. Statistical significance was set at a probability level of 0.05.

3 Results

3.1 The Self-rating Analysis

The mean scores of AAA self-rating form among gender and age showed in Table 1. Mean scores of self-rating is higher in male drivers (26.8) than in female drivers (12.8). From interpretation of score denoted that the lower score (0 to 15) is the safer driver, and female drivers are fine status. Furthermore, Table 2 revealed the decision condition of drivers' ability based AAA self-rating form. There are five and four male drivers in caution and unsafe condition and three female drivers in caution condition, respectively. Drivers are engaging in some practices that need improvement to ensure safety.

These drivers should be re-education for driving safety. There are 33% male drivers really upset affecting their driving, and less half male drivers always get eye exams and check the medications affecting their driving ability. Administrative staff should be noticed the drivers beginning to experience some natural age-related changes and drivers adjust their driving habits to keep driving safely. Checking visual ability regularly and consulting for taking medications or drugs could be provided in company.

Table 1. The mean scores of AAA self-rating form among gender and age

Variables		Mean	Standard deviation	Min	Max
Gender	Male	26.8	15.6	5	60
	Female	12.8	9.5	0	27
Age	<30	32.5	7.7	27	38
	31–54	24.0	18.1	3	70
	>55	11.0	9.8	0	22

Table 2. Meaning of scores based on self-rating form among gender and age

			Gender		Age		
Levels	Scores	Decision	Male	Female	≤ 30	31–54	≥ 55
Level 1 safe driving	0 to15	Qualified	2	4	0	4	2
Level 2 caution	16 to 34	Unqualified	5	3	1	5	2
Level 3 unsafe driving	35 and over	Unqualified	4	0	1	3	0

3.2 Hand-Eye Coordination Tests, Balance Ability and Hand Grip Strength

Table 3 shows the results of the hand-eye coordination, balance ability, and hand grip strength tests by gender and age. The results of MANOVA showed that the gender effects had found (Pillai's trace = 0.97, $F(12, 1) = 13.2$, $p < 0.05$, partial $\eta^2 = 0.96$) on 0.95 of statistical power (alpha = 0.05, two-tail). Female drivers needed more time to accomplish tracing tasks. In contrast, male drivers made more mistakes than female drivers while tracing circular ($F(1, 12) = 7.35$, $p < 0.05$) and square paths using the left hand ($F(1, 12) = 5.2$, $p < 0.05$). For the electrical mirror tracer test, female drivers made a greater number of mistakes and took longer to accomplish these tasks than male drivers ($F(1, 12) = 6.76$, $p < 0.05$). Balance ability on either foot did not differ significantly between genders. However, hand grip strength differed significantly between genders ($F(1, 12) = 21.66$, $p < 0.001$).

There were no significant differences in number of mistakes made while tracing straight, circular and square paths. However, elderly drivers spent more effort to accomplish these tasks. In contrast, younger drivers made more mistakes than older drivers in tracing the square path. Drivers required more time to accomplish mirror tracings for both hands with incrementally advancing age. There was no significant

Table 3. Results of hand-eye coordination tests, balance ability and hand grip strength among gender and age

Tests	Variables		Gender		Age		
			Male	Female	≤ 30	31–54	≥ 55
Dynamic stability tests	Straight path	Right hand mistakes	2.1	0.9	2.3	1.6	1.5
		Left hand mistakes	2.1	1.3	2.5	1.8	1.6
		Right hand time	4.6	6.6	2.5	5.7	5.6
		Left hand time	3.4	4.3	2.2	3.3	5.7
	Circular path	Right hand mistakes	3.9	2.4	2.3	3.2	4.3
		Left hand mistakes	5.2	1.8	7.0	3.1	5.1
		Right hand time	4.6	6.4	2.8	5.1	6.7
		Left hand time	5.1	4.5	3.2	4.4	6.8
	Square path	Right hand mistakes	7.5	4.0	7.5	6.3	5.5
		Left hand mistakes	11.2	7.3	17.0	8.7	9.3
		Right hand time	16.5	18.4	11.1	16.8	21.1
		Left hand time	19.3	16.4	14.5	16.1	26.2
Electrical-mirror tracer		Right hand mistakes	42.3	56.7	50.8	36.4	75.9
		Left hand mistakes	26.8	77.8	27.8	47.0	47.3
		Right hand time	94.5	140.4	71.1	102.4	153.4
		Left hand time	62.8	132.7	33.8	98.3	84.6
Balance ability		Right foot	14.7	15.0	14.1	14.0	17.6
		Left foot	11.6	14.3	12.2	13.5	13.5
Hand grip strength		Right hand (kg)	33.3	21.2	32.5	28.0	30.0
		Left hand (kg)	38.0	23.0	36.0	31.8	33.5

difference between men and women in accomplishing the circular tracing tasks. In contrast, the time to accomplish the task increased with age. There were no significant differences in balance ability or hand grip strength between age groups.

4 Discussion

Driving is one of the most complex tasks that humans perform on a regular basis, placing significant demands on sensation, perceptual, cognitive, and motor capabilities. However, age-related declines in these capabilities negatively affect driving performance. This study examined multi-dimensional abilities for professional bus drivers. Self-rating of driver behavior is the first tool for evaluation of driving performance. The study applied the AAA self-rating evaluation to examine the performance of urban bus drivers. This simple self-rating form was easily completed by the drivers and showed the condition of the driver relative to driving safety. Results of the analysis showed that male bus drivers had markedly higher unsafe scores than female drivers.

Gender and age are significantly associated with the capabilities of hand-eye coordination, balance ability and hand grip strength. Although typically a male occupation, bus driving is undergoing a process of feminization [8]. Thus, buses are not only providing the service of transportation, but also need to be comfortable and safe for both passengers and driver. Tse et al. [9] consolidated the key research on the occupational health of urban bus drivers since the 1950s. Stressors for bus drivers include poor cabin ergonomics, rotating shift patterns and inflexible running times. Greater attention to salient moderating and mediating variables in the stressor–strain relationship is featuring in more recent research. Based on our results, this study suggest that professional driving license testing should consider the different among gender and age of the drivers and should set different visual ability assessment criteria for each group.

5 Conclusion

Recommended of this study were conducted the self-rating evaluation, hand-eye coordination, balance ability and hand grip strength in the process of licensing exams test to assess driving qualification, renew driving license particularly. Bus drivers should also regularly take these ability tests to assess their driving quality and ability. In this way, an employer can accurately assess drivers' multi-dimensional abilities and verify whether they meet the demands of their jobs. On the other hand, some assistant facilities should be provided. For example, power-assisted steering wheel for female drivers, camera or blind area detection alarm for elderly drivers.

References

1. Fonda, S.J., Wallace, R.B., Herzog, A.R.: Changes in driving patterns and worsening depressive symptoms among older adults. J. Gerontol. B Psychol. Sci. Soc. Sci. **56**, S343–S351 (2001)
2. McKnight, A.J., McKnight, A.S.: Multivariate analysis of age-related driver ability and performance deficits. Accid. Anal. Prev. **31**, 445–454 (1999)
3. Janz, N.K., Musch, D.C., Gillespie, B.W., Wren, P.A., Niziol, L.M.: Evaluating clinical change and visual function concerns in drivers and nondrivers with glaucoma. Invest. Ophthalmol. Vis. Sci. **50**(4), 1718–1725 (2009)
4. Weinreb, R.N., Khaw, P.T.: Primary open-angle glaucoma. Lancet **363**(9422), 1711–1720 (2004)
5. Gentzler, M.D., Smither, J.A.A.: Literature review of major perceptual, cognitive, and/or physical test batteries for older drivers. Work **41**(S1), 5381–5383 (2012)
6. Wood, J., Chaparro, A., Hickson, L.: Interaction between visual status, driver age and distracters on daytime driving performance. Vis. Res. **49**(17), 2225–2231 (2009)

7. Owsley, C., McGwin Jr., G., Ball, K.: Vision impairment, eye disease, and injuries motor vehicle crashes in the elderly. Ophthal. Epidemiol. **5**, 101–113 (1999)
8. Cunha, L., Nogueira, S., Lacomblez, M.: Beyond a man's world: Contributions from considering gender in the study of bus drivers' work activity. Work **47**(4), 431–440 (2014)
9. Tse, J.L.M., Flin, R., Mearns, K.: Bus driver well-being review: 50 years of research. Transport Res. F-Traf. **9**, 89–114 (2006)

Research on Detection Algorithm of Roadway Intersection Rule Detection Based on Big Data

Meirun Zhang[1,2,3], Fumin Zou[1,2,3](✉), Yanling Deng[2], Xinhua Jiang[3], Lvchao Liao[3], and Yun Chen[3]

[1] Beidou Navigation and Smart Traffic Innovation Center of Fujian Province Fuzhou, Fuzhou 350118, Fujian, China
laoniu8706@126.com
[2] Fujian University of Technology, Fuzhou 350108, Fujian, China
fmzou@fjut.edu.cn
[3] Fujian Key Laboratory for Automotive Electronics and Electric Drive, Fujian University of Technology, Fuzhou 350118, Fujian, China
achao@fjut.edu.cn

Abstract. At present, the research of road network change is hot in the study of the new road, the road one-way limit line and the construction caused by the reduction of road lanes, etc. Based on the change rule of road network intersection structure and steering rules and the improved map matching algorithm, this paper proposes a trajectory feature analysis method based on vehicle steering angle. At the same time, the mathematical model of geospatial data is constructed. Based on the existing network topology information, the massive traffic trajectory data are mined and analyzed, and the detection and recognition of the change of the intersection Lane rules are realized.

Keywords: Highway intersection · Steering rule · Recognition algorithm

1 Introduction

With the continuous progress of urbanization, the traffic in urban areas is tending to or has reached saturation. Traffic management departments often implement various road optimization adjustment and road limit line measures, in order to improve the intersection efficiency and the overall road network capacity. For example, many roads have been banned from crossing, banning U-turn and so on. As a result, the traffic information of the road network changes. However, due to the lag of the updating of electronic map, the updating of these information is not timely or insufficient coverage. And it often brings inconvenience to the public. Therefore, we need more timely and effective traffic network change information. This paper analyzes the characteristics of the trajectory of the floating vehicle on the basis of the trajectory data of the floating vehicle. Which can identify

J.-S. Pan et al. (eds.), *Advances in Smart Vehicular Technology, Transportation, Communication and Applications*, Smart Innovation, Systems and Technologies 86,
https://doi.org/10.1007/978-3-319-70730-3_3

the current steering rules of the intersection, thus detecting the section of the change of the limit line, alleviating the urban traffic pressure and improving the efficiency of the public travel.

2 Vehicle Road Matching Algorithm Based on Angular Distance Weighting Model

The track data of floating vehicle contains a series point sequence set, which is arranged in sequence according to the GPS receiving time. The set point in each element includes a plurality of zodiac, respectively: the vehicle terminal number, latitude and longitude, direction, speed, data transmission time and vehicle type. In this paper, we derive the experimental data from the real-time trajectory data of 150 thousand floating vehicles collected by Fujian Traffic Information Center (Table 1).

Table 1. Track data contents of floating vehicles

Field name	Examples of data	Content description
MDID	2482866	Vehicle terminal umber
AREAID	0	Vehicle owned area
LONGITUDE	119.354637	Longitude
LATITUDE	26.032568	Latitude
SPEED	20	Vehicle speed information
DIRECT	4	Vehicle direction
GPS–DATE	2015/23 0:00:51	Data transmission time
CAR–TYPE	41	Vehicle type
STATUS	1	Vehicle condition

We construct this weight model on the basis of the vehicle trajectory data. Its principal component consists of two pieces: the angle difference and the distance difference between the vehicle and the road. In the whole evaluation process, different weights are set for the relative importance of the two section, and then we obtain the synthetic matching similarity of each section and matching road section.

2.1 Angle Similarity

The angular difference in the weight model refers to the difference between the direction value in the vehicle GPS data and the geographical orientation of the candidate section. The weight formula is expressed as:

$$W_a = \lambda F(\Delta\delta) \tag{1}$$

In this formula, λ is the angle weighting coefficient, $F(\Delta\delta) = \cos(\Delta\delta)$, $\Delta\delta$ is the angle between the direction value of the vehicle trajectory data (Eight values, each representing a direction) and the geographical orientation of the road.

2.2 Distance Similarity

The distance in the weight model refers to the spatial distance between the latitude and longitude coordinates of the vehicle GPS data and the candidate section. The weight formula is expressed as:

$$W_d = \beta / F(d) \tag{2}$$

In this formula, β is represented as distance weighting coefficient, $F(d)$ is the shortest distance from the matching point to the candidate section of the vehicle. Assumed that the longitude and latitude coordinates of the points to be matched are (x, y), the shortest distance is calculated as follows as:

$$F(d) = \frac{x(lat_{to} - lat_{from}) + y(lon_{to} - lon_{from}) + lon_{to}lat_{from} - lon_{from}lat_{to}}{\sqrt{((lat_{to} - lat_{from})^2 + (lon_{to} - lon_{from})^2}} \tag{3}$$

The minimum distance is the vertical distance of the matched points to the candidate section. When the pedal falls on the road when matching, the shortest distance is what we ask for $F(d)$. Otherwise, they are calculated separately. The smaller the $F(d)$ values, the higher distance similarity of candidate segment is. Conversely, the lower the distance similarity will be.

2.3 Weighted Similarity

The weighted similarity of candidate segments can be expressed by angle similarity and distance similarity:

$$W_s = W_a + W_d = \lambda F(\Delta\delta) + \beta F(d) \tag{4}$$

As shown in the formula, λ and β are respectively the weight of angle and distance, and they satisfy the constraint condition of $\lambda + \beta = 1$. Angle similarity has a relatively greater impact on vehicle road matching. Through the matching of ten thousand vehicle trajectories data and mathematical induction, we can draw the following conclusions. In this formula, we set the angle weight R as 0.6, distance weigh B as 0.4. When $\Delta\delta \to 0$, $F(d) \to 0$, the matching degree of candidate $W_s \to max$ sections, the likelihood of matching points to the candidate segment is greatest.

3 Design of Road Steering Change Detection Algorithm

3.1 Trajectory Feature Recognition Based on Vehicle Steering Angle

In order to identify the intersection turning rules in the outlet network, the vehicle trajectory characteristics can be determined by analyzing the vehicle trajectory characteristics and the vehicle movement trends. When the floating vehicle that conforms to this particular track characteristic and movement state reaches 1/4 of the total data, the steering rules of the road section and intersection can be judged.

Design of Trajectory Feature Extraction Algorithm. The track characteristic recognition algorithm based on vehicle steering angle mainly tracks and analyzes the change of steering angle of track data points of floating vehicles. Because the trajectory of a single floating vehicle can not fully represent the vehicle's running trend and state, so we make an algorithm, this algorithm is based on a 22 point trajectory by 3 adjacent receiving chronological ordering of the steering angle calculation, by using the change of steering angle calculation formula, assign to the state of the vehicle.

In order to describe the feature extraction algorithm explicitly, the definition is as follows:

Definition 1 (Change of steering angle). *The steering angle refers to the angle based on the trajectory of the vehicle ordered before and after the two track point the direction of travel form, to the north of* 0°.

Definition 2 (Steering undetermined table). *The storage table of intersection steering rules is used as storage format, which is used to record the change of steering angle.*

As the limited and one-sided of single track point information representation, the vehicle trajectory feature extraction algorithm is mainly for the spatial information of 3 adjacent track points before and after the vehicle is calculated. In a cross section of a road network, the path of the vehicle on the road is shown in Fig. 1. The four curves, which are connected by a single locus of the vehicle received by the time series, represent the four running states of the vehicle: straight, left turn, right turn, and u-turn.

Assume that $p_1, p_2, ..., p_n$ are the continuous trajectory point data of a vehicle over a period of time. Each point of p_i contains several attributes: time t_i, speed d_i, direction $[lon_i, lat_i]$, and coordinates. In the formula, lon_i represents the vehicle longitude coordinates, lon_j latitude coordinates, $Agl(lon_i, lat_i)$ is the angle formed by the northward direction and the vector $\overrightarrow{p_i p_j}$ which is composed by point p_i and point p_j.

$$\triangle\Theta = Agl(p_{i-1}, p_i) - Agl(p_i, p_{i+1}) \tag{5}$$

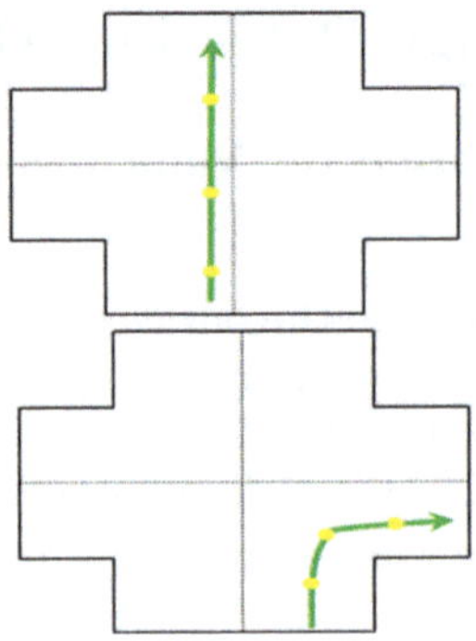

Fig. 1. Vehicle running state at intersection

In the formula, $\triangle\Theta$ is composed of the difference between the vector formed by the angle of track points of the vehicle and the north direction. That is, the steering variation during vehicle operation. $\triangle\Theta$ values vary, the vehicle's driving conditions are different, the specific circumstances shown in Fig. 2.

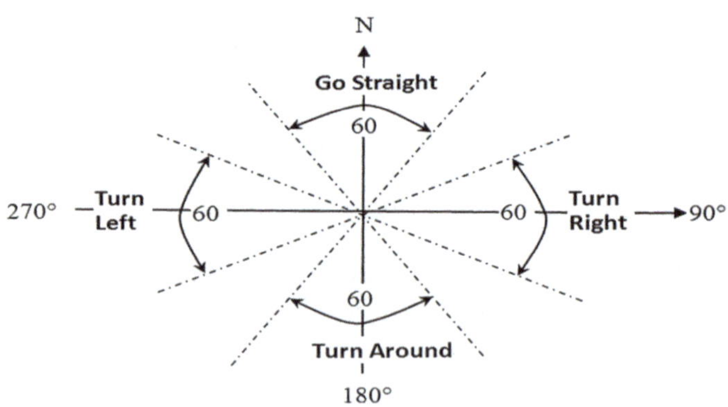

Fig. 2. The basis of judging between $\triangle\Theta$ and vehicle running state

As shown in the diagram, the vehicle running state is divided into three types: the left turn of the vehicle, the right turn of the vehicle and the U-turn of the vehicle. The left turn and right turn are mainly judged by the change of the steering angle. In contrast, The vehicle 's U-turn is more complex.

When the vehicle turns around, not only the change of the steering angle needs to meet the corresponding conditions, but also the relative distance has some characteristics.

$$|\overrightarrow{p_{i-1}, p_{i+1}}| = |\overrightarrow{p_{i-1}, p_i} + \overrightarrow{p_i, p_{i+1}}| < max(\overrightarrow{p_{i-1}, p_i}, \overrightarrow{p_i, p_{i+1}}) \qquad (6)$$

In this formula, p_{i-1}, p_i, p_{i+1} represent the three contiguous points of the vehicle. When the vehicle is moving straight or left and right, the $|\overrightarrow{p_{i-1}, p_{i+1}}|$ at both ends must be greater than $|\overrightarrow{p_{i-1}, p_i}|$ and $|\overrightarrow{p_i, p_{i+1}}|$, which represent respectively as the distance from the starting point to the middle point and the distance from the middle point to the stopping point. The relative distance between them is growing positively. When the vehicle is in a U-turn, the running track of the vehicle will turn back. In this process, its relative driving distance is a negative growth state.

For a floating vehicle running in the network, if its running state meet to GPS trajectory point criterion in the sequence, then we can judge the steering and steering section of the vehicle. At the same time, according to the data storage format of the designed steering rules, we can increase the number of turns in the database of the steering rules by combining nodes and sections. As shown in this formula, if $\{p_{i-1}, p_i, p_{i+1}\}$ is a sequence of floating car points that satisfies the left turn condition in the table, and can be used to determine that the vehicle is moving from the current section to another.

Effective Determination of Steering Rules. The positioning system of a floating vehicle is prone to error when it is positioned under the influence of the outside environment such as the dense building area or the weather. If only the trajectory points of the voucher car are used to judge the actual turning rules of the intersection, the credibility of the results obtained is very low. In order to reflect the actual turning rules of the road network and deal with several floating vehicles at the same time, a mathematical statistic model is needed to determine the effectiveness of the turning rules.

$$C_{n,r_i} \geq M, \frac{C_{n,r_i,r_j}}{C_{n,r_i}} \geq \eta \tag{7}$$

In this formula C_{n,r_i} represents the sum of the number of turns in the N network node with r_i as the same starting section. Among them, the experience value of M is 100. C_{n,r_i,r_j} indicates the number of times a vehicle moves from the initial section r_i to the r_i section at the node N connection. $\frac{C_{n,r_i,r_j}}{C_{n,r_i}}$ represents the proportion of specific steering, and the ratio of the empirical value η is 5%. As shown in Table 2 of the steering rules, at a crossing, if the ratio of a particular turn to the sum of all turns exceeds the set threshold A, then the steering rule can be determined to be valid; otherwise, the decision is invalid.

Table 2. Steering rules

Node	Initial section	Turning section	Statistic times	Proportion (%)	Effectiveness
482	6215	6216	495	49.06	TRUE
482	6215	6218	2	0.2	FALSE
482	6215	6233	512	50.74	TRUE

3.2 Change Detection of the Turning Rules for the Intersection

Whether the road network information changes or not depends on whether the information of the road map, the road network information and the vehicle data detected by the floating vehicle are consistent. The same goes for the change of rules at the intersection. The detection of the change of lane limit can be transformed to the comparison between the electronic map and the steering rule of the actual road network. The algorithm firstly extracts the rule list of the database stored in the database by node ID, and then traverses all the nodes. In this process, it can extract all the steering rules data of this node and process them one by one, and determine whether it is valid or not. If this validity is true, then the network topology information table obtained by searching the primary key map for querying the electronic map is consistent with the actual result, and that means the information has not changed. However, if the result is different, the change in the steering rules is indicated, and the change result is recorded and stored in a temporary list.

The pseudo code for the implementation of this algorithm is represented as follows:

Algorithm 1. Change detection of the turning rules for the intersection

Input: Two related tables in database the table about turning rules in the electric map and real road information from the float car data

Output: New table to store the changed information about the turning rules

```
Set table < T − result > to restore the changed turning rules of road
Select * from < T − rule > order by Node-ID
for each node N of the< T − rule >: do
    for each rule R of the node N: do
        if validity is True then
            query mapped rule from map database R
            if R is equal to R then
                No change, next rule of the node N
            else
                Insert < T − result >where node = N and rule = R
            end if
        else
            No comparison, next rule of the node N
        end if
    end for
end for
Return table < T − result >
```

The detection link of road steering limit change is mainly based on the original electronic map spatial data table and the actual road network steering rule table, and it needs to be queried through the primary key index. At the same time, it creates a new table that meets the requirements of the test to store the contrast results. The time complexity of the algorithm is expressed as (n), and the algorithm is scalable.

4 Conclusion

In this paper, an improved matching algorithm based on weight model and topological relation is proposed for real-time road change detection. We can use in different types of vehicle road limit line dynamic detection, including Road ban, left turn, right turn, straight line, U-turn, and one-way limit line and speed limit detection. The algorithm in this paper can be used for the change detection of the road steering limit line, and can build a more objective and effective verification method for the large area of the road network limit line rules change, so as to improve the verification efficiency and reduce the workload.

Acknowledgments. This work is supported by National Natural Science Foundation of China (No. 61304199), Fujian Science and Technology Department (No. 2014H0008), Fujian Transportation Department (No. 2015Y0008), Fujian Education Department (No. JK2014033, JA14209,JA15322), and Fujian University of Technology (No. GYZ13125, 61304199, GY-Z160064). Many thanks to the anonymous reviewers, whose insightful comments made this a better paper.

References

1. Deng, Y.L., Zou, F.M.: An algorithm for identifying the change of road restriction based on K-means clustering. In: 2016 International Conference on Mathematical, Computational and Statistical Sciences and Engineering, pp. 94–104. DEStech Publications Press, Shen Zhen (2016)
2. Bae, Y., Lee, W.H.: Automatic road extraction from remote sensing images based on a normalized second derivative map. IEEE Geosci. Remote Sens. Lett. **12**, 1–5 (2015)
3. Xiong, Y.E., Xu, Z.L.: Methods of intersection dynamic subsections based on floating car data. Traffic Inf. Saf. **27**, 38–43 (2009)
4. Zhuang, L.J., He, Z.C.: A large-scale floating car data-based algorithm of turning rule extraction at intersections. J. Wuhan Univ. Technol. (Transp. Sci. Eng.) **37**, 1084–1088 (2013)
5. Yang, L., Zou, Z.J.: Study of Urban intersection lane structure based on floating car tracking data. Sci. Surv. Mapp. **39**, 76–80 (2016)
6. Tan, X.S., Wang, J.: Detection of road intersections using floating car data. Geogr. Geo-Inf. Sci. **31**, 34–38 (2015)
7. Li, X., Zhang, J.S.: An extraction algorithm of track features based on trend set of heading angle variable. J. Geo-Inf. Sci. **17**, 1172–1178 (2015)

Predicting the Travel Time in Using Recurrent Neural Networks: A Case Study of Fuzhou

Luming Li[1,2,3](✉) and Xinhua Jiang[1,3]

[1] School of Information Science and Engineering, Central South University, Changsha, China
{liluming,xhj}@csu.edu.cn

[2] School of Intelligent Transportation, Hunan Communication Polytechnic, Changsha, China

[3] Fujian Key Laboratory for Automotive Electronics and Electric Drive, Fujian University of Technology, Fuzhou, China

Abstract. Travel time plays an important role in many ITS application such as traffic control and trip guidance. However the travel time information is acquired after the real driving. This makes travel time prediction an important way to estimate the real-travel time before actual traveling. In this paper, we focus on predicting the travel time of a road segment using deep learning methods. In our work, the historical travel time information collected by Fuzhou taxicab is extracted. Different recurrent networks architectures are applied. Experimental result shows that the deep learning models considering the temporal relation work well on travel time prediction.

Keywords: Travel time prediction · Recurrent network · LTSM

1 Introduction

In recent years, intelligent transport system has become a fundamental infrastructure for building a smart city. Working as a key part of the intelligent transport system, travel time prediction is growing in importance for many applications such as route planning [16], navigations [2], traffic monitoring [13], route pattern recognition [15] and passenger pickup location recommendation [17] etc. Due to its nonlinear feature, travel time prediction is still a challenging task. In this paper, we focus on predicting travel time using deep learning method. First, the travel time information from a real trajectory dataset is extracted. Then, serval recurrent neural network architectures are applied to predicting the travel time of road segments in Fuzhou city. Finally, the model's performance is evaluated.

The remainder of the paper is organized as follows. In Sect. 2, some of the related works are reviewed. The detail of the model is presented in Sect. 3. Training strategy and experiments are stated in Sect. 4. Finally, we conclude our work in Sect. 5.

J.-S. Pan et al. (eds.), *Advances in Smart Vehicular Technology, Transportation, Communication and Applications*, Smart Innovation, Systems and Technologies 86,
https://doi.org/10.1007/978-3-319-70730-3_4

2 Related Works

Travel time prediction problem have been studied by many researches using various methods. Reza et al. [11] designed an Integrated Moving Average Model (ARIMA) based method to predict the short-term travel time. The method generally assumes the historical data are fitted with some time serious model and provides interpretable parameters with a simple structure. Tak et al. [12] predicted the travel time using a multi-level k nearest method with data fusion. Moridpour et al. [10] proposed a SVM based model to predict the bus's travel time. In their model, a Genetic Algorithm (GA) is used to determine the optimal set of model parameters. Zhang et al. [7] proposed a gradient boosting method to improve the travel time prediction of the urban link. The proposed STGBRT model combines simple tree together which could uncover the underlying patterns in travel time data to enhance the prediction accuracy. Kumar et al. [7] predicted the bus travel time using a spatial-temporal discretization approach. In their work, the traffic flow and density was first described in terms of speed in the form of a partial differential equation. Then, a speed based equation was discretized using the Godunov scheme. Finally the prediction was made through a Kalman filter based model.

Apart from the models and algorithm mentioned above, neural networks is wildly applied in travel time prediction since it has a great advantage in modeling the complex nonlinear problem. Mazloumi et al. [9] applied neural networks in the context of predicting bus travel time over different sections. The proposed framework consists with two neural networks. The first network predicts the average travel time given a certain set of input, whereas the second neural network predicts the variance imposed by training data noise. Wang et al. [14] developed a space-time delay neural network model to integrate the spatial-temporal correlation of road traffic network. Their test result shows the STDNN's prediction ability outperforms the ARIMA based models. Zhang et al. [18] proposed a Takagi–Sugeno–Kang Fuzzy Neural Network (TSKFNN) approach to predict freeway corridor travel time with an online computing algorithm. Their test results show that using count, speed, and occupancy together as input produces the best prediction results.

With the improvements of computing capability, the deep learning methods are reporting huge success in many research domain including the area of ITS. Huang et al. [6] explored the deep belief networks to traffic flow prediction. Lv et al. [8] applied a deep architecture model SEAS to learn the future traffic flow and reported the SEAS has the better performance than SVM based model. Chen et al. [1] using a long-short term memory to learn and predict the patterns of traffic conditions and reported successes in comparison with multilayer perceptron model, decision tree model and SVM. Obviously, exploring deep learning in the travel time prediction is of great significance.

3 Methods

In this paper, serval long short term memory based architectures are explored to our travel time prediction application. In this section, the problem is described formally. Then the architectures which are used to predict the travel time are introduced.

3.1 Problem Definition

Travel time prediction can be treated as a sequence processing problem. Giving a series of acquired time stamped vector $\{X_1, X_2, X_3 \ldots X_t\}$, a prediction model aim to predict the future travel time p_{t+1} for a specified road segment r.

3.2 Model Architecture

Long Short Term Memory (LSTM) neural network was first proposed by S. Hochreiter [5]. It has a complex structure called LSTM cell in the network's hidden layer. LTSM has been successfully applied in many sequence data processing problem like natural language processing, speech recognition etc. Since the time stamped travel time data is also a sequence data, the LSTM model can be directly applied to the travel time prediction. LSTM has many variations such as peephole LSTM, Convolutional LSTM etc., the architecture we choose is a typical structure of LSTM with forget gate.

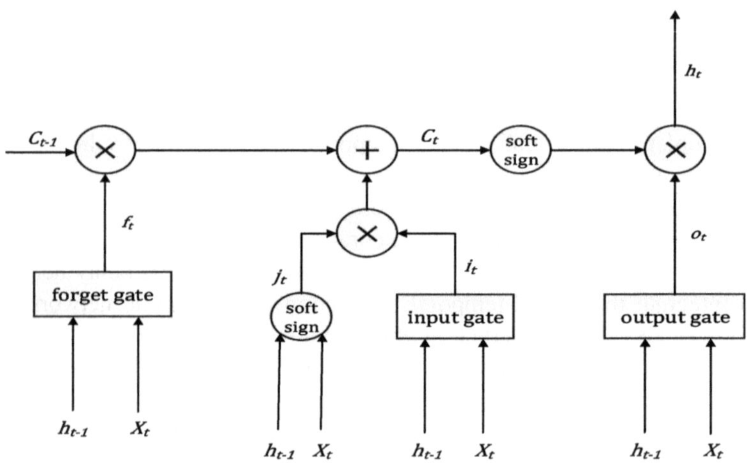

Fig. 1. The LSTM architecture

As it shows in Fig. 1. The LSTM architecture, the main component of LSTM's hidden layer is a gate controlled cell structure. The output h_t is configured by three gates: forget gate, input gate and the output gate. The outputs of the three gates are f_t, i_t and o_t. The C_t and C_{t-1} are two consecutive cell states. Forget gate decide whether to forget the formal cell state C_{t-1}. Input gate controls the entering of the outside data X_t. Output gate gathering all the information and compute the hidden layer output. The precise model is described below:

$$f_t = \sigma\left(\theta_X^f X_t + \theta_h^f h_{t-1} + b_f\right) \tag{1}$$

$$j_t = softsign\left(\theta_X^j X_t + \theta_h^j h_{t-1} + b_j\right) \tag{2}$$

$$i_t = \sigma\left(\theta_X^i X_t + \theta_h^i h_{t-1} + b_i\right) \tag{3}$$

$$o_t = softsign\left(\theta_X^o X_t + \theta_h^o h_{t-1} + b_0\right) \tag{4}$$

$$C_t = f_t \odot C_{t-1} + j_t \odot i_t \tag{5}$$

$$h_t = softsign(C_t) \odot o_t \tag{6}$$

Here the θ_X^* stands for the X_t's weights on component. θ_h^* are the weights matrix for the previous hidden output h_{t-1}. $\odot$ Is the Hadamard product. σ stands for the sigmoid function.

$$\sigma(x) = \frac{1}{1 + e^{-x}} \tag{7}$$

In order to avoid saturation and faster the training speed, the soft sign is used as the activation function to instead the hyperbolic tangent activation. The soft sign is described below.

$$softsign(x) = \frac{x}{1 + |x|} \tag{8}$$

When h_t is calculated, a liner regression layer is used to generate the final prediction.

$$p_{t+1} = \theta h_t + b \tag{9}$$

One of the goal in this work is to evaluate the gate's impotency in travel time prediction problem. This can be done by removing the different the forget gate and the input gate to see the performance of modified model. First, we want check whether the model work well if it has long term memory. By removing the Eq. (1) and changing Eq. (5) to the Eq. (10), the prediction model is capable to "remember" all the data.

$$C_t = C_{t-1} + j_t \odot i_t \tag{10}$$

By removing model's input gate, we makes the X_t always enter into LSTM cell. This change was made based on one assumption that when prediction the travel time, the recent information has the most impact. For LSTM without the input gate, Eq. (4) is removed and C_t is calculate through:

$$C_t = f_t \odot C_{t-1} + j_t \tag{11}$$

4 Experiments

4.1 Data Preparation

We use the Fuzhou Taxicab Trajectory Dataset. The original data is collected through the taxicab's GPS devices in every 30 s. For each taxicab, speed, direction, location and timestamp are generated. The data from Jun 1st 2015 to Jun 31st 2015 is toked; the time period in our study ranges from 7:30 A.M to 9:30 P.M. The data covers the urban area of Fuzhou city which is the area of approximate 1400 km^2. The road map of the Fuzhou City is shown in Fig. 2.

To generate the travel time information, a digital map of Fuzhou city is used. A grid based map matching method is applied to match each taxi to the road segments of Fuzhou city. Then for each 10 min, timestamp, number of taxi on the road, and average travel time for each road segment are calculated. On some road segments, there are data missing due to there are no taxis go through the specific road in a time period. In these case, the missing data are filled with the historical average value (Fig. 2).

Fig. 2. The main road networks of Fuzhou city

4.2 Model Training

Model training's goal is to figure out the all the weights and biases which can minimize the difference of the predicted results and the real travel time. In order to evaluate the model's effectiveness, a cost function is designed below:

$$\mathrm{J}(\Theta, \mathrm{B}) = \frac{1}{2}(p_{t+1} - x_{t+1})^2 \tag{12}$$

Where m is the number of the predicted result; Θ. and B stand for all the weight matrices and biases.

In our work, the dimension of the input vector is set to 4, which are the timestamps, the road's travel time, the travel time of upstream road's and downstream road's. The output is a one-dimensional scaler which stands for the travel time of the specific road segment in future 10 min. Glorot's initialize strategy is used to initialize the model [4]. According to the Gers's idea, the forget gates bias b_f is set to 1 [3]. A stochastic gradient descent (SGD) for the backup propagation through time (BPTT) algorithm is designed to training the model, the training process is described below:

```
Input:
Training dataset X_train, Dimeson of hidden layer dh, Number
of times to iterate through the complete dataset epoch,
Initial learning rate α.

Output:  Weight matrices Θ, Biases B
Method:
nepoch ← 100, α ← 0.005, ecnt ← 10
for  dh in {4,5,6,7,8}
initialize Θ and B with the uniform distribution
for epoch in 1 to 100
   perform forward propagation to compute p_{t+1}
   perform BPTT to compute ΔΘ, ΔB
   Θ ← Θ − αΔΘ
 B ← B − αΔB
   if  epoch%10 == 0
        calculate  cost J
        if J is not decreasing
              α ← α * 0.8
         end if
    end if
if α < 10^-3  break
end for

save Θ and B for dh
end for
```

4.3 Experimental Analysis

To evaluate the performance of the prediction model, 20 road segments are chosen. The evaluation matrices are percentage error (MAPE), and mean squared error (RMSE) which are defined below, where n is the total number of prediction.

$$MAPE = 1 - \frac{\sum_n \frac{|p_{t+1} - x_{t+1}|}{x_{t+1}}}{n} \tag{13}$$

$$RMSE = \sqrt{\frac{\sum_n |p_{t+1} - x_{t+1}|}{n}} \tag{14}$$

The test result are shown in Fig. 3. The MAPE and RMSE for Test Result, where the LSTM-* states for the LSTM without corresponding gate unit.

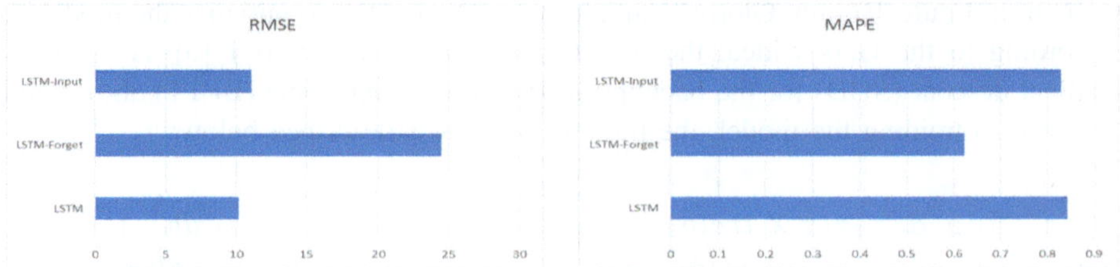

Fig. 3. The MAPE and RMSE for test result

The result shows that the recurred learning based architecture performs well in our dataset. The LSTM models shows the best performance, the model without the forget gate performs worse. Which means in all the component in a LSTM cell the most important component is the forget gate. In our case, the LSTM-Input's accuracy is very close to the model without the input gate. This approved the assumption that in travel time prediction problem, the newly coming data has the greatest impact.

5 Conclusions

In this paper, serval recurrent network architecture are applied to predicting the travel time of the roads in Fuzhou city. Travel time information is extracted from a real trajectory dataset generated by Fuzhou taxicab. 20 road's travel time are predicted. The experimental results show that recurred learning based architecture can worked well in predicting the travel time in Fuzhou city. It also shows that the input gate in the LSTM model is less important than the forget gate unit in our task.

Acknowledgement. This work is supported by Natural Science Foundation of Hunan Province, China (2017JJ5002).

References

1. Chen, Y., Lv, Y., Li, Z., Wang, F.-Y.: Long short-term memory model for traffic congestion prediction with online open data. In: 2016 IEEE 19th International Conference on Intelligent Transportation Systems (ITSC), pp. 132–137. IEEE (2016)
2. Dobbelstein, D., Henzler, P., Rukzio, E.: Unconstrained pedestrian navigation based on vibro-tactile feedback around the wristband of a smartwatch. In: Proceedings of the 2016 CHI Conference Extended Abstracts on Human Factors in Computing Systems, pp. 2439–2445. ACM (2016)
3. Gers, F.A., Schmidhuber, J., Cummins, F.: Learning to forget: continual prediction with LSTM. Neural Comput. **12**, 2451–2471 (2000)
4. Glorot, X., Bengio, Y.: Understanding the difficulty of training deep feedforward neural networks. In: AISTATS, pp. 249–256 (2010)
5. Hochreiter, S., Schmidhuber, J.: Long short-term memory. Neural Comput. **9**, 1735–1780 (1997)

6. Huang, W., Hong, H., Li, M., Hu, W., Song, G., Xie, K.: Deep architecture for traffic flow prediction. In: International Conference on Advanced Data Mining and Applications. Springer, pp. 165–176 (2013)
7. Kumar, B.A., Vanajakshi, L., Subramanian, S.C.: Bus travel time prediction using a time-space discretization approach. Transp. Res. Part C Emerg. Technol. **79**, 308–332 (2017)
8. Lv, Y., Duan, Y., Kang, W., Li, Z., Wang, F.Y.: Traffic flow prediction with big data: a deep learning approach. IEEE Trans. Intell. Transp. Syst. **16**, 865–873 (2015). doi:10.1109/TITS.2014.2345663
9. Mazloumi, E., Rose, G., Currie, G., Moridpour, S.: Prediction intervals to account for uncertainties in neural network predictions: methodology and application in bus travel time prediction. Eng. Appl. Artif. Intell. **24**, 534–542 (2011)
10. Moridpour, S., Anwar, T., Sadat, M.T., Mazloumi, E.: A genetic algorithm-based support vector machine for bus travel time prediction. In: 2015 International Conference on Transportation Information and Safety (ICTIS). IEEE, pp. 264–270 (2015)
11. Reza, R., Pulugurtha, S.S., Duddu, V.R.: ARIMA model for forecasting short-term travel time due to incidents in spatio-temporal context. In: Transportation Research Board 94th Annual Meeting, vol. 15-5553 (2015)
12. Tak, S., Kim, S., Jang, K., Yeo, H.: Real-time travel time prediction using multi-level k-nearest neighbor algorithm and data fusion method. In: Computing in Civil and Building Engineering (2014), pp. 1861–1868 (2014)
13. Wang, F., Hu, L., Zhou, D.D., Sun, R., Hu, J.J., Zhao, K.: Estimating online vacancies in real-time road traffic monitoring with traffic sensor data stream. Ad Hoc Netw. **35**, 3–13 (2015). doi:10.1016/j.adhoc.2015.07.003
14. Wang, J., Tsapakis, I., Zhong, C.: A space–time delay neural network model for travel time prediction. Eng. Appl. Artif. Intell. **52**, 145–160 (2016)
15. Wang, T.B., Zhang, D.Q., Zhou, X.S., Qi, X., Ni, H.B., Wang, H.P., Zhou, G.: Mining personal frequent routes via road corner detection. IEEE Trans. Syst. Man Cybern. Syst. **46**, 445–458 (2016). doi:10.1109/tsmc.2015.2444416
16. Wang, Z., Zlatanova, S., Moreno, A., van Oosterom, P., Toro, C.: A data model for route planning in the case of forest fires. Comput. Geosci. **68**, 1–10 (2014). doi:10.1016/j.cageo.2014.03.013
17. Yuan, N.J., Zheng, Y., Zhang, L., Xie, X.: T-finder: a recommender system for finding passengers and vacant taxis. IEEE Trans. Knowl. Data Eng. **25**, 2390–2403 (2013)
18. Zhang, Y., Ge, H.: Freeway travel time prediction using Takagi–Sugeno–Kang fuzzy neural network. Comput.-Aided Civ. Infrastruct. Eng. **28**, 594–603 (2013)

Driving Behavior Motivation Model Research Based on Vehicle Trajectory Data

Yun Chen[1,2,3], Xin-hua Jiang[3(✉)], Lyuchao Liao[1,2], Fu-min Zou[1,3], and Mei-run Zhang[2,3]

[1] Beidou Navigation and Smart Traffic Innovation Center of Fujian Province Fuzhou, Fuzhou 350118, Fujian, China
chenyun@smail.fjut.edu.cn

[2] Fujian University of Technology, Fuzhou 350118, Fujian, China
achao@fjut.edu.cn

[3] Fujian Key Laboratory for Automotive Electronics and Electric Drive, Fujian University of Technology, Fuzhou 350118, Fujian, China
xhjiang@fjut.edu.cn

Abstract. The development of floating car technology provides a new method for studying driving behavior motivation. Massive vehicle trajectory data often contains rich potential semantic information such as traffic patterns and driving behavior habits. To overcome the nonstationarity of the spatial distribution of the original trajectory data, this paper designs a semantic mining method of traffic trajectory data to construct complex behavioral modeling of traffic trajectory data based on LDA latent semantic topic model, so that to depict the driving behavior interest patterns and explore the cognitive mechanism of driving behavior. Research results show that the driving motivation model constructed can estimate the probability of the potential driving motivation topic in each spatial grid, and then the driving trajectory will be transformed from random probability events into driving intention with intrinsic certainty.

Keywords: Traffic trajectory · Driving behavior · Latent semantic analysis

1 Introduction

The continuous development and application of floating car technology provides a new way to study driving behavior motivation, floating car trajectory data forms into traffic trajectory data [1]. Massive trajectory data often contains rich potential semantic information such as traffic pattern and driving behavior habits, fully exploiting these potential and hidden semantic information can provide lots of valuable decision support for urban road planning and traffic management optimization. By digging out trajectory data, the semantic correlation between driving behavior and traffic network can be learned, so that to reveal

J.-S. Pan et al. (eds.), *Advances in Smart Vehicular Technology, Transportation, Communication and Applications*, Smart Innovation, Systems and Technologies 86,
https://doi.org/10.1007/978-3-319-70730-3_5

public traveling patterns and apply to driving route recommendation, which has become one of popular researches [2].

At present, related researches are focused on the analyzing and processing with the original trajectory data, as the actual trajectory data is affected by the geographical feature of road network and its own intrinsic properties, the spatial distribution of the data has characteristics of non-stationary, so that these methods that directly perform statistical analysis and measurement on original trajectory data are often easy to produce error, especially in the case of high spatial non-stationary of the data, which lead the trajectory differences with characteristic significance are swamped by a mass of small differences without characteristic significance, resulting in the difficulties in digging out the potential common behavior from massive trajectory data.

In order to characterize the driver's behavior motivation, the semantic mining modeling of traffic trajectory data needs to span the low-level primitive time-space features and high-level behavior semantic features, namely the model not only reflects the spatiotemporal statistical features of original trajectory data, but also fully digs out the internal semantic behavior features from massive trajectory data, so that to solve the wide gap problem between the original low-level motion feature and high-level semantic behavior analysis. Therefore, this paper explores to construct a traffic driving cognitive model based on the potential semantic topic distribution to characterize the driving behavior as a specific semantic topic probability sequence.

2 Problem Description

2.1 A Summary of Driving Behavior Motivation Research

The driving trajectory is characterized by uncertain, and the traffic trajectory data possess the characteristic of wide space coverage and low coverage strength, whose spatial distribution are usually nonuniform [3], which means that the data are usually concentrated in some areas, while the rest areas are very few or even without data coverage.

Each driver has a specific driving trajectory temporal and spatial distribution mode, which contains the driver's interest and preference information. Digging these information is also the core content of intelligent traffic information service such as driving OD forecasting, traffic condition recommendation and individual behavior analysis. Therefore, this paper proposes two assumptions:

(1) If a user's driving trajectory frequently appears in a certain area, the user is more concerned about this region.
(2) If a user frequently stays in a certain area for a long time, the user is more familiar with this region.

In this paper, the region that the driver is more familiar and concerned is defined as the driver's driving interest preference area [4]. Analyzing the temporal and spatial distribution and aggregation property of trajectory data is necessary to extract the driver's interest area. The most direct method is the statistical calculation with vehicle data of the spatial grid, which is easy influenced by abnormal and noisy data, and is difficult to ensure the robustness of data distribution, and is also hard to find the intrinsic behavior feature of trajectory data. Therefore, this paper try to explore to construct a traffic driving cognitive model based on the potential semantic topic distribution, so that to describe the driving behavior as a specific semantic topic probability sequence.

2.2 Driving Behavior Motivation Modeling Based on LDA

The aim of potential semantic analysis is to discover the potential semantic relations between objects in the data set, this relationship cannot be described simply by frequency characteristic [5], it needs to be analyzed with structural knowledge in the semantic space.

To analyze the semantic relation in the text and reduce the influence of high frequency words, *Hofmann* [6] put forward a probabilistic latent semantic analysis (PLSA) method to study the inherent semantic structure knowledge among text data. Although PLSA is easy to extract the semantic information from training data, it is difficult to understand new data. Then *David* [7] put forward a topic model based on the LDA, which used a potential random variable with K dimension subject to Dirichlet distribution to represent the documents' topic probability and simulate the process of document generation.

Similar to the document generation process, the semantic knowledge of the driving trajectory is also combined in an agnostic way. Based on LDA latent semantic topic model, this paper explores the complex behavior modeling of traffic trajectory data, the core idea is to view each track as multinomial distribution of a number of driving behavior motive topics, and each driving motive is a probability distribution of spatial grid. Similarly, the driving motive model can be regarded as a generation model of the trajectory events, namely, each track selects some driving motive topic through the certain probability, and then accesses some spatial grid with certain probability, generating the relevant topic models that describe the inherent driving behavior motive of trajectory data.

Based on the hierarchical structure of LDA model, we divide the traffic trajectory data into trajectory data set layer, single track layer and spatial grid layer. The entire traffic trajectory data is a collection set of M traffic tracks and denoted by $D = \{tr_1, tr_2, ..., tr_M\}$, a single track is a sequence of N spatial grids, that is $Tr =< g_1, g_2, ..., g_N >$, and g_i is the No. i space grid passed by the trajectory data sequence (Fig. 1).

Given a trajectory data set D, each track is represented by LDA model as a potential driving intention motive sequence whose combination ratio is generated by the *Dirichlet* distribution. Implicit driving motivation is defined as the polynomial distribution of spatial grid in the trajectory data set D to represent the relevant properties between spatial grids, these potential driving motives are shared by all tracks and each track has a specific driving intention motive probability distribution. LDA model is determined by the parameter (α, β) of the

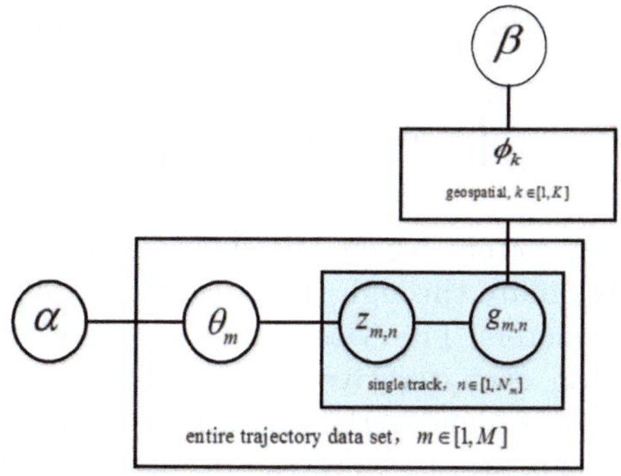

Fig. 1. LDA algorithm display.

entire trajectory data layer, α reflects relative strength of the implied driving motions in the trajectory data set, β representing the probability distribution of all implied driving motivations themselves. The random variable θ indicates the proportion of the implied driving motives in the single track layer and z represents the driving motive proportion of each single spatial grid distributed by single track, and g is the spatial grid vector representation of trajectory data. Marking interest spatial position of entire trajectory as $\{g_1, g_2, ..., g_N\}$.

Based on the probability graph model of LDA, we can further derive the joint distribution of all variables [8]:

$$p(\vec{g}_i, \vec{z}_i, \vec{\theta}_i, \varphi) = \prod_{n=1}^{N_i} p\left(g_{i,n}|\vec{\varphi}_{z_{i,n}}\right) \bullet p\left(z_{i,n}|\vec{\theta}_i\right) \bullet p\left(\vec{\theta}_i|\vec{\alpha}\right) \bullet p\left(\Phi|\beta\right) \quad (1)$$

Where, φ_k is the probability distribution of spatial grid of the driving motive k; θ_m represents the topic probability distribution of the No. m document; k and m as the parameters of polynomial distribution are used to generate driving behavior topics and spatial grids respectively; N_m represents the length (data size) of the No. m track; $w_{m,n}$ and $z_{m,n}$ represent the No. n spatial grid and driving motive in the No. m track respectively. And α and β, as the parameters of *Dirichlet* distribution, generally take the fixed value with symmetrical distribution and scalar representation; θ_m and φ_k obey the *Dirichlet* distribution, and their distribution function is:

$$Dir\left(\mu \mid \alpha\right) = \frac{\Gamma(\alpha_0)}{\Gamma(\alpha_0), ..., \Gamma(\alpha_k)} \prod_{k=1}^{N} \mu_k^{\alpha_k - 1} \quad (2)$$

Where, $0 \leq \mu_k \leq 1, \sum_{k=1}^{K} \mu_k = 1, \alpha_0 = \sum_{k=1}^{K} \alpha_k, \Gamma$ is *gamma* function. For each grid g in the entire geographic space, the appearance probability in the *No. i* track is:

$$p(w_{i,n} = g|\vec{\theta}_i, \Phi) = \sum_{k=1}^{K} p(w_{i,n} = g|\vec{\varphi}_k) \bullet p(z_{i,n} = k|\vec{\theta}_i) \quad (3)$$

For the entire traffic trajectory dataset, the likelihood function is:

$$p(W|\theta,\Phi)=\prod_{i=1}^{M}p(\vec{w}_i|\vec{\theta}_i,\Phi)=\prod_{i=1}^{M}\prod_{n=1}^{N_i}p(w_{i,n}|\vec{\theta}_i,\Phi) \tag{4}$$

The driving model parameters based on LDA cannot be calculated directly, using the way of approximate through the *GibbsSampling* algorithm to deduce the model optimal parameters. The *GibbsSampling* algorithm performs iterative process in each dimension, each iterative process mainly selects one of dimensions and perform sample process under the circumstances of the other dimensions constrained, repeating this process until the result is converged (Fig. 2).

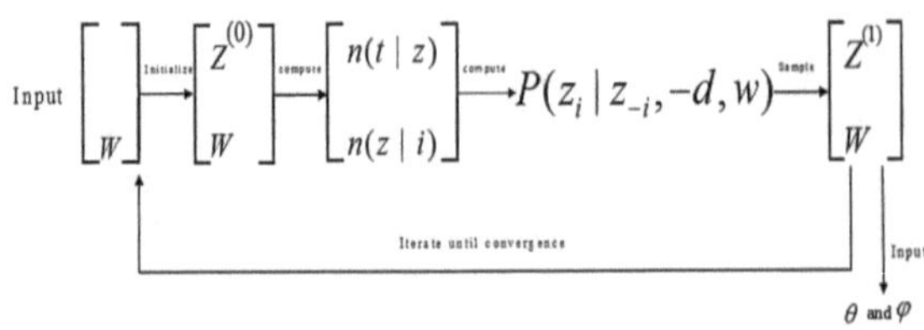

Fig. 2. Driving motivation modeling flowchart

In the driving motivation analysis process, we first make fractal for driver's trajectory based on driving motive analysis, trajectory fractal is to divide the tracks into segments based on the potential driving intention so that each sub-segment of the track corresponds to a specific driving intention. Trajectory fractal makes each sub-track as close as possible to a single driving intention. Then, the driving motive is initialized, that is to construct the driving motive set. And then randomly assigning the each spatial grid to a driving motivation $z^{(0)}$, counting up the spatial grid number that belongs to driving motivation z in each track Tr and appearance frequency of spatial grid g in each motivation z. And computing $p(z_j \mid z_{-j}, Tr, g)$ iteratively through *GibbsSampling*, namely, estimating the probability distribution that belongs to various driving motive of the current spatial grid through other spatial grid driving motive probability distribution until the result is converged, then outputting the driving motive model parameter θ, φ and the final potential driving motive topic $z_(i,n)$ of each spatial grid.

3 Experiment

3.1 Experimental Environment

This experiment uses a rack-mounted server (Intel Xeon E52650 dual CPU, 128 GB memory), the operating environment is Windows 2008 R2 (64 bit), the algorithms are implemented by using R language and Python.

3.2 Experimental Data Set

The experimental data set used is a driver's trajectory data including a total of 128 days from May 25, 2015 to September 29, 2015, the number of the total trajectory data is 685,577.

3.3 Personal Driving Behavior Motive Mining Experiment

According to the extracted single-driver trajectory data set, constructing single-driver motive mining dataset, the experiment uses finite precision (the precision is about 1 km) to connect latitude and longitude to form string, so that to achieve labeling processing of two-dimensional space, forming the road network space grid dictionary table. Based on driving trajectory fractal method, the daily driving tracks are divided into some sub-segments, and then analyzing the *Dirichlet* distribution property of the tracks so that to generate the driver's motive topics according the method mentioned above.

The maximum iteration number of this experiment is set to 5000, the number of topics generated is set to 10, and the priori values of trajectory data set layer parameters (α, β) are both set to 0.02, 10 driving motives are generated by training the model based on experimental data set. Further, principal component analysis is performed for the 10 driving motive topics, and project each topics to two-dimensional space composed by two principal components, and the driving motive topic distribution diagram (Fig. 3 Left) can be obtained, from the graph's distribution relation, we can analyze the neighborhood relation between the various topics, and we can also generate the spatial heat relationship in the global schema. Fig. 3 Right shows the top 30 spatial hotspots for the driver that are sorted by topic significant.

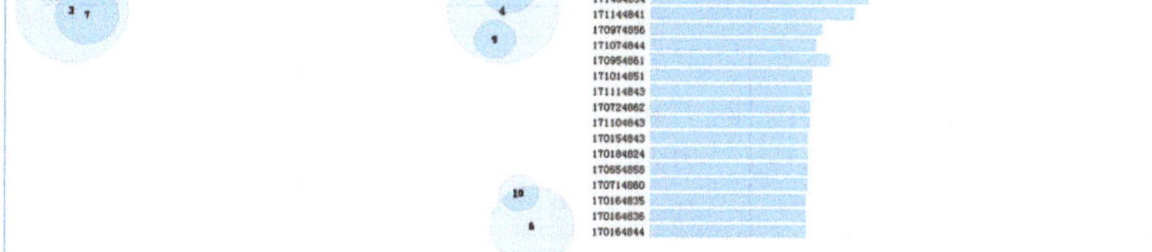

Fig. 3. Driving motivation topic global distribution diagram

3.4 Experimental Results and Analysis

For each generated driving motivation topic, there are different access probabilities and heat at each spatial location, heat information (top 30) is shown in Fig. 4 when choosing the "Driving motivation topic 5".

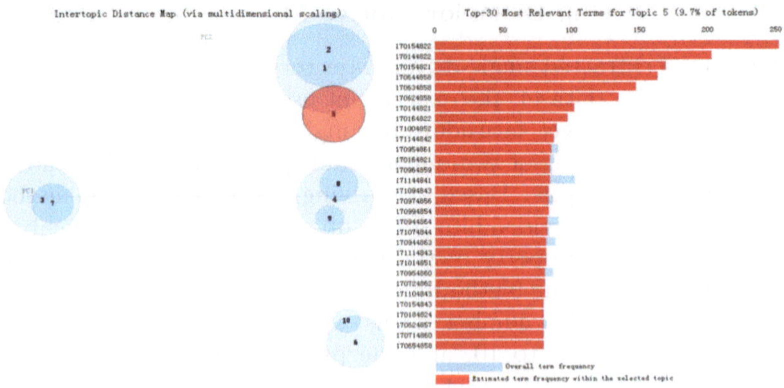

Fig. 4. Spatial probability distribution diagram of driving motivation

The statistical analysis for the first 8 driving motivation topic is shown in Table 1, the statistical analysis result shows that the drive has a specific driving interest topic. For example, in this experiment the drivers driving motivation topic mainly includes: "topic 1", "topic 3", "topic 4", "topic 6", "topic 9" and "topic 10". As shown in the table above, different spatial positions have their own topic probability feedback. Taking the most significant position in the global topic distribution (spatial grid ID is 170154822) as example, calculating its probability distribution through the driving motivation model, and named the grid as "G1" probability (Table 1), it can be found that this spatial position mainly reflects the "topic 3" and "topic 5" driving motivation tendency.

Table 1. Driving motivation topic statistical analysis

Items	Topics							
	Topic 1	Topic 2	Topic 3	Topic 4	Topic 5	Topic 6	Topic 7	Topic 8
Data quantity	33673	2554	21236	24187	2884	13920	6635	5038
Total probability	12.75	3.72	16.24	21.38	6.23	10.97	5.21	2.83
G1's probability	3.57	0	1.25	39.12	0	7.29	0	0

In order to understand the practical significance of each driving motivation topic more intuitively, the spatial grid distribution of each motive topic is projected onto the map. Top 9 motivation topics are listed in Fig. 5. It is easy to find that although in system model, each spatial grid is independent identically distributed, it is not randomized in each topic, and is characterized by significant regularity, which indicates that the driving motivation topic has inherent tendency. The practical significance of this driver's driving motivation topic can be obtained. For example, for driving topic 2, the active route is Zhangzhou City-Dongshan County through No. 324 national highway and No. 201 provincial road.

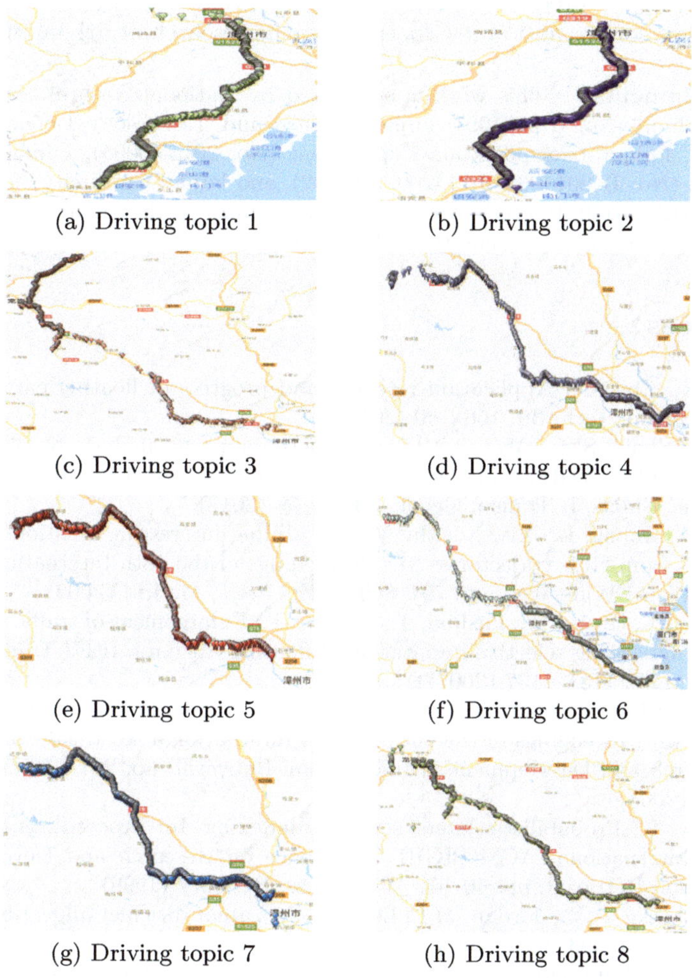

(a) Driving topic 1 (b) Driving topic 2 (c) Driving topic 3 (d) Driving topic 4 (e) Driving topic 5 (f) Driving topic 6 (g) Driving topic 7 (h) Driving topic 8

Fig. 5. Driving behavior motive subject diagram

4 Conclusion

It can be seen from the experimental analysis results that the probability of the driver's potential driving motivation topics reflected by each spatial grid can be estimated by the driving motivation model. Therefore, during the driving process, the driving motivation topic probability will gradually converge with the grid number increasing passed by. And then the random probability events are transformed into driving motive intention with inherent certainty, which will provides important predictive support for applications such as driving route recommendation service and a new method for the research of driving motivation.

Acknowledgments. This work is supported by National Natural Science Foundation of China (No. 61304199), Fujian Science and Technology Department (No. 2014H0008), Fujian Transportation Department (No. 2015Y0008), Fujian Education Department (No. JK2014033, JA14209, JA1532), and Fujian University of Technology (No. GYZ13125, 61304199, GY-Z160064). Many thanks to the anonymous reviewers, whose insightful comments made this a better paper.

References

1. Zheng, K., Zhu, D.: Application research and progress of floating car technology. Mod. Electron. Tech. **39**, 156–160 (2016)
2. Fang, Z., Shaw, S.-L., Wei, T., Li, Q.: Spatiotemporal analysis of critical transportation links based on time geographic concepts: a case study of critical bridges in Wuhan, China. J. Transp. Geogr. **23**, 44–59 (2012)
3. Zheng, Y., Zhang, L., Xie, X., Ma, W.-Y.: Mining interesting locations and travel sequences from GPS trajectories. In: Proceedings of the 18th International Conference on World Wide Web, pp. 791–800. IEEE Press, Madrid (2009)
4. Kumagai, M., Hiruta, T., Fushiki, T., Yokota, T.: Enlargement of traffic information coverage area using selective imputation of floating car data. IEEJ Trans. Electron. Inf. Syst. **127**, 1131–1137 (2007)
5. Dupret, G.: Latent concepts and the number orthogonal factors in latent semantic analysis. In: Proceedings of the 26th Annual International ACM SIGIR Conference on Research and Development in Information Retrieval, pp. 221–226. ACM Press, Toronto (2003)
6. Hofmann, T.: Probabilistic latent semantic indexing. In: Proceedings of the 22nd Annual International ACM SIGIR Conference on Research and Development in Information Retrieval, pp. 50–57. ACM Press, Berkeley (1999)
7. Blei, D.M., Ng, A.Y., Jordan, M.I., Lafferty, J.: Latent dirichlet allocation. J. Mach. Learn. Res. **39**, 993–1022 (2003)
8. Heinrich, G.: Parameter estimation for text analysis. Technical report, Global Grid Forum (2004)

A Ranging Algorithm for Mobile Vehicles Based on Kalman Filter

Junmin Wang[1,3], Fumin Zou[1,3(✉)], Maolin Zhang[2,3], Lyuchao Liao[1], and Yaxue Pang[1,3]

[1] Beidou Navigation and Smart Traffic Innovation Center of Fujian Province Fuzhou, Fuzhou 350108, Fujian, China
wjmfjut@163.com
[2] Fujian Key Laboratory for Automotive Electronics and Electric Drive, Fujian University of Technology, Fuzhou 350108, Fujian, China
mailzml@163.com
[3] Fujian University of Technology, Fuzhou 350108, Fujian, China
fmzou@fjut.edu.cn

Abstract. In order to detect the behavior of illegal vehicle jumping and provide a simple and reliable basis for it, this paper presents an algorithm based on Kalman filter for measuring the distance between two cars. This paper mainly studies the ranging algorithm based on the constant size of the license plate area. Meanwhile, by introducing the Kalman algorithm to predict the position of the license plate in the next frame of the picture and to improve efficiency. In order to verify the reliability and accuracy of the algorithm, this paper has carried out static and dynamic experiments. According to the experimental results, we can find that the algorithm can meet the requirements of ranging and have good accuracy.

Keywords: Vehicle jumping detection · Monocular distance measurement · Kalman filter

1 Introduction

With the development of intelligent vehicle assistant driving technology, video ranging technology is rapidly entering the field, used to distance between the vehicle and the surrounding objects. At present, machine vision ranging includes monocular ranging [1] and binocular ranging [2] or multi-field [3] vision ranging technology. For binocular and multidimensional vision technology, to match the characteristic of picture is a most important obstacle that hinder the technology to be developed rapidly. However, other affecting factors include real-time requirements and cost of technology. The Zhao [4] put forward a real-time ranging algorithm based on image depth information. The Wu [5] proposed an improved distance measurement method based on vehicle shadow. The Liu [6]

J.-S. Pan et al. (eds.), *Advances in Smart Vehicular Technology, Transportation, Communication and Applications*, Smart Innovation, Systems and Technologies 86,
https://doi.org/10.1007/978-3-319-70730-3_6

proposed a fast image segmentation method based on difference image stack and ranging method.

This paper proposes a ranging algorithm based on the information of license plate images. The size of most license plates is $440 * 140\,\mathrm{m}^2$ in China. Thus, we can calculate the measure the distance according to this characteristic. In this paper, the Kalman predictor is used to predict the region of the target license plate, which can reduces the number of license plate detection and recognition time. The results of the experiments show that the method can satisfy stability and real-time requirements.

2 Ranging Model and Algorithm Design

The general frame of the algorithm is showed as Fig. 1, which includes five modules. The video capture module obtains pictures by the car DVR. The image preprocessing module is to provide more effective information for distance ranging. The positioning module identifies the size of the license plate with a rectangle. The ranging module is used to calculate the distance and output the result. The prediction module is used to predict the position of the license plate in the next frame.

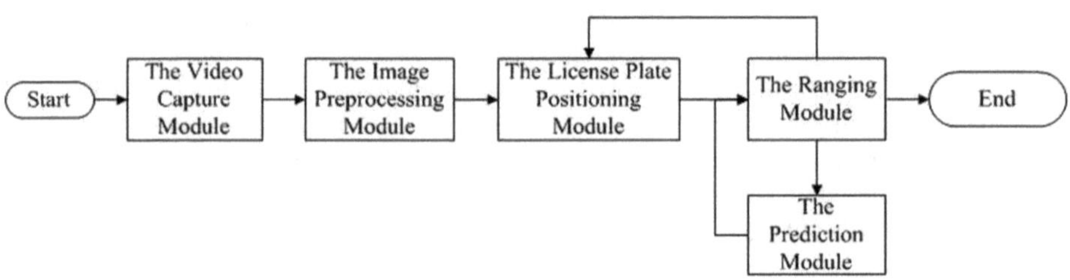

Fig. 1. General frame of ranging

2.1 Analysis of Algorithm

The imaging geometric model of the camera contains the relationship between the positions and spatial coordinates of all the information. The model is generally described by the pixel coordinates system (u-v), the image coordinates system (x-y), the camera coordinates system (X_c, Y_c, Z_c) and the world coordinates system (X_w, Y_w, Z_w). Assumed that the physical dimensions of each pixel in the u-axis and v-axis directions are d_x and d_y, the image coordinates system listed as Fig. 2.

The intersection point of the camera optical axis and the image plane is defined as the origin of the coordinates system, in which the x-axis is parallel to the u-axis and the y-axis to the v-axis. The point (U_0, V_0) is the coordinates of point O_1 in coordinates system u-v. The d_x represents the physical size of each pixel on the horizontal x-axis and the d_y represents the physical size of each pixel

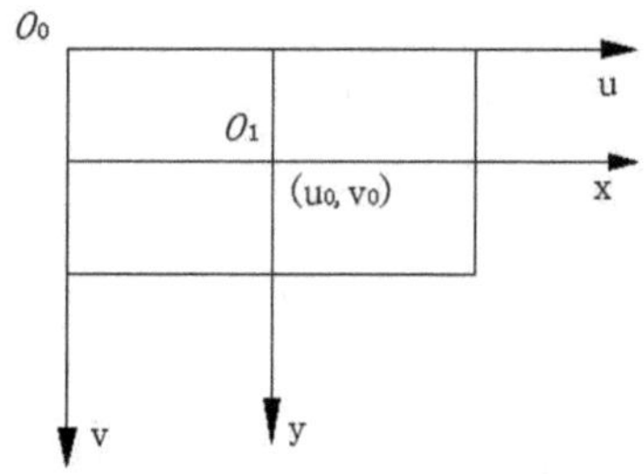

Fig. 2. General frame of ranging

on the horizontal y-axis. The relationship between the pixel coordinates system and image coordinates system as follows:

$$u = \frac{x}{dx} + u_0 \tag{1}$$

$$v = \frac{y}{dy} + v_0 \tag{2}$$

The relationship can be expressed as follows:

$$\begin{bmatrix} u \\ v \\ 1 \end{bmatrix} = \begin{bmatrix} \frac{1}{dx} & 0 & u_0 \\ 0 & \frac{1}{dy} & v_0 \\ 0 & 0 & 1 \end{bmatrix} \begin{bmatrix} x \\ y \\ 1 \end{bmatrix} \tag{3}$$

2.2 The Linear Model of Camera Imaging

The relationship between the coordinates system of the camera and the world can be showed as Fig. 3. The point O is the optical center. The X_c-axis and Y_c-axis are parallel to the x-axis and y-axis of the image coordinates system respectively. The Z_c is the optical axis of the camera. The intersection of the optical axis and plane of the image is point O_1. The camera coordinates system are make up of the point O, X_c-axis, Y_c-axis and Z_c-axis.

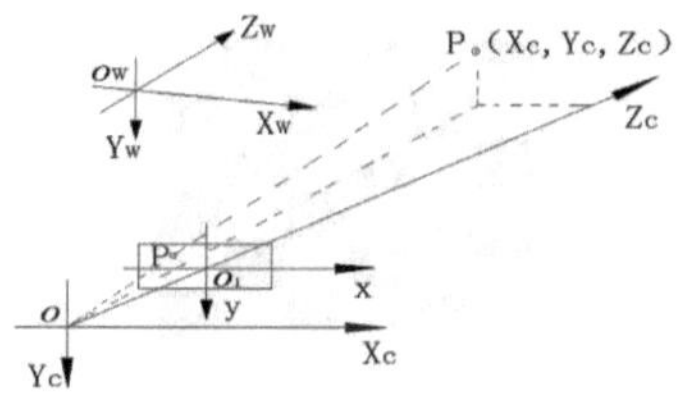

Fig. 3. Camera coordinate system and world coordinate system

The world coordinates system can be converted to the camera coordinates system, and the transition relation is shown in the (5):

$$\begin{bmatrix} X_c \\ Y_c \\ Z_c \\ 1 \end{bmatrix} = \begin{bmatrix} \boldsymbol{R} & \boldsymbol{T} \\ 0^T & 1 \end{bmatrix} = M_1 \begin{bmatrix} X_w \\ Y_w \\ Z_w \\ 1 \end{bmatrix} \tag{4}$$

The $\boldsymbol{R}$ is a orthogonal unit rotation matrix. The $\boldsymbol{T}$ is a translation vector. The M_1 is a internal parameter of the camera. The conversion relationship between the camera coordinates system and the image coordinates system can be expressed as follows:

$$Z_c \begin{bmatrix} x \\ y \\ 1 \end{bmatrix} = \begin{bmatrix} f\,0\,0\,0 \\ 0\,f\,0\,0 \\ 0\,0\,1\,0 \end{bmatrix} \begin{bmatrix} X_c \\ Y_c \\ Z_c \\ 1 \end{bmatrix} \tag{5}$$

Inputting the (4) and the (5) into the (3), we can get the model as follows:

$$Z_c \begin{bmatrix} u \\ v \\ 1 \end{bmatrix} = \begin{bmatrix} \alpha_x & 0 & u_0 & 0 \\ 0 & \alpha_y & v_0 & 0 \\ 0 & 0 & 1 & 0 \end{bmatrix} \begin{bmatrix} R & T \\ 0^T & 1 \end{bmatrix} \begin{bmatrix} X_w \\ Y_w \\ Z_w \\ 1 \end{bmatrix} = M_2 M_1 \begin{bmatrix} X_w \\ Y_w \\ Z_w \\ 1 \end{bmatrix} \tag{6}$$

The $\alpha_x = f/dx$, $\alpha_y = f/dy$. The M_2 is a external parameters of camera. We can calculate the value of $M_2 * M_1$ is by the method of camera calibration.

2.3 Camera Calibration

By the process of camera calibration, we can get the three-dimensional position and orientation of camera in the world coordinates system [7]. As a general rule, the size of the actual object in the image is inversely proportional to the distance from the camera. The essay use checkerboard target calibration method to calibrate the camera. The checkerboard calibration template is shown in Fig. 4. The size of each black or white square is $10 * 10\,\text{mm}^2$.

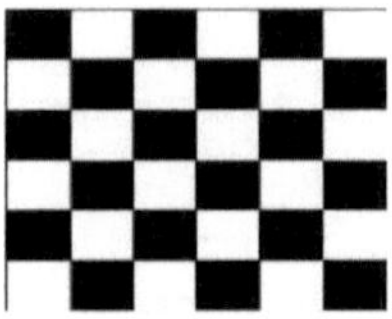

Fig. 4. Checkerboard calibration template

By getting the checkerboard calibration template from different distance and make sure that only the template is included for each shooting. Then, calculate the image area of the template obtained at distances. Based on the (6), we can simplify the camera linear model into the (7) according to the paper [8]:

$$d = m\sqrt{s} \tag{7}$$

The m is obtained by test on different distances, such as 100 mm, 150 mm, 200 mm and 250 mm. According to results of the camera calibration test, we can get the arithmetic mean of m is 1.298 (Table 1).

Table 1. Test results of camera calibration

No.	d (mm)	S (mm^2)	m (s)
1	100	5935.408	1.297
2	150	13211.785	1.306
3	200	24300.023	1.282
4	250	36643.223	1.305

We define N_a as the pixel size of the entire image and N_b as the size of the rectangle that identified by location module. At the same time, S_a represents the cross-sectional area of the plane where the license plate is located; S_b is $440 * 140\,\text{mm}^2$. The relationship as the (8):

$$\frac{N_a}{N_b} = \frac{S_a}{S_b} \tag{8}$$

And,

$$S_a = 61600 * \frac{N_a}{N_b}(mm^2) \tag{9}$$

Based on the (9) and the value of m, we can calculate the actual distance from the camera to the front plate by followed the (10).

$$d = 1.298 * \sqrt{61600 * \frac{N_a}{N_b}(mm^2)} \tag{10}$$

2.4 Kalman Predictor

The Kalman predictor is one of the common methods to estimate the target, and further improve the efficiency of the algorithm by estimating the license plate position, reducing the range of detection, recognition and matching. The basic principle of the Kalman predictor is to use the measured value to correct the estimated state and finally obtain a reliable state estimate.

$$\boldsymbol{X}(k+1) = \boldsymbol{A}\boldsymbol{X}(k) + \boldsymbol{B}\boldsymbol{W}(k) \tag{11}$$

$$\boldsymbol{Y}(k) = \boldsymbol{H}\boldsymbol{X}(k) + \boldsymbol{V}(k) \tag{12}$$

In this paper, the k represents the sequence number of each frame in the video information. The matrix $\boldsymbol{X}$ is the system state vector. The matrix $\boldsymbol{W}$ is the noise vector of the system. The matrix $\boldsymbol{Y}$ is the observation vector. The matrix V is the observed noise vector, $\boldsymbol{A}$ is the state transition matrix, $\boldsymbol{B}$ is the noise driving matrix, and $\boldsymbol{H}$ is the observation matrix. The state prediction equation and the updated equations can be obtained according to Kalman's prediction theory. The prediction equations as follows:

$$\boldsymbol{X}^{'}(k+1) = \boldsymbol{A}(k+1,k)\boldsymbol{X}^{'}(k|k) \tag{13}$$

$$\boldsymbol{P}(k+1|k) = \boldsymbol{A}(k+1,k)\boldsymbol{P}(k|k)\boldsymbol{A}^{'}(k+1,k) + \boldsymbol{Q}(k) \tag{14}$$

$$\boldsymbol{P}(k+1|k+1) = [\boldsymbol{I} - \boldsymbol{K}(k+1)\boldsymbol{H}(k+1)]\boldsymbol{P}(k+1|k) \tag{15}$$

The updated equations as follows:

$$\boldsymbol{K}(k+1) = \boldsymbol{P}(k+1|k)\boldsymbol{H}^{T}(k+1)\left[\boldsymbol{H}(k+1)\boldsymbol{P}(k+1|k)\boldsymbol{H}^{T}(k+1) + \boldsymbol{R}(k+1)^{-1}\right] \tag{16}$$

$$\boldsymbol{X}^{'}(k+1|k+1) = \boldsymbol{X}^{'}(k+1|k) + \boldsymbol{K}(k+1)\left[\boldsymbol{Y}(k+1) - \boldsymbol{H}(k+1)X^{'}(k+1|k)\right] \tag{17}$$

$$\boldsymbol{P}(k+1|k+1) = [\boldsymbol{I} - \boldsymbol{K}(k+1)\boldsymbol{H}(k+1)]\boldsymbol{P}(k+1|k) \tag{18}$$

Where, the $\boldsymbol{I}$ is the unit matrix and the $\boldsymbol{P}$ is the variance matrix. When the violation occurs, the speed of the vehicle is relatively low. Therefore,in the experiment referred to in this paper, the vehicle in the video information can be moved almost uniformly. The horizontal and vertical coordinates of the center of the identified rectangular box are defined as the initial measurements of the Kalman predictor. In addition, the relevant parameters are defined in this experiment as follows: $\boldsymbol{A} = 1$, $\boldsymbol{B} = 1$, $\boldsymbol{H} = 1$, Q = 0, $\boldsymbol{R}$ = [4,0;0,1], $\boldsymbol{P}(0)$ = [2,0;0,1].

3 Ranging Experiment and Result Analysis

In order to verify the accuracy and real-time performance of the algorithm, the paper arranges vehicle experiments in both static and dynamic status. The experimental procedures are as follows: Step1: Gets information about each frame image. Step2: Preprocess the image. Step3: Matching license plate in the given area and marking the license plate area with a rectangle. Step4: Calculate the distance, and output the result. Step5: Based on the results of Step3, estimate the area where the license plate may appear. The area is centered on the centroid of the rectangle identified by the above frame, while the length of the long side is 1.5 times the previous one. Step6: If we can match the license plate in step5, proceed to step4. If the algorithm can not match the license plate, proceed to step3.

3.1 Static Vehicle Experiment and Result Analysis

The environment of the experiments is showed in the Fig. 5. The distance between the license plate and the rear is 5 cm. Then, mark the position of the distance against the rear, 1 m, 1.5 m, 2 m, 2.5 m and so on. The camera used in the experiment is a Logitech C270 camera with a maximum resolution of 1280 ∗ 720. Thus, the value of N_a is 921600. We install camera at the fixed height of 1.2 m and an inclination angle of 87° to conduct two groups of experiments. The ranging algorithm is implemented by OpenCV-3.1.0 visual processing function library. And, we can found the algorithm the can satisfy the requirements of static ranging according to the Table 2.

Fig. 5. Experiment environment

Table 2. Experiment result of dynamic distance ranging

NO	Test results (cm)	Real distance (cm)	Error (%)	Running time (s)
1	98.26703	105	6.41	0.39395
2	227.303452	205	10.87	0.341132
3	320.364197	305	5.03	0.233536
4	409.206177	405	1.03	0.228833
5	489.061584	505	3.16	0.221364

3.2 Dynamic Vehicle Experiment

To verify the algorithm, we have a dynamic vehicle experiment. The video information is about 25 frames per second. Figure 6 is test done on the campus road. Figure 7 is test on the urban center road. Compared with the urban center road, the speed of the vehicle on campus is slower, and the surrounding environment is relatively simple. Figures 6 and 7 have been marked out the license plate, and the results are displayed in the image.

The dynamic experiment shows that the monocular ranging algorithm proposed in the paper can be used for dynamic vehicle distance measurement. The red data on the left corner of the chart represents the time that consumed on the distance measurement. It can be found that the time is in the range of 0.1 to 0.3 s. So it can effectively meet the real-time requirements of ranging.

(a) (b)

Fig. 6. Distance ranging on campus road

(a) (b)

Fig. 7. Distance ranging on urban center road

4 Conclusion

The paper proposes a monocular ranging algorithm for measure the distance. The efficiency of the algorithm is improved by the Kalman predictor. Generally, the speed and distance of the vehicle are relatively small. Thus, according to the results of static experiment and dynamic experiment, the proposed algorithm can meet the requirement of vehicle jumping distance detection, and has good robustness and accuracy.

Acknowledgments. This work is supported by Fujian Science and Technology Department (No. 2014H0008), Project of Science and Technology Plan of Fujian Provincial Transportation (No. 2015Y0008), Universities Scientific Research Special of Fujian Provincial (No. JK2014033), Fujian University of Technology (No. GY-Z13125), and Scientific Research Startup Fund for Doctor of Fujian University of Technology (No. 61304199). Many thanks to the anonymous reviewers, whose insightful comments made this a better paper.

References

1. Zhao, T., Guo, L., Zhang, L.: Spatial positioning algorithm based on monocular vision. J. Northwest. Polytech. Univ. **01**, 47–51 (2009)
2. Zheng, X., Zhao, J., Liu, M.: Identification and location of tomato based on binocular stereo vision. Comput. Eng. **22**, 155–156 (2004)
3. Zhang, L., Mao, Q., Feng, D.: 3D face modeling based on multi-camera. Comput. Dev. Appl. **04**, 2–3 (2005)

4. Zhao, S., Sheng, H., Wang, H.: A real time ranging algorithm based on monocular vision. J. Suzhou Univ. **08**, 2–3 (2016)
5. Jun, W., Li, W., Geng, L.: Forward vehicle detection and ranging based on monocular vision. Comput. Eng. **02**, 26–32 (2017)
6. Liu, W., Chen, L., Ma, X.: High-speed target monocular pose ranging method based on color image. Chin. J. Sci. Instrum. **03**, 675–682 (2016)
7. Guo, C., Hong, Y.: Automatic camera calibration method using checker-board target. Comput. Eng. Appl. **12**, 176–179 (2016)
8. Han, Y., Zhang, Z., Dai, M.: Monocular vision ranging method for target ranging. Opt. Precis. Eng. **05**, 1110–1117 (2011)

The Evaluation of the Video Codec Performances on the Driving Recorder with Different Vehicle Speeds

Shi-Huang Chen[1(✉)], Zhi-Quan Huang[1], Wen-Kai Liu[2], and Jui-Yang Tsai[2]

[1] Department of Computer Science and Information Engineering, Shu-Te University, Kaohsiung 824, Taiwan
shchen@stu.edu.tw, hhzq4216@gmail.com

[2] Smart Network System Institute, Institute for Information Industry, Taipei 105, Taiwan
{wkliu,jytsai}@iii.org.tw

Abstract. Driving recorder has become one the necessary equipment for the current automotive. In Taiwan, there are various brands driving recorder with different capabilities, and almost all of the automotives are installed at least one set of driving recorder. The main function of driving recorder is to record the road video in case of clarify for the responsibility of an accident or unexpected situations. Therefore, the design of driving recorder will focus on the higher recorded video quality as well as higher compression ratio in order to take up less storage space. This paper mainly discusses the different video codec technologies and evaluates their performance on the driving recorder applications. The video codec technologies mentioned in this paper includes the early MPEG-4, currently H.264, and the newest H.265. This paper simulates these three different video codecs, i.e., MPEG-4, H.264, and H.265 to evaluate their coding performances on 6 road videos captured from different vehicle speeds. Then this paper will find the appropriate video codec for the specific vehicle speed as well as road conditions. The experimental results could provide a well reference for the selection of driving record video codec.

Keywords: Driving recorder · Dash cam · MPEG-4 · H.264 · H.265

1 Introduction

Driving Recorder or called Dash cam is a type of video camera for vehicles and is growing in popularity in worldwide, specially, in China and Russia. This is a video camera installed in vehicles in order to record what is happening on the roads. Driving recorder could be used for simple entertainment or to securing evidence in the event of accidents [1, 2]. According to data from retail analysts GfK, the sales of driving recorders increased by 395% in 2016, growing faster than any other consumer electronic product including smart fitness bands, tablets and digital cameras.

There are different types of driving recorder models that vary in size and functionality. Basically their primary design is to record events that occur outside the vehicle. In market place, there are three types of driving recorder: (1) Front facing video camera

J.-S. Pan et al. (eds.), *Advances in Smart Vehicular Technology, Transportation, Communication and Applications*, Smart Innovation, Systems and Technologies 86,
https://doi.org/10.1007/978-3-319-70730-3_7

or 'one video cam' product. This type of driving recorders typically mount on the front windscreen or dashboard whilst facing forward and capturing events and everything within sight. (2) 'Dual video cams' front and rear driving recorder. In addition to the front facing camera, there is also a secondary camera which mounts at the back of the vehicle to provide coverage of all events happening behind the vehicle. (3) Four video cams driving recorder. In addition to the front and rear cameras, there are also one affixed to the left and right side of the vehicle. From experience, commercial vehicles such as truck transport or major logistic companies often tend to use multi-video cams products.

Therefore, the key component of driving recorder is the video codec. There are 3 developed video codecs could be used in driving recorder today. They are MPEG-4, H.264, and H.265. All of these three types of video codecs are belong to hybrid codec structure. The principle encoding processes are video data transform, motion estimation, motion compensation, and entropy coding for bit-stream. The decoding processes are the reverse of the encoding. MPEG-4 is the most common streaming format and it consists of many parts, of which only MPEG-4 Part II is used for video coding. MPEG-4 Part II calls on video encoders such as DivX or XviD in order to encode the video, while audio is typically carried in MP3 format. Modern updates to MPEG-4 are now using H.264 as well. H.264 is a more advanced technology for compressing video. It analyzes foreground (changing parts) and background (static parts) in video and avoids redundant information. The compression rate is significantly higher than MPEG-4. H.264 relies on x264 for encoded video (as well as others, such as DivX or XviD), and audio is often encoded using AAC or MP3 audio codecs depending on the size and quality. H.264 is touted as 1.5 to 2 times as efficient as basic MPEG-4 compression, which leads to smaller file sizes and seamless playback on more devices. In fact, H.264 is now included in the MPEG-4 codec (MPEG-4 Part 10: Advance Video Coding, known as MPEG-4 AVC).

High Efficiency Video Coding (HEVC) or H.265 is a new standard for video compression developed by the ISO and ITU-T in 2013. HEVC/H.265 has the potential to deliver better performance than earlier standards such as H.264/AVC. HEVC promises a 50% storage reduction as its algorithm uses efficient coding by encoding video at the lowest possible bit rate while maintaining a high image quality level. HEVC has the same basic structure as previous standards such as MPEG-4 and H.264/AVC with various incremental improvements such as [3]:

- More flexible partitioning, from large to small partition sizes,
- Greater flexibility in prediction modes and transform block sizes,
- More sophisticated interpolation and de-blocking filters,
- More sophisticated prediction and signaling of modes and motion vectors,
- Features to support efficient parallel processing.

Even though HEVC/H.265 is already finalized, it is still not popular. One reason is that HEVC/H.265 comes with the trade-off requiring almost 10x more computing power. This means the cost of driving recorder using HEVC/H.265 codec is greater than the one using H.264/AVC or MPEG-4 codec. This paper will discuss the encoding performances of MPEG-4, H.264, and H.265 in six different types of vehicle speed and to find the most suitable video codec for driving recorder. The following sections are organized as follows: Sect. 2 describes the experimental method as well as the experimental results, and Sect. 3 gives a conclusion of this paper.

2 Experimental Method and Results

2.1 Simulation Environment

This paper makes use of JVC GZ-MG20TW hard disk digital camera to capture the vehicle driving videos. The JCV GZ-MG20TW comes with F1.8 lens and 68 million pixels CCD. The original videos will be encoded using JVC exclusive lossless video codec with frame size of 720 × 480 and frame rate of 29.97 FPS. This paper takes video frames from urban and highway regions, respectively. In the urban regions, this paper records the driving video frames at three different vehicle speeds, i.e., 40,50, and 60 km/h. Whereas in the highway region, this paper records the driving video frames at three different vehicle speeds, i.e., 90,100, and 110 km/h. Each video clip contains 300 video frames, approaching 10 s video length.

The simulation video codec used in this paper is VidCoder. VidCoder is a kind of free software [5]. In this paper, VidCoder is carried out by a PC with Intel Core i5 2.2 GHz CPU, 8G RAM, and Window 10 64-bit OS. VidCoder supports MPEG-4 encoding with ffmpeg library. VidCoder performs H.264 and H.265 encoding via x264 and x265 libraries, respectively. This paper using Peak Signal-to-Noise Ratio (PSNR) to evaluates the encoding performances. The definition of PSNR is given as [4]

$$\mathrm{PSNR} = 10\log\frac{255\times 255}{MSE}(dB) \tag{1}$$

where MSE (mean square error) is defined as [4]

$$\mathrm{MSE} = \frac{\sum_{I=1}^{M}\sum_{j=1}^{N}\left(f_{i,j} - f'_{i,j}\right)^2}{M\times N} \tag{2}$$

In Eq. (2), the variables M and N are the frame size, e.g., $M = 720$, and $N = 480$. $f_{i,j}$ and $f'_{i,j}$ are the luminance pixel values at (i, j) position in the original and decoded video frames, respectively.

2.2 Experimental Results

Tables 1, 2, 3, 4, 5, and 6 show the MPEG-4, H.264, H.265 encoding performances at vehicle speed equals to 40, 50, 60, 90, 100, and 110 km/h, respectively.

Table 1. The MPEG-4, H.264, H.265 encoding performances at vehicle speed = 40 km/h

MPEG-4		H.264		H.265	
CP. Ratio	PSNR (dB)	CP. Ratio	PSNR (dB)	CP. Ratio	PSNR (dB)
5.31	39.29	5.91	39.7	8.44	39.74
7.43	37.89	8.83	38.57	12.42	38.77
9.78	36.76	12.99	37.46	17.99	37.82
12.24	35.89	18.76	36.36	25.71	36.87
14.69	35.18	26.89	35.24	36.51	35.91
17.68	34.56	37.86	34.16	51.64	34.94

Table 2. The MPEG-4, H.264, H.265 encoding performances at vehicle speed = 50 km/h

MPEG-4		H.264		H.265	
CP. Ratio	PSNR (dB)	CP. Ratio	PSNR (dB)	CP. Ratio	PSNR (dB)
5.16	39.63	5.64	40.05	8.09	40.25
7.08	38.25	8.40	38.93	11.80	39.27
9.20	37.15	12.30	37.81	17.00	38.30
11.32	36.28	17.75	36.70	24.54	37.34
13.47	35.58	25.56	35.57	34.87	36.36
16.07	34.97	35.83	34.46	49.29	35.37

Table 3. The MPEG-4, H.264, H.265 encoding performances at vehicle speed = 60 km/h

MPEG-4		H.264		H.265	
CP. Ratio	PSNR (dB)	CP. Ratio	PSNR (dB)	CP. Ratio	PSNR (dB)
4.85	39.20	5.51	39.64	7.69	39.72
6.77	37.77	8.08	38.46	11.10	38.70
8.61	36.62	11.68	37.28	15.74	37.68
10.63	35.72	16.65	36.10	22.22	36.67
12.71	34.98	23.65	34.93	31.43	35.66
15.13	34.35	33.22	33.76	43.99	34.63

Table 4. The MPEG-4, H.264, H.265 encoding performances at vehicle speed = 90 km/h

MPEG-4		H.264		H.265	
CP. Ratio	PSNR (dB)	CP. Ratio	PSNR (dB)	CP. Ratio	PSNR (dB)
4.42	38.92	5.35	39.33	7.15	39.51
6.05	37.47	7.74	38.09	10.06	38.44
7.84	36.31	11.08	36.86	14.14	37.38
9.67	35.40	15.70	35.62	19.75	36.31
11.45	34.65	22.25	34.41	27.60	35.24
13.51	34.01	30.82	33.19	38.70	34.17

Table 5. The MPEG-4, H.264, H.265 encoding performances at vehicle speed = 100 km/h

MPEG-4		H.264		H.265	
CP. Ratio	PSNR (dB)	CP. Ratio	PSNR (dB)	CP. Ratio	PSNR (dB)
4.30	39.10	5.04	39.59	6.78	39.76
5.83	37.64	7.21	38.33	9.52	38.68
7.56	36.47	10.28	37.09	13.27	37.62
9.30	35.55	14.49	35.84	18.42	36.55
10.93	34.79	20.45	34.61	25.73	35.48
13.08	34.15	28.45	33.38	35.99	34.41

Table 6. The MPEG-4, H.264, H.265 encoding performances at vehicle speed = 110 km/h

MPEG-4		H.264		H.265	
CP. Ratio	PSNR (dB)	CP. Ratio	PSNR (dB)	CP. Ratio	PSNR (dB)
4.18	39.01	5.08	39.40	6.74	39.62
5.65	37.53	7.23	38.13	9.38	38.53
7.30	36.34	10.21	36.85	12.95	37.45
8.96	35.40	14.34	35.58	17.82	36.36
10.66	34.64	20.17	34.30	24.70	35.27
12.56	33.98	28.17	33.05	34.17	34.18

Figure 1 shows the encoding execution time of H.265, H.264, and MPEG-4 with different PSNR. It follows from Tables 1, 2, 3, 4, 5, and 6 and Fig. 1, the sorting of encoding performance is H.265 > H.264 > MPEG-4. The performance gap between these three video codecs will be extended when the compression ratio is increase. Besides this outstanding bit-rate reduction, when compared to H.264 and MPEG-4,

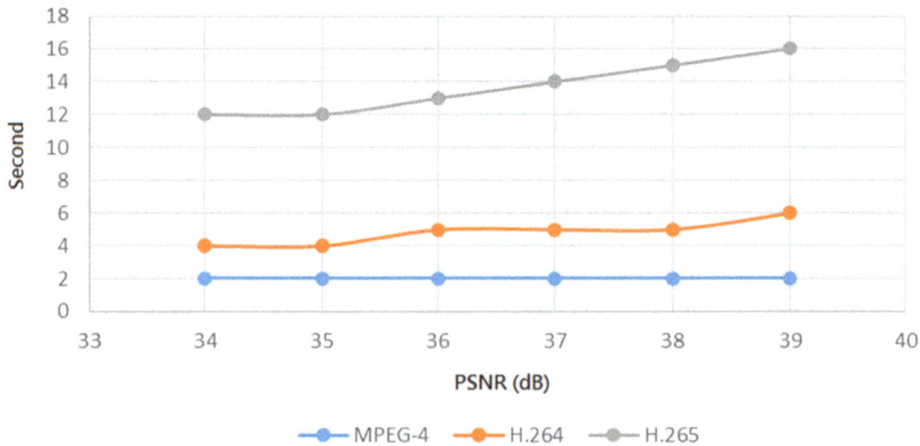

Fig. 1. The encoding execution time of H.265, H.264, and MPEG-4.

H.265 delivers a better visual quality with similar file size or bit-rate. However, the encoding execution time of H.265 is significantly longer than H.264 and MPEG-4. In this simulation, the average encoding time of MPEG-4 is within 2 s, H.264 is about 5 s, and H.265 reaches 15 s. It means that the computing power of H.265 will greater than those of H.264 and MPEG-4. Figure 2 shows the compression ratio and PSNR of MPEG-4, H.264, H.265 at vehicle speed equals to 40, 50, 60, 90, 100, and 110 km/h, respectively. The compression ratio of these three video codecs is similar to the statistical analysis of the PSNR values, whichever regardless of the vehicle speed.

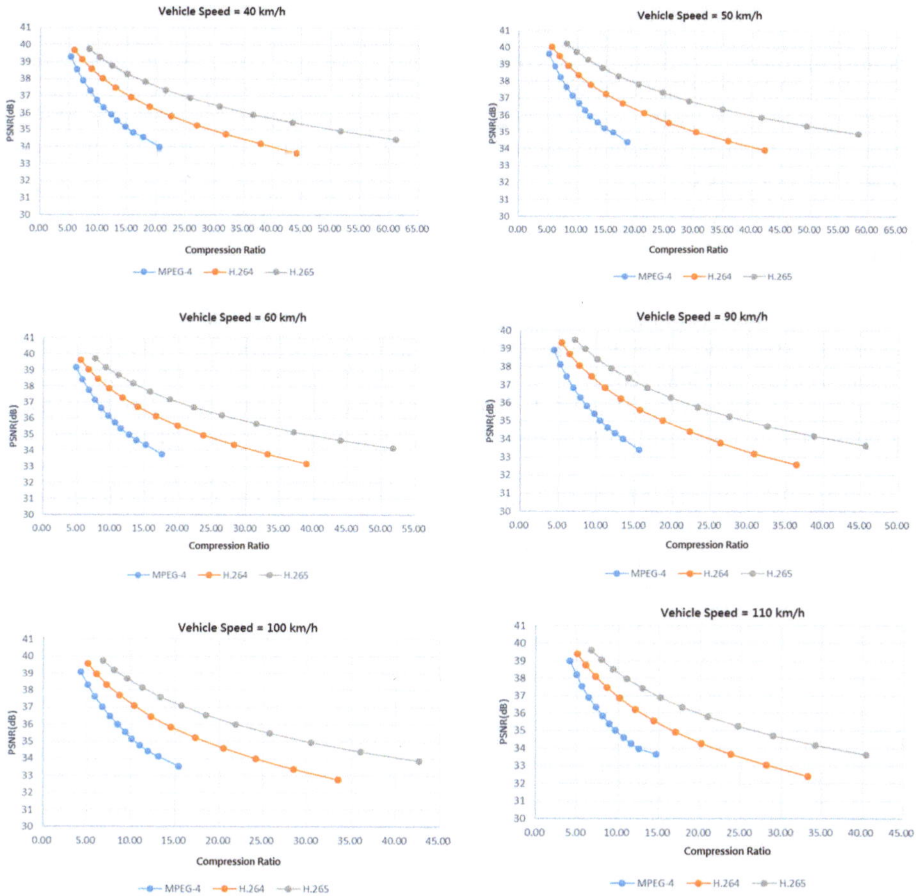

Fig. 2. The compression ratio and PSNR of MPEG-4, H.264, H.265 at vehicle speed = 40, 50, 60, 90, 100, and 110 km/h.

3 Conclusion

This paper simulates three different video codecs, i.e., MPEG-4, H.264, and H.265 to evaluate their coding performances on 6 types of driving videos captured from different vehicle speeds and road conditions. According to the experimental results, H.264 is the most suitable video codec for driving recorder at the present time. H.264 has the better trade-off between decoding video quality and encoding execution time. However, H. 265 has unbeatable better visual quality at a very low bit-rate. Once hardware manufacturers are starting to produce new IC to support H.265 codec, it is obviously H.265 will be the main video encoder for driving recorder in the future.

Acknowledgement. This study is conducted under the "Advance Smart IoV System Development Project" of the Institute for Information Industry which is subsidized by the Ministry of Economic Affairs of the Republic of China. This study is also supported in part by the Ministry of Science and Technology, Taiwan, under Grant MOST 105-2221-E-366-003-MY3 and MOST 106-2622-E-366-002-CC3.

References

1. Fleming, B.: Advances in automotive electronics. IEEE Veh. Technol. Mag. **9**(2), 4–19 (2014)
2. Wang, K., Huang, W., Tian, B., Wen, D.: Measuring driving behaviors from live video. IEEE Intell. Syst. **27**(5), 75–80 (2012)
3. Sullivan, G.J., Ohm, J.-R., Han, W.-J., Wiegand, T.: Overview of the High Efficiency Video Coding (HEVC) Standard. IEEE Trans. Circuits Syst. Video Technol. **22**(12), 1649–1668 (2012)
4. Gonzalez, R.C., Woods, R.E.: Digital Image Processing, 2nd edn. Prentice Hall, Upper Saddle River (2002)
5. VidCoder. http://vidcoder.codeplex.com/

Intelligent Railway Technology

Railway Train Working Diagram Plan System

Shaoquan Ni[1,2,3], Miaomiao Lyu[1,2,3(✉)], and Hongxia Lv[1,2,3]

[1] School of Transportation and Logistics, Southwest Jiaotong University, Chengdu 610031, China
lvmiaomiao@swjtu.cn

[2] National Railway Train Diagram Research and Training Center, Southwest Jiaotong University, Chengdu 610031, China

[3] National and Local Joint Engineering Laboratory of Comprehensive Intelligent Transportation, Southwest Jiaotong University, Chengdu 610031, China

Abstract. Research background and significance of railway train working diagram plan system is introduced, and system research process including six stages is proposed, and point out that at present, the train working diagram planning system research is in the fifth stage. Then, the train working diagram system 4.0 developed at present is introduced, and effect of the system are proposed. Furthermore, system key technology and main innovation are analyzed, including the unified algorithm applying for both single lines and double lines, the general data model and computers-supported cooperative work method for train working diagram plan system. Then, based on the constantly changing railway network construction and transportation market environment, the challenges of high-speed railway transportation organization and freight transportation organization based on limit delivery time are analyzed. Finally, research prospects of the system is put forward.

Keywords: Railway · Train working diagram · System

1 Research Background and Significance

Train working diagram connects the railway transportation network as a whole, it is the basis of railway transportation organization, and more, fully embodies the centralized use of railway resources, that is, whole railway network organization likes a game of chess.

This research was supported by the National Natural Science Foundation of China (Project No. 61403317, 61703351), National Key Research and Development Program of China (2017YFB1200702, 2016YFC0802208), Science and Technology Plan of Sichuan province (Project NO. 2017ZR0149,2017RZ0007), and Science and Technology Plan of China Railway Corporation (Project No.: 2016X006-D), and the Fundamental Research Funds for the Central Universities(2682017ZDPY04, 2682017CX022, 2682017CX018).

J.-S. Pan et al. (eds.), *Advances in Smart Vehicular Technology, Transportation, Communication and Applications*, Smart Innovation, Systems and Technologies 86,
https://doi.org/10.1007/978-3-319-70730-3_8

It is difficult to plan train working diagram because of characteristics that newly built high-speed railway lines and the existing lines are intertwined into a network, both passenger trains and freight trains run on the same railway line, transport capacity and demand contradictions prominent, etc. The traditional manual plan method and the early plan method by a stand-alone computer can't meet the needs of the train working diagram plan (TWDP) demand under the complex conditions of railway network transportation organization [1].

The only way for the railway to enhance its rapid reaction capability to cope with rapid changing market, is in-depth researching TWDP theory and method, and developing the railway TWDP system and improving the level of TWDP informatization and intelligence.

2 System Research and Application Review

2.1 System Research Process

Based on development law and process, six stages of TWDP system are shown as bellow [2].

Stage 1: Focus on solving specific problems by specific method, that is, according to the characteristics of different sections of different railway bureau, study the specific solution to solving the distinct individualized specific problem;

Phase 2: Committed to the study of the versatility and adaptability of the system, that is, trying to solve the train working diagram planning problem of different sections and different lines using the same system;

Phase 3: Establish a system to plan the train working diagram in the scope of railway bureau, the system is effective for single railway lines and double lines, automatic blocks and semi-automatic blocks, so, it is general, more functional, effective for the whole railway network, more, it can improve the efficiency of the planning train working diagram;

Stage 4: Starting from the goal of planning the national railway network train working diagram as a whole, focus on the establishment of the whole passenger train working diagram planning system to create the conditions for optimization of passenger train working diagram, and to improve the quality and efficiency of TWDP;

Stage 5: Study the technological means for plan the whole railway network train working diagram, establish the system supporting the remote network TWDP. Focus on elaborate, intelligent and systematic system construction technology, achieve to plan the train working diagram and other related plans cooperatively, establish an integrated management information system for internal railway, achieve the integration of information collection, management, query and comprehensive decision support for TWDP based on the computer network, and provide technical support for realizing the railway TWDP and management centralized, unified, efficient.

Phase 6: Achieve to plan the train working diagram automatically based on whole railway network, greatly improve the automation and efficiency and the level of information management degree of TWDP system.

At present, TWDP system research is in the fifth stage.

2.2 Characteristic of Train Working Diagram System 4.0

(1) Upgrade the system based on 64-bit operation system, and solve the problem of data management capacity constraints in version of system 3.0, and it is beneficial to improve planning and dynamic management, and to realize train working diagram multi-program comparison.
(2) Construct train working diagram planning and management integrated information platform, including technical data management, all plan-making related trains operation, for example train operation plan, vehicle (EMU) scheduling plan, locomotive turnover plan, vehicle distribution plan, station operation plan, and train operation traction calculation, etc., train operation diagram version management, train operation information sharing, train operation simulation and verification, furthermore, consider the information exchange and sharing demand with other railway information system.

2.3 Summary of Research and Development Process

A suitable research and development process for planning train working diagram by computer in mainland is gradually explored, that is from the pursuit of establishment train working diagram automatic and optimization planning system, to strive for building decision-making and human-computer interaction system for practical use and ensuring the quality of plan, and from plan train working diagram by a stand-alone computer in the scope of local area, to collaborative making the whole railway network train operation diagram based on computer network.

2.4 Consequence of Planning Train Working Diagram by Computer

(1) Realize the revolution from traditional way of planning train working diagram manually to planning computerized;
(2) Realize the revolution of planning train working diagram by networking computers-supported cooperative work instead of planning by a stand-alone computer.

2.5 Effect of Planning Train Working Diagram by Computer

(1) Improve the efficiency of planning train working diagram greatly and shorten the time;
(2) Improve the quality of TWDP effectively by changing plan mode from individual closed plan to computers-supported cooperative work.

3 System Key Technology and Main Innovation

3.1 Key Technology Points

(1) The unified algorithm applying for both singer line and double line TWDP is proposed, laid the technical foundation for planning whole railway network train working diagram.
(2) The general data model of the train working diagram is established, laid the technical foundation for the data management of the railway network train working diagram.
(3) The computers-supported cooperative work method is built [3], laid the technical basis for networking TWDP.

3.2 Main Innovation

(1) Develop the whole railway network TWDP system, realize TWDP by computer instead of by manual.
(2) Establish of the computers-supported cooperative work technology, make a breakthrough of networking TWDP.

Study a series of TWDP system key technologies such as the shared control structure, the cooperative mechanism, the concurrent control method and the access control strategy, and establish sets of computers-supported cooperative work technology for TWDP system. Then, propose a hybrid structure with a combination of concentration and distribution, and proposes two kinds of cooperation modes for TWDP, they are "cooperation under centralized control" and "equal cooperation" [4]. System concurrency control hybrid control strategy and technical solution which are integrated by improved full-order centralized control method, the locking method and concurrency control method based on operation transform. An improved role-based [5] train operation diagram access control method is proposed by improving the regional attributes and virtual rights of the subject and object.

(3) Establish a computers-supported cooperative work mode, break the traditional way of closed work, and achieve revolutionary changes in TWDP.

As a result of data synchronization, planners can make and adjust detailed TWDP based on the entire railway network, break the traditional mode in make TWDP on two minutes grid based on solidified long distance passenger train rough plan.

(4) Propose TWDP intelligent solution based on hierarchical decision-making and satisfactory optimization.

The hierarchical decision model and satisfactory optimization model are established, and TWDP method based on state space method and heuristic search technique is constructed.

(5) Construct the planning technology applying for both single lines and double lines, break difficulties of planning network train working diagram by computer.

Different train operation organization modes apply for different railway lines, railway has two modes including two-way traffic in single-line and one-way traffic in double-line, how to design unified algorithm suitable for both modes is the primary problem of planning network train working diagram.

(6) Establish a set of integrated information platform including all related train operation plans, realize the system integration and innovation, improve the refinement degree of TWDP, and it is beneficial to comprehensive optimization of TWDP.
(7) Establish data management technology based on the entire railway network TWDP.

Put forward data structure design method of TWDP system based on the railway interval, and establish the data management method based on the user role. And, system has a high degree of automation in system data consistency maintenance, and supports all train operation data reporting, download, and realizes the data dynamic management. TWDP system is tightly coupled system, all the data is one-time read into the computer memory, and has high operating efficiency.

(8) Upgrade TWDP system based on 64-bit operating system, greatly enhance the system's processing power.

With the expanding of railway network, system memory capacity is a technical bottleneck. "TWDP system 4.0" is developed on 64-bit operating system, solve the problem of data management capacity constraints based on 32-bit system, improve the system's ability to adapt to the development of the railway, and the system's life cycle can be greatly extended.

4 Challenges Based on High-Speed Railway Network Operations and Railway Logistics

Based on the constantly changing railway network construction and transportation market environment, the challenges of high-speed railway transportation organization and freight transportation are shown below.

4.1 High-Speed Railway Transportation Organization

(1) Transportation organization model

High-speed railway passenger flow consists of both types including passenger only travelling on one high-speed railway line and passengers travelling across several lines. At present, long distance trains are mainly adopted for cross-line passenger flow, the train organization mode is the same as the traditional way. High-speed railway train running fast, so travelling time on the railway network is greatly reduced, in order to operate more trains to meet the demand of passengers, high-speed railway transportation organization organizer should develop new transportation organization model that is direct transportation combined transfers connection for cross-line passenger flow in the future.

(2) Planning mechanism

The key of the high-speed rail transportation organization is to coordinate the relationship between the railway transportation organization and the transportation marketing, establish a new mechanism for the comprehensive, rapid and accurate response to the change of the transportation market is very important. In the future, high-speed railway operation plan should be made based on the market-oriented, integrated design ideas.

(3) Planning technical characteristics
① Synergy integration

High-speed railway operation plan should be made based on market-oriented, and rational use of transportation capacity resources, and passenger flow forecasting and train plan, train working diagram, EMU scheduling should be designed integrated.

② Dynamic

Railway passenger flow has seasonal, volatility characteristics, passenger flow has sharp ups and downs fluctuating in one year, one quarter, one week, or even within a day, planners must make regular train working diagram, weekend train working diagram and peak period train working diagram by season and year according to passenger flow rule.

③ High precision

In order to ensure the high quality of high-speed railway service level, high-speed trains must have a high punctuality rate, which requires operation plan must be highly reliable and implementable, plan must be made comprehensively, accurately and detailed.

④ Intelligent

Newly built high-speed railway and existing regular railway has been integrated, complex railway network has already been constructed. In order to adapt quickly to market changes, it is necessary and urgent to realize high-speed railway operation plan intelligent, in order to shorten the make period of high-speed railway operation plan, and improve plan's quality.

4.2 Freight Transportation Organization Based on Limit Delivery Time

(1) Background of the challenge

The traditional railway transportation is centered on the internal operation efficiency of railway organizations, and there is no good balance between customer needs and internal railway organization, and also no time to attend to delivery time which is the goods owner most concerned about. With the development of globalization and modern logistics industry, the demand for delivery goods quickly becomes strong, the transportation market requires service concept of transportation organizer from "producer as the center" to "consumer as the center", railway transportation must meet the goods owner' needs.

(2) Solution

Based on limit delivery time freight transportation organizations should be "market-oriented, customer-centric", basic plan must be adjusted according to recent railway freight market demand and the dynamic changes of railway transportation resources, the plan based on this formation is the implementation plan. The implementation plan is adopted to guide the daily scheduling command work, and it should be adjusted by daily operation plan to ensure the precision of implementation plan, and the demand of freight limit delivery time transportation can be achieved by the way.

The implementation plan must include freight full transportation process, and must be made when approve the goods owner's transportation request, so it can provide real-time tracking, reliable, timely delivery transportation services for goods owners. To this end, existing transport organization method must be improved and redesign. The key is to establish the concept of railway freight transportation organization based on limit time delivery, to study its transportation organization method and carry out the transportation organization process reengineering, and to speed up the construction of intelligent railway transportation information system.

5 Research Prospects

Based on the current research and development of the system, combined with the railway network establish and operation organization revolution, we will study the following aspects, including optimization of long distance train operation scheme which is crossing some railway bureau' scope, optimization of window maintenance design scheme under the railway network condition, cooperative intelligent optimization of train plan and train working diagram, cooperative intelligent optimization of train operation plan and EMU scheduling, optimization of high speed railway periodic train working diagram under condition of railway network, optimization technology of the railway freight train operation diagram based on the limit delivery time.

References

1. Ni, S., Lu, H., Zhang, J., Chen, T.: Research on concurrency control method of train working diagram compilation system based on group coordination. Sciencepaper Online **4**(10), 738–741 (2009)
2. Ni, S.: The Train Working Diagram Making System of Railway. Southwest Jiaotong University, Chengdu (2013)
3. Yu, L.: Study of Common Integrated Framework Supporting Group Work. Chinese Academy of Science, Beijing (2001)
4. Shi, M., Xiang, Y., Yang, G.: Computer Supported Cooperative Work Theory and Application. Electronic Industry Press, Beijing (2000)
5. Ye, C.: Study on Attribute Restricted Delegation Model in Role Based Access Control. Chongqing University, Chongqing (2005)

Railway Passenger Service Mode on "Internet+"

Tao Chen[1,2,3], Zhen-ning Zhou[4], Jie-ru Zhang[1,2,3], and Shao-quan Ni[1,2,3(✉)]

[1] School of Transportation and Logistics, Southwest Jiaotong University, Chengdu 610031, Sichuan, China
nishaoquan@126.com

[2] National Railway Train Diagram Research and Training Center, Southwest Jiaotong University, Chengdu 610031, Sichuan, China

[3] National United Engineering Laboratory of Integrated and Intelligent Transportation, Chengdu 610031, Sichuan, China

[4] Yue-hai Railway Limited Liability Company, Hankou 570125, Hainan, China

Abstract. The development of "Internet+" railway passenger service mode has important significance under the new situation of economic development. At first, this paper uses the service chain theory to analyze railway passenger service demand, content and elements comprehensively. Then, the existing railway passenger service mode of the content, characteristics and existing problems are discussed. At last, for realizing the goals for providing a full range of passengers services, extension services, advanced services, an new railway passenger service mode based on "Internet+" is proposed, and railway passenger service information platform technology architecture is put forward based on cloud computing. Thus, a reference for the railway passenger service to achieve "Internet+" can be proposed.

Keywords: "Internet+" · Railway passenger service · Mode · Service chain · Cloud computing

1 Introduction

The "Internet+" was first proposed by contemporary ecological thinker Mr. Zhang Ronghuan in 2007. It refers to the application of Internet thinking and Internet technology to various traditional industries. And we should give full play to the optimization and integration of the Internet in the allocation of traditional industry resources, improve the innovation and productivity of traditional industries based on

This research was supported by the National Natural Science Foundation of China (Project No. 6140 3317, 61703351), National Key Research and Development Program of China (2017YFB1200702, 2016YFC0802208), Science and Technology Plan of Sichuan province (Project NO. 2017ZR0149, 2017RZ0007), and Science and Technology Plan of China Railway Corporation (Project No. 2016X 006-D),and the Fundamental Research Funds for the Central Universities (2682017ZDPY04, 26820 17CX022, 2682017CX018).

J.-S. Pan et al. (eds.), *Advances in Smart Vehicular Technology, Transportation, Communication and Applications*, Smart Innovation, Systems and Technologies 86,
https://doi.org/10.1007/978-3-319-70730-3_9

the Internet platform. Since 2015, the "Internet+" action plan has been promoted to the national strategic height, and all industries have become increasingly pay attention to internet [1].

Under this circumstance, the "Internet+" also has gradually been applied on railway passenger service, and the building of 12306 domestic passenger service website is more outstanding. This website has improved the convenience and satisfaction of the passenger ticket purchase, and obtained better benefits. But in general, the Internet of passenger service is still not high. In terms of passenger product development, promotion of passenger service mode, improvement of passenger service quality, passenger service business development and other aspects, there is still a large gap between passengers' expectations. A complete Internet passenger service system has not yet been established. And the development of railway passenger service based on "Internet+" has been proved to be the innovation and upgrading of traditional passenger service mode. At the same time, as the internal demand to improve its own competitiveness, its development model needs to be further studied.

This paper focuses on the goal of the Internet of passenger travel service. Firstly, the service chain theory is used to study the content, scope and mechanism of passenger service. Then, this paper combines the "Internet+" technology with the railway passenger service chain, and puts forward the system, content and mode of "Internet+" railway passenger service mode, to promote the railway passenger service implementation "Internet+" for reference.

2 Railway Passenger Service Chain Model

Traditional passenger travel chain demand analysis, customer relationship analysis and other methods mainly focus on customer demand as the center to promote and optimize the product structure and production mode of passenger service enterprises. Because of the focus on the passenger service enterprise itself, it is not conducive to the expansion of the service depth and breadth of passenger service.

The service chain aims to maximize the overall and optimal service of the passengers, and organize various agencies or elements related to the service to form a complete network chain structure [2]. The service chain is essentially the complete service of all service organizations participating in the travel of passengers. Due to the focus on other service organizations and elements including core passenger service enterprises, therefore, it is easier to explore and develop active services, full-service services, extension services and advance services on the basis of integrating all service resources, and more suitable to play the characteristics of "Internet+".

Analyze the relationship between passenger demand, passenger service and service elements by constructing a service chain model, the model construction method proposed by Ref. [3] was established, and the railway passenger service chain was constructed as shown in Fig. 1. Railway passenger service can be divided into pre-service, interim service and post-service. Pre-service refers to the services that passengers need before the bus, including information consultation, policy consultation, travel planning, booking tickets, insurance and other services. The interim service refers to the services that passengers need during their trip, including short haul, station service, train service, business services, etc. Post-service refers to the service required by the passenger after the end of the ride, including the service evaluation, service remediation, etc. The railway enterprise is the core service provider in this service chain, including government supervision departments, public transport departments, intermediary agents, insurance companies, bank credit, business services enterprises and other service agencies. These service agencies cooperate with each other to provide complete travel services for passengers.

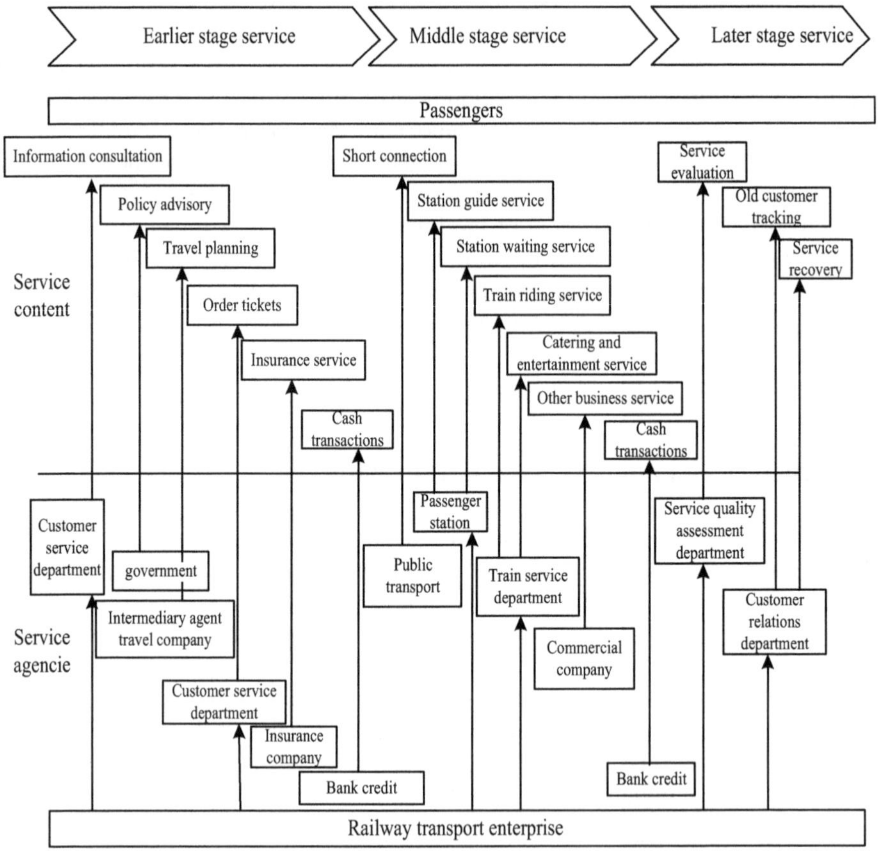

Fig. 1. Rrailway passenger service chain model

3 "Internet+" Railway Passenger Service Mode

3.1 Current Railway Passenger Service Model Analysis

The current railway passenger service model has the following problems: (1) The railway passenger service scope is still limited to the traditional services such as the order ticket service, station service and train service, and other extension services. (2) Despite the introduction of 12306 online booking system, the railway passenger service mode is mainly based on human services and equipment guidance. The degree of automation and autonomation of passenger service need to be further improved. (3) There is a lack of personalized service for travelers. (4) The business model of the free business model mainly lacks the overall planning and connection. The traditional passenger railway passenger service model is shown in Fig. 2.

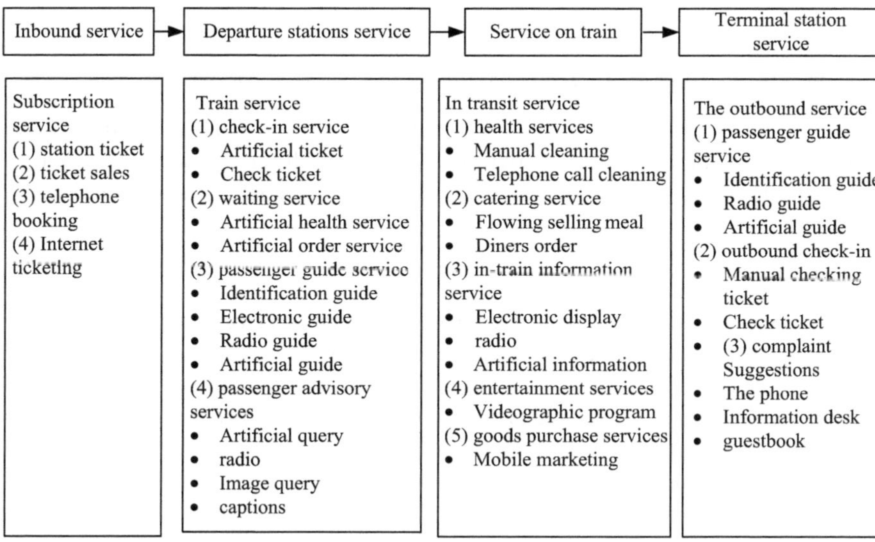

Fig. 2. Current railway passenger service mode

3.2 "Internet+" Railway Passenger Service Model Characteristics

The railway passenger service chain model shows the full demand of the passenger, and also shows the network connection between the service providers. "Internet" technology and platform can support and realize the construction of railway passenger service chain. The railway passenger service model under "Internet+" shall include the following characteristics:

(1) Cross-border integration. The "Internet+" can remodel the different areas and industries in the railway passenger service chain, and the cross-border integration will make the railway passenger service base more solid. Integration can also refer to the integration of identity, the passenger even turn into the designer of the service.
(2) Reshape the structure. Information revolution, economic globalization, Internet and other technologies are constantly improving the original industrial structure. As a new kind of innovative thinking and technology, "Internet+" will also reshape the traditional railway passenger service structure.
(3) Respect human nature. The starting point of the "Internet+" model should be respect for human nature. Only in this way, Internet technology can truly promote the innovation and progress of railway passenger service. The "Internet+" railway

passenger service model can utilize cloud computing and big data mining to find and meet the personalized needs of passengers.

(4) Connect everything. The "Internet+" model makes it easier to connect people, people and things, greatly reducing the cost of communication. The "Internet+" railway passenger service model should also make full use of Internet platforms and Internet of things technology to enhance the connectivity of passengers and services.

3.3 Design of "Internet+" Railway Passenger Service Model

Relying on the Internet platform, cloud computing, big data mining, Internet of things technology, etc., aiming at the shortcomings of the traditional model of railway passenger service, based on the passenger service chain, design "Internet+" railway passenger service mode.

(1) Pre-trip service

① Create customized products for passenger traffic service and passenger service
Using passenger service platform and large data analysis technology, analyze different travelers' favorite tourism products, customize personalized railway tourism products for tourists, and make travel plans for every passenger. Travelers can choose their favorite travel plans based on their own preferences. The following activities are jointly arranged by railway companies and tour companies, hotels, and short-haul operators, including hotel accommodation, scenic spot tickets, round-trip tickets and so on. Carry out the railway online tourism business and make full use of the internal resources of the railway, including railway passenger transport products, railway hotel resources and tourism resources. At the same time, it combines external resources to create special railway travel products and expand passenger service content.

② Introduce a variety of purchase scheme base on market
According to market demand and passenger demand data, a variety of passenger ticket forms such as annual ticket, monthly ticket, special discount ticket and membership ticket are introduced at any time to enhance the attraction of railway travel. In order to facilitate the journey of the passengers, carry out the joint tickets with public transport, and travel companies. Enhance the passenger ticket service level through online selection of seats.

③ Provide customized passenger service according to passenger demand
Booking catering online means that passengers can book their meals online while booking tickets online. Travelers can also choose not to order catering services, and only choose to order online during the trip. Besides, hotel booking service, entry and exit taxi rental can be booked online. According to the passengers' demand, special passengers like minors, old people, sick passengers and so on will be escorted. Passengers can compare online and buy insurance at any time.

(2) Station and train service in passenger travel

① Using Internet of things technology to build smart station [4]
Make full use of Internet technology to build digital smart station. Use perception, navigation, positioning and other wireless Internet technologies to upgrade the station's facilities and improve the quality of passenger service. When the passengers arrived at the station, simple layout of the station should be pushed to passengers' mobile phone. This will allow passengers to easily understand the location of the waiting room, the entrance passage, the ticket office, the shopping area, the toilet, etc. It is easy to find the location of the ticket office or waiting room through the navigation of the mobile phone, and convenient for passengers. Advertise station client inside, outside the station and on the train. Or post WeChat qr codes in bold locations, and scan the qr code to download the client. By paying attention to these things, travelers can use their mobile phones to quickly query the information of the car and the delay time of the trip.

② Use a smart ticket to assist travelers throughout the journey
The intelligent identification chip is installed in the passenger ticket, and each passenger ticket will temporarily store the information of the passengers' ride. The read-write recognition device can be placed in the various areas of the station and train, passengers can easily enter the station by placing the ticket on the sensor area for identification. Passengers will receive all kinds of prompt guidance information, train information and problem solving at the same time. Tickets will automatically show the seats or seats reserved for passengers when they get on the train. The client will send the station name and the delay time according to the train ride to passengers, and then push the destination information accordingly. When the passengers get to their destination, the tickets will send a notice of impending arrival to give travelers a rough idea of where they will arrive. When the passengers arrive, collect the tickets and empty the passengers' data.

③ o2o business services
The railway can develop the convenience store function of the APP, which can be operated on the Internet or on the mobile phone client according to the need, and browse or order the goods on the station or train. After that, the service personnel will directly send the items to the passengers, and the passengers can choose to pay for the goods online or on delivery. The business around the station can also integrated into the railway online shopping mall. Passengers can order online, consume or take within the station.

(3) Passenger later period service

① Push the customer service evaluation list through mobile phone client to set up rewards or gifts. Regularly visit regular customers and give gifts. According to the customers' return visit data, the passenger products are analyzed and evaluated, and provide data support for train operation plan and passenger service products.
② Set up 24 h of online consultation service and accept service dispute. Push service disputes to customer service department leaders, and quickly develop service recovery plan for passengers. Supervise the progress of service disputes.

③ Regularly push changing travel schedule, ticket price information, service information, travel recommendation information, etc.

The "Internet+" railway passenger service model is shown in Fig. 3.

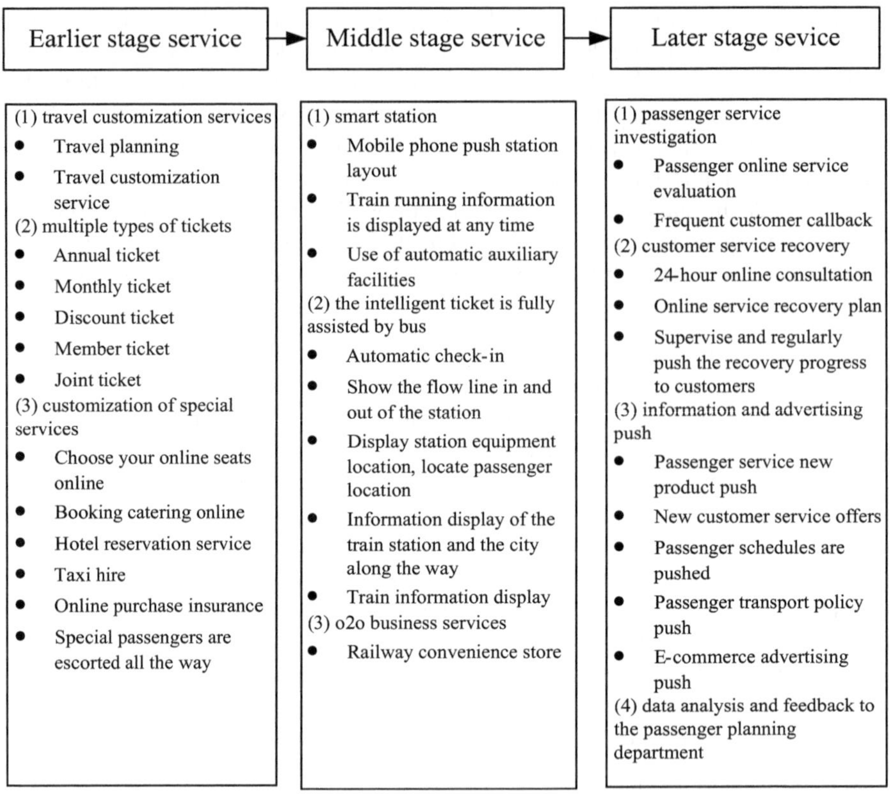

Fig. 3. "Internet+" railway passenger service mode

3.4 The Overall Technical Framework of the "Internet+" Railway Passenger Service

The cloud computes a huge computing processor over the network and hands it over to multiple servers and returns the results back to the user. It can calculate the large amount of data that the user sends out in a short time, the computation speed and the stability have very good performance. In addition, the SOA architecture system in cloud computing can easily combine loosely coupled services to provide better and faster responses. The number of railway service passengers is large, and the number of data that demands response at the same time is large, besides, the service chain involves many enterprises and institutions. Therefore, it is fast and convenient to use cloud computing to frame railway passenger service information platform.

According to the "Internet+" railway passenger service model, a railway passenger service information platform based on public cloud is established as the core system of various business joint operation. And develop various passenger service application systems (order tickets, travel plans, quality feedback, etc.) on this information platform. Railway passenger service information platform provide a unified public business interface for big business, supply chain (such as tourism, hotels, catering companies, etc.) The third party business system of the bank government agency to cost savings by renting railway passenger service platform application system. The other information systems in the railway connect to the railway service information platform through the interface and exchange data with the information platform. The overall technical architecture of the cloud-based railway passenger service platform is shown in Fig. 4.

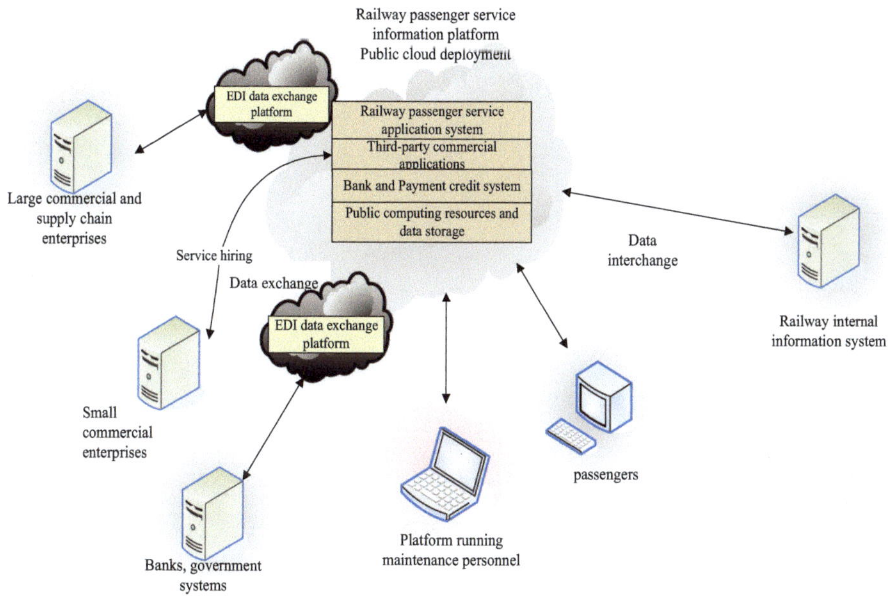

Fig. 4. The overall technical framework of the cloud-based rail passenger service platform

4 Conclusion

The development of "Internet+" railway passenger service model is of great significance for the railway to adapt to the economic development under the new situation. This paper analyzes the demand, content and constituting elements of railway passenger service by using the theory of service chain, and discusses the characteristics and shortcomings of the existing railway passenger service model. In order to provide passenger service, extension service and advance service, this paper proposes a new model of railway passenger service based on Internet technology and the realization of technical architecture. It provides reference for the realization of "Internet+" of railway passenger service.

References

1. The State Council: National Issue (2015) 40, Guidance notes on actively promoting "Internet+" action. The State Council, Beijing (2015)
2. Liu, Q., Yin, H.: Construction and implementation of service chain model. Technol. Prog. Countermeasures **03**, 23–26 (2011)
3. Hu, Z., Ning, X.: Concept, model and application of service chain. Bus. Res. **07**, 111–114 (2003)
4. Shan, X., Wang, F., Li, F., Li, P.: The research and design of railway "Internet+" passenger service. Railway Comput. Appl. **11**, 5–8 (2015)

Design of High Speed Train Operation Plan Simulation System

Dingjun Chen[1], Xinyi Lin[2], Hui Zhang[1], and Kaiteng Wu[3(✉)]

[1] School of Transportation and Logistics, Southwest Jiaotong University, Chengdu 610031, Sichuan, China
[2] School of Information Science and Technology, Southwest Jiaotong University, Chengdu 610031, China
[3] Key Laboratory of Numerical Simulation of Sichuan Province, Neijiang 610031, China
wukaiteng@263.net

Abstract. In view of the existing high-speed railway train operation plan system cannot accurately, comprehensively and intuitively reflect the transport equipment operating load and ability to adapt to the situation, cannot reflect the train group operation, operation diagram, station work plan between the complex interactive relationship, a train operation simulation module is designed in this paper which through the construction of high-speed railway traffic related electronic map and technical standards database, computer simulation on the train diagram has been made. Then, a high-speed railway operation plan simulation system is developed, which includes train planning, traction calculation and train operation simulation. Finally, simulation results show that high-speed railway operation plan simulation system can accurately simulate the whole process technical operation of the train group operation, operation process, generate statistical analysis results, thus we can get real-time visual inspection of the adaptability of operating equipment ability in the operation scheme, and provide strong support for the work of high-speed train running organization. It is of great significance to improve the level of high-speed railway traffic organization.

Keywords: High-speed railway · Trains · Simulation · System

1 Introduction

Through the train diagram, the railway system links the transportation and production activities of the whole road network into a unified whole. Train diagram is the comprehensive plan of railway transportation work, the basis of railway traffic organization,

---

This research was supported by the National Natural Science Foundation of China (Project No. 6140 3317, 61703351), National Key Research and Development Program of China (2017YFB1200702, 2016YFC0802208), Science and Technology Plan of Sichuan province (Project NO. 2017ZR0149, 2017RZ0007), and Science and Technology Plan of China Railway Corporation (Project No. 2016 X006-D), and the Fundamental Research Funds for the Central Universities (2682017ZDPY04,268 2017CX022, 2682017CX018) and key laboratory of Numerical Simulation of Sichuan Province (20 17KF001).

J.-S. Pan et al. (eds.), *Advances in Smart Vehicular Technology, Transportation, Communication and Applications*, Smart Innovation, Systems and Technologies 86,
https://doi.org/10.1007/978-3-319-70730-3_10

and the tool to coordinate the railway departments and units to carry out production activities according to certain procedures. The quality of its preparation has a direct impact on the safety, efficiency and service level of the railway transportation.

With the development of economy and society, the railway system needs to improve its adaptability to market changes, thus the period of train diagram scheduling tends to shorten. The train diagram working system is developing and improving continuously, which lays a good foundation for the optimization of train diagram [8]. However, the existing train operation diagram does not verify the capability of the transport equipment, thus it may affect the feasibility of the operation diagram.

With the development of computer information technology and railway operation management technology, the simulation of railway operation has been proved to be an effective tool for analysis and has been widely used. At present, in the field of railway simulation the research work abroad mainly focuses on the capability simulation of railway network, optimization simulation of passenger line network, simulation of vehicle structure, etc., and obtained certain research results. Azadeh proposed a railway system with integrated simulation, multivariate analysis and multiple decision analysis model in [1], which is based on data packet analysis in computer simulation and it selects optimal options when considering a large number of inputs and outputs. In [2] Abirl simulated the Spanish railway signal and the robustness of train diagram, and derived its impact on the carrying capacity. Feng Li studied how to deal with the train scheduling problem on the single-track railway in [3], got the improved simulation model to reduce the running time, and designed an algorithm for obtaining a more effective operation strategy based on the train global information. Ping Gao in [4] proposed to determine the importance level of railway locomotive vehicle using Monte Carlo simulation technology. Domestic research on railway simulation includes train operation status, station operation system simulation and so on. Chen Gang studied simulation method of the operation system of marshalling station as well as modeling technology, research structure of simulation system and simulation algorithm and developed a dynamic simulation software for marshalling station operation system in [5]. Pan Jinshan et al. in [6] have established the simulation model based on the GIS technology, the railway line and the station basic geographic information data, and on this basis carried on the train movement simulation. The model of railway station yard diagram model was studied by Lu Gongyuan in [7], and established a railway station arrival and departure system simulation model based on timed colored Petri nets.

In this paper, a train operation simulation integration module is designed by constructing the relevant electronic map and technical standard database of high-speed railway, which put the train operation planning, the traction calculation, the train operation simulation in an organic whole, thus enhanced the fine level and the management level of train operation plan.

2 Architecture of Train Operation Plan Simulation System for High-Speed Railway

The overall architecture of the train operation plan simulation system of high-speed railway is mainly composed of "three systems" and "three levels". The three systems include: standard specification system, operation management system and security guarantee system. The three levels include: computer network platform layer, data resource layer and application layer. In the architecture, the structure and function of each layer are relatively independent, and the next layer provides support and services for the upper layer (Fig. 1).

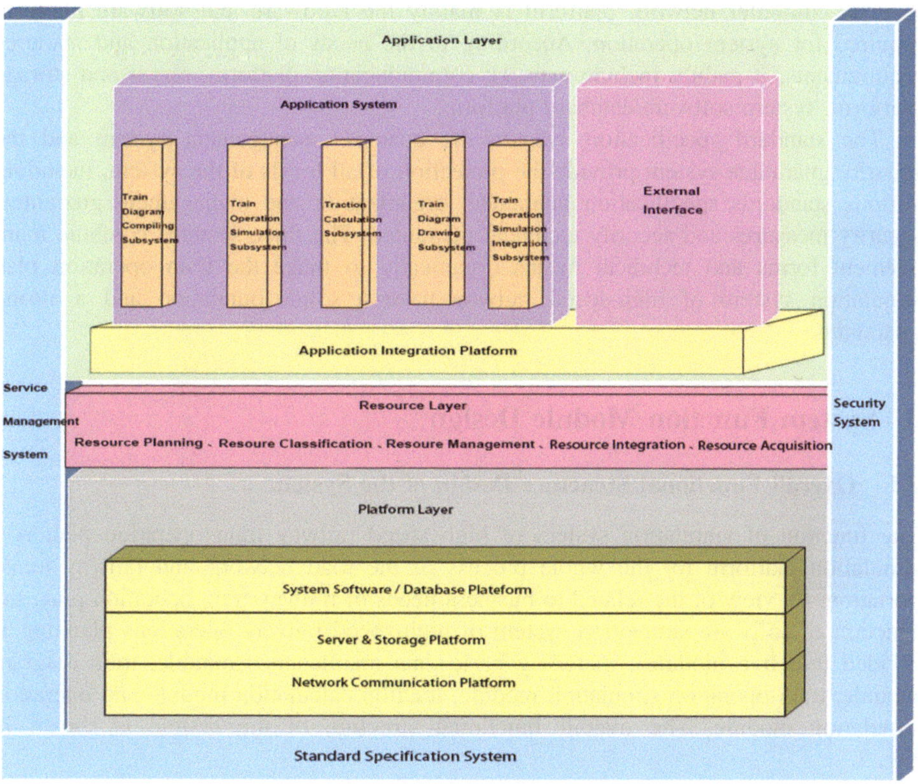

Fig. 1. Framework of overall system

Application layer is a collection of related business logic application system of high-speed railway operation plan. According to the needs of the business, various application systems are integrated, including train diagram preparation subsystem, train simulation subsystem, operation drawing subsystem, traction calculation subsystem, etc. In the process of system construction, a unified application integration platform is established, which can improve the sharing degree of each business data, simplify the business process, automate various processing processes and optimize the existing business processes.

The core task of the data resource layer is to utilize and develop the resources involved in the high-speed railway map. In Accordance with the actual requirements of the business system, resources construction is carried out under the guidance of the standard specification system.

The computer network platform is mainly the hardware and software platform required for system operation. According to the needs of application and resource requirements, it mainly include network communication platform, server and storage platform, system software/database platform.

The standard specification system, the business management system and the security guarantee system provide the protection of all levels of the system, including various standards specification guarantee, organization and management guarantee, security measures and security technical guarantee. The three systems combine management forms and technical means organically to make the train operation plan simulation system of high-speed railway have a solid foundation and a strong guarantee.

3 System Function Module Design

3.1 Overall Functional Structure Design of the System

The function of simulation system of high speed railway train operation plan is a simulation platform for the whole process of the road network and railway transportation. IN view of the related to business needs of railway train operation plan, the function of software simulation system of high-speed railway operations planning is divided into five modules, including basic data maintenance module, train diagram module, train operation simulation module, traction calculation module and operation simulation module. The overall functional structure of the system is shown in Fig. 2.

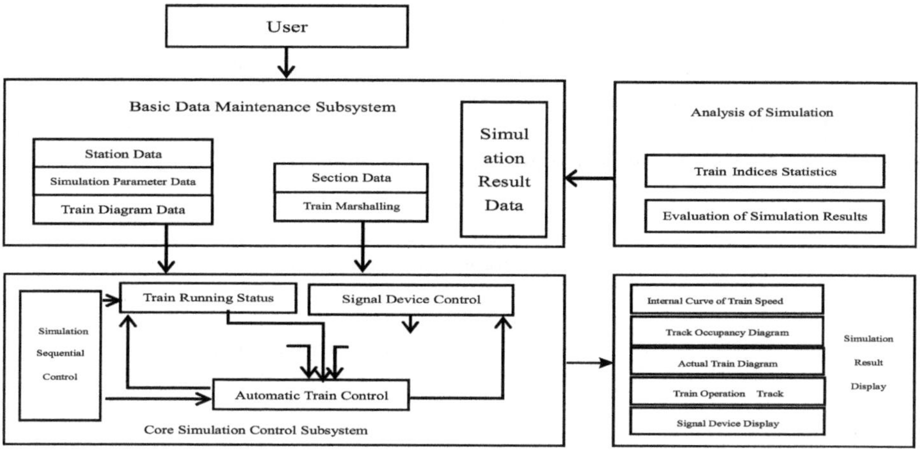

Fig. 2. System functional structure diagram

3.2 Design of Basic Data Maintenance Module

The basic data maintenance module is to maintain all the basic data needed for the simulation system, including train group data, train diagram data, traffic line equipment data, etc., it also contains the train running simulation data for simulation playback and simulation results analysis. Its function is shown in Fig. 3 below.

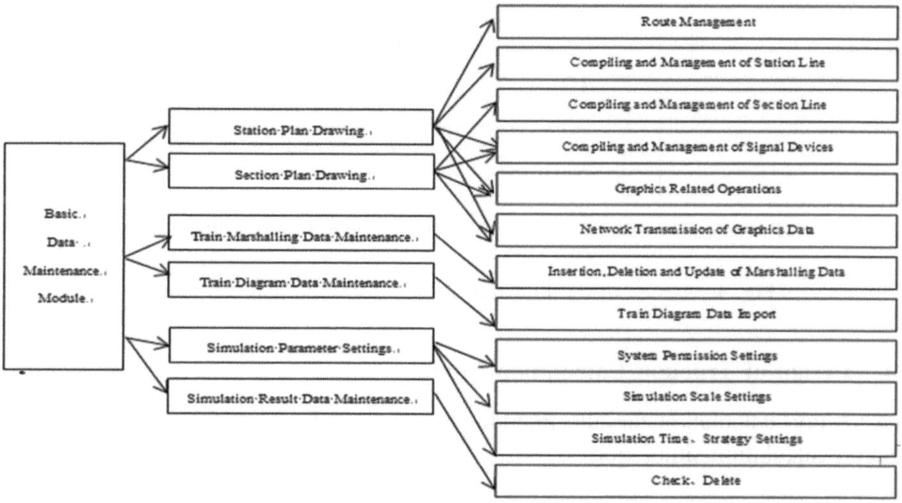

Fig. 3. Design of basic data maintenance module function

3.3 Design of Train Diagram Module

The train diagram module is used to realize the management of the relevant information of the train operation plan and the optimization of it, and the adjustment of man-machine interaction, the train running plan drawing, the train timetable generation and such function as the statistical analysis of the index (Fig. 4).

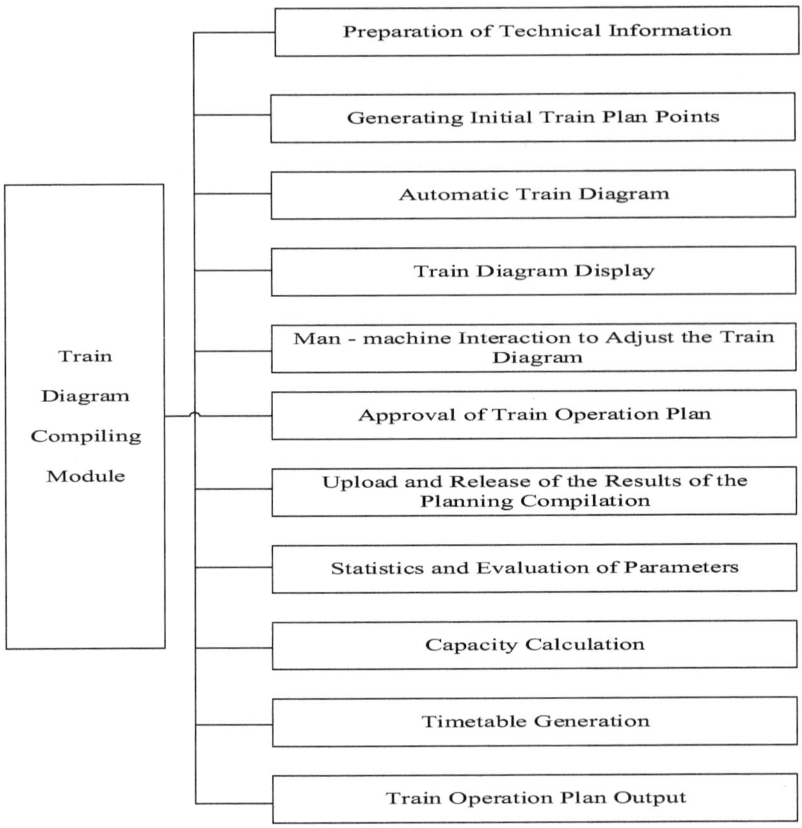

Fig. 4. Design of the function of train operation diagram module

3.4 Design of Traction Calculation Module

The traction calculation module includes two main functions: data management and traction calculation simulation.

(1) Data management: This section manages and maintains all the basic data required by the simulation system, including the line data, the motor group data, marshalling data and parameter data. Its functional design is shown in Fig. 5 below.

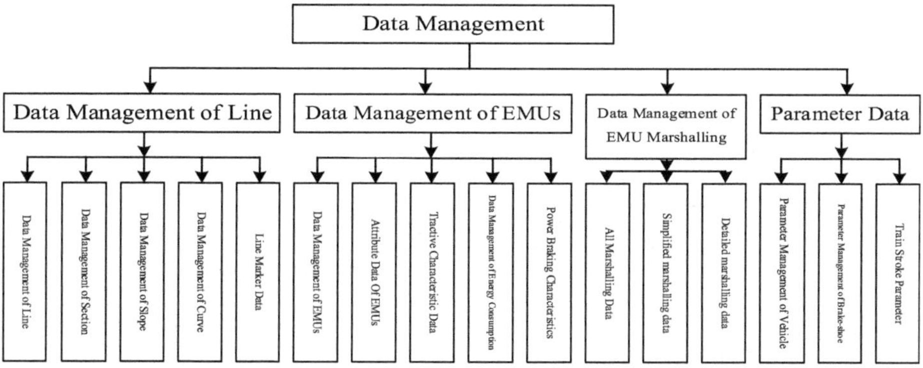

Fig. 5. Functional design of data management

(2) Simulation of traction calculation: This is the core part of the train operation simulation system. It realizes the control of the train operation process and its function design is shown in Fig. 6:

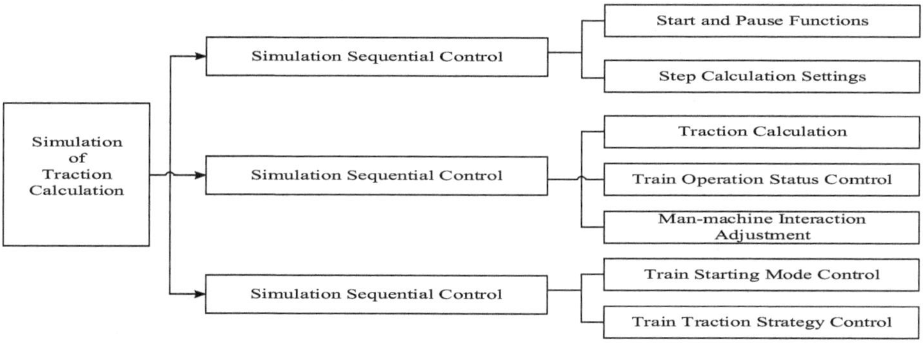

Fig. 6. Functional design of traction calculation simulation

3.5 Design of Operation Simulation Module

The operation simulation module is the core part of the simulation system of the high-speed railway train operation plan, including the train running simulation control and the simulation result output, which can realize the control of the train operation process.

(1) Train operation simulation control: Train operation simulation control function design is shown in Fig. 7.
(2) Simulation result output: Simulation results output mainly includes simulation process display, simulation result analysis and simulation result printing. The functional design included in this section is shown in Fig. 8.

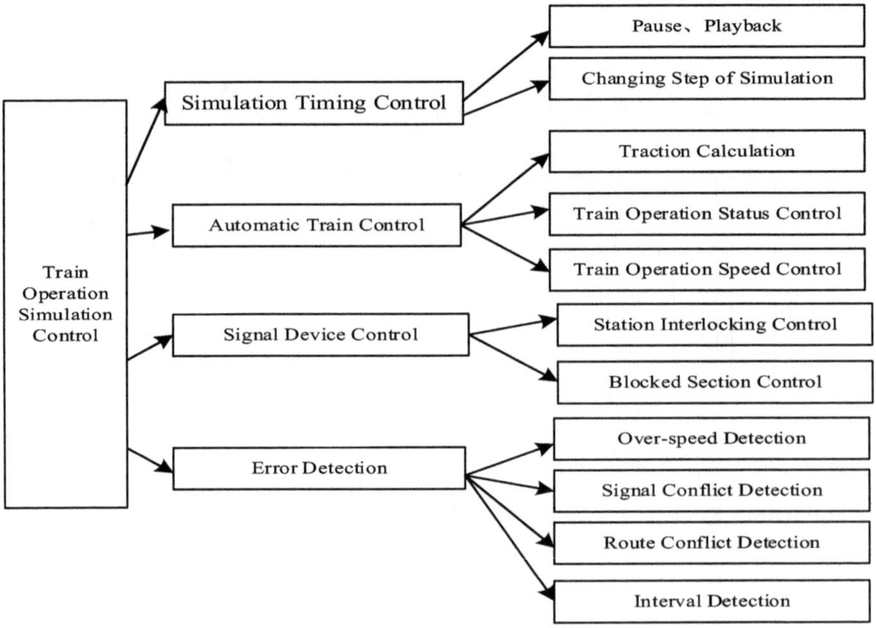

Fig. 7. Functional design of train operation simulation control

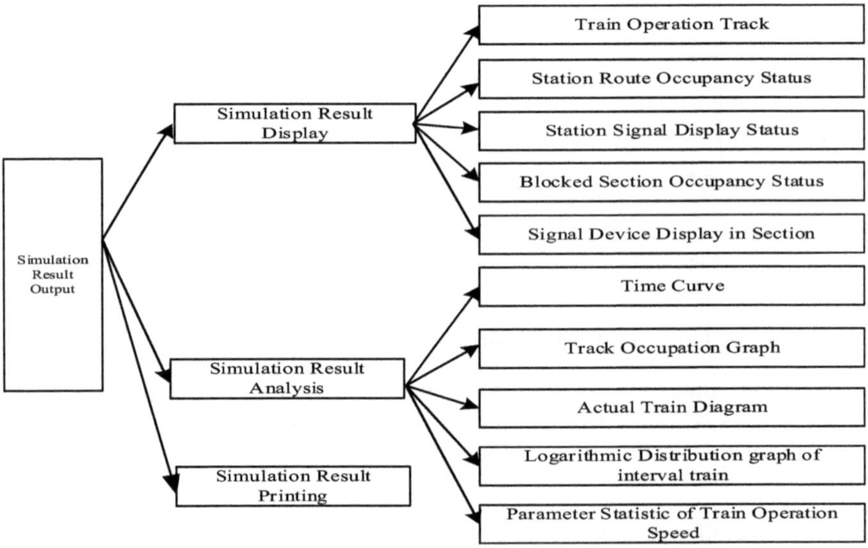

Fig. 8. Functional design of simulation result output

3.6 Design of Integrated Simulation Module for Train Operation

The integrated simulation module for train operation is a centralized linkage display of train diagram, traction calculation and train operation simulation on the premise of the functional simulation of them. The subsystem is divided into three parts, each part of the content shown differs from the other. The train diagram part is the dynamic operation display of road network and the train running on it. The traction calculation part is aimed at the traction curve display of the selected train, which belongs to micro display. The train operation simulation shows the whole road network diagram, including the line and the section, and realizes the real-time simulation of the train group according to the traction calculation results.

4 Simulation Results

The train operation plan simulation system of high speed railway adopts Visual C++ as the developing tool. The hardware platform is composed of database server, application-side machine, ink-jet color plotter and laser printer and other equipment. Application-side machine configuration requirements: CPU 2.6G and above, 4 core, memory 8.0G and above, hard disk 500G and above, network card 1000M and other necessary configuration. The module of ink-jet color plotter and laser printer are determined according to the actual conditions. The operating system should be Win7 X64 Ultimate.

Taking the actual data of the Chengdu-Dujiangyan express railway line as an example, the main operating mileage of Chengdu-Dujiangyan railway express railway is 67 km, with 12 stations; Lidui branch line is 6 km long and has 3 stations; Pengzhou branch line is 20 km long and has 6 stations. A total of 19 pairs of EMU are in operation. The simulation results of the train operation plan of high speed railway are shown in Fig. 9. The simulation results show that the simulation system can simulate the train running behavior of the railway line (including the line and the section) and the train running state on the line in real time. According to the calculation results of

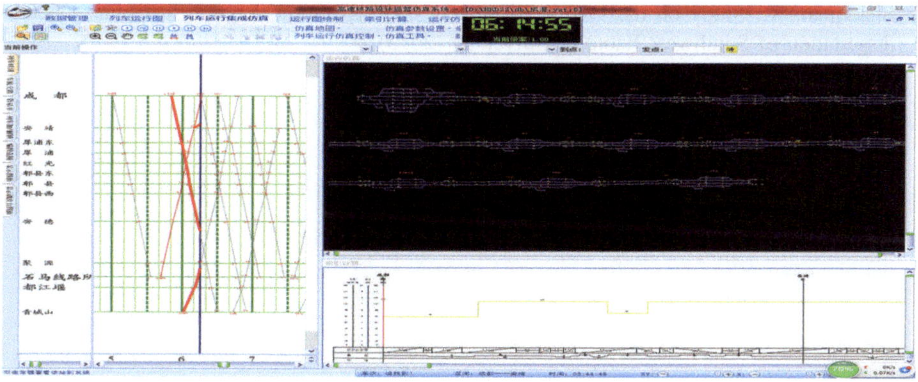

Fig. 9. The simulation result interface of the Chengdu-Dujiangyan railway

traction calculation, the real-time simulation of all train groups running on the line is realized. At the same time, the train diagram is proved to meet the station's technical operation and equipment capability, and the feasibility of the train diagram is improved.

5 Conclusion

In this paper, a simulation system of high-speed railway operation plan is designed and developed, which includes train operation planning, traction calculation and train operation simulation. Experiments show that the system can accurately simulate the train group to run the whole process of technical operations, the operation process, generate statistical analysis results, get the real-time test of the capacity adaptability of a variety of operating equipment in the operating program.

References

1. Azadeh, A., Ghaderi, S.F., Izadbakhsh, H.: Integration of DEA and AHP with computer simulation for railway system improvement and optimization. Appl. Math. Comput. **195**, 775–785 (2008)
2. Abril, M., Barber, F., Ingolotti, L., et al.: An assessment of railway capacity. Transp. Res. Part E **44**, 774–806 (2008)
3. Li, F., Gao, Z., Li, K., Yang, L.: Efficient scheduling of railway traffic based on global information of train. Transp. Res. Part B **42**, 1008–1030 (2008)
4. Gao, P., Wu, S., Cheng, Z.: Improved criticality analysis of railway locomotive components by simulation. In: Annual Reliability and Maintainability Symposium, RAMS 2007, pp. 494–498 (2007)
5. Chen, G.; Dynamic simulation of railway marshalling station operation system. Doctoral dissertation of Southwest Jiaotong University (2006)
6. Jinshan, P., Shaoquan, N., Hongxia, L.: Research on train operation simulation system based on GIS. Railway Comput. Appl. **16**(3), 1–4 (2007)
7. Gongyuan, L., Qiyuan, P.: Research on the framework of railway station operation simulation system. Railway Transp. Econ. **34**(1), 27–32 (2012)
8. Shaoquan, N.: Study on the system of railway train operation diagram in China. Doctoral dissertation of Southwest Jiaotong University (2013)

Service Quality Evaluation of Railway Freight Transportation Network Based on Bayes Theory

Xuepeng Zhang[1], Shengdong Li[2(✉)], Xiucong Li[2], Xinyi Lin[2], and Xiaoyuan Lv[2]

[1] Chengdu Railway Bureau, Chengdu, China
[2] Southwest Jiaotong University, Chengdu, China
Shengdongli@my.swjtu.edu.cn

Abstract. With the continuous improvement of the railway freight network, the effective evaluation can ensure that the railway remains competitive in the competition of freight market. Based on the analysis of the influencing factors of railway freight transportation network service quality, this paper constructs the event tree model according to the logical relationship among the factors, and then maps it into Bayesian network model. By calibrating the probability value of the basic events, the Bayesian network probability calculation method is used to obtain the top event probability value, which can determine the grade of the railway freight network service quality. Finally, a case study of Chengdu-Chongqing region is made was carried out to verify the effectiveness of the method.

Keywords: Railway transportation · Freight transport network · Service quality · Bayesian network · Evaluation

1 Introduction

Railway freight transportation plays an important role in national economic activities. With the development of social economy, the demand for quality of railway freight service has also been improved, and the transportation of cargo has become more and more timely and personalized. With the continuous reform of railway freight transportation in China, the construction of railway freight transportation network is becoming more and more perfect. The railway freight network mainly consists of two parts: the cargo transportation infrastructure network and the related operation service. The evaluation of the quality of the freight network service is beneficial to provide the basic basis for the reform and development of the freight transport, and further enhance the service level of the railway freight service network.

J.-S. Pan et al. (eds.), *Advances in Smart Vehicular Technology, Transportation, Communication and Applications*, Smart Innovation, Systems and Technologies 86,
https://doi.org/10.1007/978-3-319-70730-3_11

Scholars at home and abroad have done a great deal of research on the service quality evaluation of transport network. However, existing research mainly focused on the field of road transport, and has done little direct research on the railway transport network. Moreover; it studies mainly the supply and demand of transport service network and the equality index. Chen [1] introduced capacity reliability as a new network performance index which includes connectivity reliability and travel time reliability, and a comprehensive methodology combining reliability and uncertainty analysis was used to analysis the performance of networks. Choiu [2] used the concept of reserve capacity of signal-controlled junctions and a projected gradient approach was presented. Chen and Kasikitwiwat [3] provide a quantitative assessment of capacity flexibility for the transportation network and two approaches based on the concept of reserve capacity and demand pattern were proposed. Zhou [4] used grey relationship analysis method and principal components analysis method to evaluation the networks. Ji [5] puts forward a railway network capacity adaptability study framework including supply and demand total amount adaptability, distribute structure adaptability, service quality adaptability and capacity flexibility, and AHP-grey fuzzy evaluation method was used.

Bayesian network is based on the Bayesian formula of the graphical probability method, which has shown a wide range of applicability in the transport of dangerous cargo on the railway, emergency rescue decision-making and other engineering computing. The influencing factors of the quality of railway freight network service is analyzed, the event tree model of network service quality is established, and the event tree model into Bayesian network model is mapped in this paper. Through the Bayesian network probability calculation method, we can obtain the network quality of service probability to determine the network quality of service level.

2 Analysis on the Influencing Factors of Network Service Quality

The service object of the railway freight transportation network is the owner of cargo, and the satisfaction of the owner's demand directly determines the service quality of the network. Therefore, the impact of quality factors of the network service in this paper is mainly considered from the perspective of the needs of the owner, including security, flexibility and reliability, rapidity, adaptability and economy. The established evaluation index system of service network is shown in Table 1.

Table 1. Evaluation index system of service network in railway freight transport

Target layer	Index layer	Category	Evaluation index
Evaluation index system of service network in rail way freight transport (Q)	Security (A)	Accident (F)	Accident rate of ten thousand freight (F_1)
			Turnover accident rate of millions of cargo (F_1)
		Accident emergency rescue capability	Accident emergency rescue capability (L)
	Flexibility and reliability (B)	Flexibility	Service flexibility (M)
		Reliability (G)	Fulfillment rate of delivery period (G_1)
			Average delivery period (G_1)
	Fastness (C)	Rapidity (H)	Carrier handling time (H_1)
			Average delivery speed of goods (H_1)
		Convenience (I)	Distribution density of freight station (I_1)
			Connectivity of road network (I_1)
	Adaptability (D)	Transport network adaptability	Adaptation coefficient of transport network (N)
		Transport capacity adaptability	Adaptability of transport capacity (O)
	Economy (E)	Social and economic (J) benefits	Economic growth rate (J_1)
			Freight transport revenue (J_2)
		Transportation cost benefits (K)	Freight transport rate (K_1)
			Transport facility cost (K_2)

3 Service Quality Evaluation of Railway Freight Transportation Network Based on Bayes Theory

3.1 Theory Foundation of Bayesian Networks

Bayesian networks, also known as Bayesian belief networks (BBN), belong to the family of probabilistic graphical models in a set of variables. Generally, the Bayesian network consists of the directed acyclic graph (DAG) and the conditional probability table (CPT).

Each data sample is represented by an n-dimensional eigenvector $X = \{x_1, x_2, \ldots, x_n\}$, which describes n measures of n attribute samples $A_1, A_2, \ldots, A_n$ respectively. Assuming that there are m classes $C_1, C_2, \ldots, C_m$, and an unknown data sample X is given, the Bayesian network can estimate the posterior probability of each class, and the basic reasoning formula is shown in formula (3).

$$P(C_i|X) = \frac{P(X|C_i) \cdot P(C_i)}{P(X)} \tag{1}$$

Since $P(X)$ is constant for all classes, it only need that $P(X|C_i) \cdot P(C_i)$ is the maximum. Among them, $P(C_i) = s_i/s$. s_i is the number of training samples in class C_i, and s is the total number of training samples. In order to simplify the calculation, the simple assumption of which the class conditions are independent can be made, that is, the attribute values are independent of each other, namely:

$$P(X|C_i) = \prod_{k=1}^{n} P(x_k|C_i) \tag{2}$$

Where, Probability $P(x_k|C_i)$ can be estimated from the training samples.

3.2 The Event Tree Model of Freight Network Service Quality

An event tree is a representation of the logical relationship between a certain event that may occur in the system and the various causes of events by using arborescence. Therefore, according to the analysis of the influencing factors in the first section, this paper constructs the event tree model of the railway freight network service quality evaluation, as shown in Fig. 1.

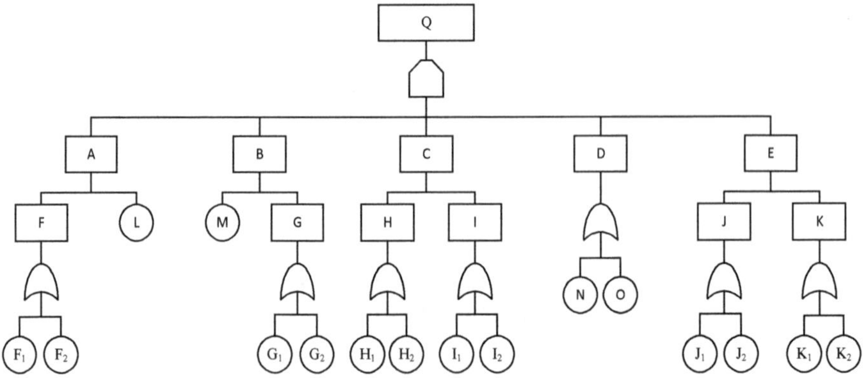

Fig. 1. Event tree model of freight network service quality evaluation

3.3 Probability Calculation Based on Bayesian Network

When the fault tree model is mapped to the Bayesian network model, the node corresponds to the event; the logic gate corresponds to the connection strength in Bayesian network, as shown in Fig. 2.

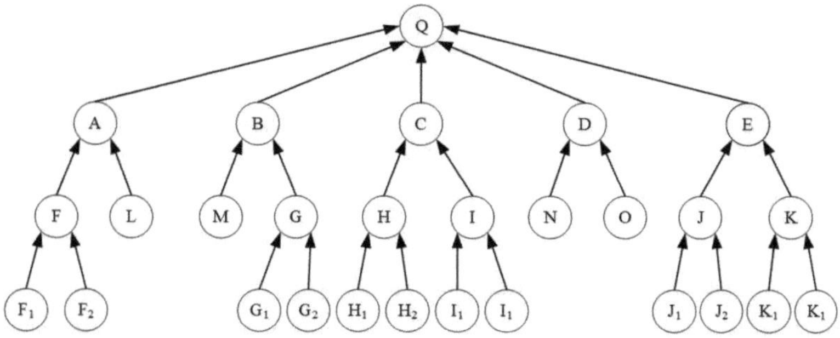

Fig. 2. Bayesian network of railway freight network service quality

Starting from the basic event node step by step, the probability value of the top event is calculated, and then the quality grade of the network system can be determined according to the probability value. If the event node contains a logical structure, the probability distribution formula is calculated as shown in Eq. (3):

$$P(S = 1|X_1, X_2, \ldots, X_n) = \begin{cases} P(X_1 = 1) \times P(X_2 = 1) \times \ldots \times P(X_n = 1), \; logic \;\; AND \;\; gate \\ 1 - P(X_1 = 0) \times P(X_2 = 0) \times \ldots \times P(X_n = 0), \; logic \;\; OR \;\; gate \end{cases} \tag{3}$$

Where, $X_i = 1$ indicates that the quality of network services is affected, and $X_i = 0$ is not affected.

Table 2. Result remarks

Remarks	Excellent	Good	General	Poor	Bad
Intervals	[0.9, 1.0]	[0.8, 0.9)	[0.7, 0.8)	[0.6, 0.7)	[0.0, 0.6)

The probability values of the network service quality are calculated, and the remarks of the probability values are shown in Table 2. Therefore, when the freight market changes, it's easy to find out the key influencing factors, and adjust the service marketing strategy and transportation organization measures timely to maintain the competitiveness of railway freight transportation.

4 Case Study

Taking the railway freight network in Chengdu and Chongqing as an example, this paper evaluates its service quality. Chengdu-Chongqing region takes Chengdu-Chongqing economic circle as the basis, with a total area of about 200,000 km^2, population 87.24 million, mainly covering 16 cities in Sichuan and 29 districts and counties in Chongqing.

According to the socio-economic and railway freight development status of Chengdu-Chongqing area, the statistical indicators are collected and the evaluation values are transformed into the [0, 1] range, as shown in the table.

In Table 3, the evaluation value of each evaluation index is the prior probability of the basic event, and the middle event probability can be obtained by formula (5): $P(G = 1|G_1, G_2) = 1 - P(G_1 = 0)P(G_1 = 0) = 1 - 0.4531 * 0.1997 = 0.9095$, the same can be obtained $P(F = 1|F_1, F_2) = 0.9640$, $P(H = 1|H_1, H_2) = 0.8799$, $P(I = 1|I_1, I_2) = 0.7046$, $P(J = 1|J_1, J_2) = 0.9149$, $P(K = 1|K_1, K_2) = 0.8439$, $P(D = 1|N, O) = 0.9868$.

Table 3. The statistics value about related evaluation index

Evaluation index	Index value	Evaluation value
Accident rate of ten thousand freight F_1	0.08	0.8147
Turnover accident rate of millions of cargo F_2	0.045	0.8058
Accident emergency rescue capability L	A	0.9134
Service flexibility M	B	0.6324
Fulfilled rate of delivery periodG_1	0.60	0.5469
Average delivery period G_2	4.5	0.8003
Carrier handling time H_1	0.67	0.7922
Average delivery speed of goods H_2	26.83	0.4218
Distribution density of freight station I_1	0.03	0.1419
Connectivity of road network I_2	0.67	0.6557
Adaptation coefficient of transport network N	8.92	0.8491
Adaptability of transport capacity O	0.83	0.9130
Economic growth rate J_1	0.07	0.6487
Freight transport revenue J_2	45.21	0.7577
Freight transport rate K_1	4.50	0.7431
Transportation facility cost K_2	126.7	0.3922

Note: the data came from "The statistical communique on national economic and social development of Sichuan Province in 2016", "The statistical communique on national economic and social development in Chongqing in 2016" and "Regional planning of Cheng Yu Economic Zone (2011–2020)".

Then we use the BNT toolkit in Matlab to compute the posterior probability of each node in the Bayesian network: $P(A = 1|F, L) = 0.9734$, $P(B = 1|M, G) = 0.9875$, $P(C = 1|H, I) = 0.9649$, $P(E = 1|J, K) = 0.9972$. So we can get $P(Q = 1|A, B, C, D, E) = 0.9128$. According to the quality remarks table, the service quality of the network is the highest.

5 Conclusions

The development of community economy makes the railway pay more attention on the integrity in the competition of the freight market. The evaluation of the service quality of the railway freight network can improve the competitiveness of the railway in the freight transportation market. Based on the analysis of the factors affecting the service quality of railway freight transportation network, this paper uses event tree model and logical relationship to combine and analyze the relationship between the factors. Through the collection of related information, calibration the prior probability of basic events, based on the fault tree model is mapped to a Bayesian network model, Bayesian conditional probability formula is used to obtain the posterior probability of network quality of service, which is a basis of quality evaluation. The Bayesian network model constructed in this paper provides a theoretical basis for the analysis and evaluation of railway freight transportation service network.

Acknowledgement. This research was supported by the National Natural Science Foundation of China (Project No. 61403317, 61703351), National Key Research and Development Program of China (2016YFC0802208), Science and Technology Plan of Sichuan province (Project No. 2017ZR0149, 2017RZ007), and Science and Technology Plan of China Railway Corporation (Project No. 2016X006-D), and the Fundamental Research Funds for the Central Universities (2682017ZDPY04, 2682017CX022, 2682017CX018).

References

1. Chen, A., Yang, H., Hong, K.L., et al.: Capacity reliability of a road network: an assessment methodology and numerical results. Transp. Res. Part B Methodol. **36**, 225–252 (2002)
2. Chiou, S.W.: Reserve capacity of signal-controlled road network. Appl. Math. Comput. **190**, 1602–1611 (2007)
3. Chen, A., Kasikitwiwat, P.: Modeling capacity flexibility of transportation networks. Transp. Res. Part A Policy Pract. **45**, 105–117 (2011)
4. Zhou, H.H.: Study on Performance Measures of LOS for the Express Freight Network System in Intermodal Transportation System. Beijing Jiaotong University (2009)
5. Ji, L.J.: Study on Adaptability between Supply and Demand in Railway Freight Network. Beijing Jiaotong University (2013)
6. Xing X.X.: Study of Combined Evaluation about Emergency Plans in Railway. Lanzhou Jiaotong University (2012)

Train Headway Calculation and Simulation System for High-Speed Railway

Guolong Gao[1,2], Jie Zhang[1,2(✉)], Wenjing Lai[1,2], and Yuchao Xin[1,2]

[1] School of Transportation and Logistics, Southwest Jiaotong University, Chengdu 610031, China
zhangjie@swjtu.cn

[2] National and Local Joint Engineering Laboratory of Comprehensive Intelligent Transportation, Southwest Jiaotong University, Chengdu 610031, China

Abstract. Train headway is very important for the operation efficiency and safety of high-speed railway. With the demand of high-speed train headway calculation and simulation, based on the principle of traction effort calculation and the control in train operation system, train headway calculation and simulation system for high-speed railway was designed. For the system, some key issues were discussed such as automatic train traction effort calculation, train headway calculation and simulation method. With this system, it will be benefit for signal-block layout and train operation optimization.

Keywords: High-speed railway · Train headway · Train traction effort calculation · System design

1 Background

High-speed railway was characterized as high speed and punctuality, comfort, safety and reliability, which developed fast in China for recent years.

For high-speed railway operation, traffic density directly affects the operational efficiency and the transport capacity. To ensure the safety of the train operation and adequate braking distance, the traffic density will be expressed as a certain train headway. Train headway is an important part of high-speed railway train operation technology standard system, its calculation process and the result are the basic theoretical basis of train operation and management.

Gill D.C used a heuristic gradient algorithm to solve the signal block layout [1]. Chang C.S. studied signal arrangement of urban rail transit system by genetic algorithm at 1999 [2]. Wong determined the inert point in train running process accurately [3].

This research was supported by the National Natural Science Foundation of China (Project No. 61403317, 61703351), National Key Research and Development Program of China (2016YFC0802208), Science and Technology Plan of Sichuan province (Project NO. 2017ZR0149, 2017RZ0007), and Science and Technology Plan of China Railway Corporation (Project No. 2016X006-D), and the Fundamental Research Funds for the Central Universities (2682017ZDPY04, 2682017CX022, 2682017CX018).

J.-S. Pan et al. (eds.), *Advances in Smart Vehicular Technology, Transportation, Communication and Applications*, Smart Innovation, Systems and Technologies 86, https://doi.org/10.1007/978-3-319-70730-3_12

Chen Rongwu et al. proposed a new algorithm for train headway based on the IEEE CTBC recommendation model [4–6]. According to above studies, this paper will discuss the system design for Train Headway Calculation and Simulation System for High-speed Railway.

2 System Demand Analysis

2.1 Business Demand

According to train headway simulation calculation process, the main application of train headway calculation and simulation system for high-speed railway could be divided into the following categories:

1. High-speed railway train operation facilities simulation data management
2. Train traction effort calculation
3. Train headway calculation
4. Simulation results output.

2.2 Data Demand

The business data is composed of rail line profile data, signal data, EMU data and train traction effort calculation parameters.

1. Rail line data: station, track, ramp, curve, bridge, tunnel and speed limit;
2. EMU data: EMU attributes and EMU traction effort, braking, energy consumption data;
3. Signal data: signal mode, block and signal;
4. Train traction effort calculation parameters: working condition data, braking arguments.

2.3 System Interface Demand

The system should have the function of outputting the results in EXCEL report, PDF report and CAD graphical report, and interface should be designed for other system of railway operation such as Signal-block Layout System, Train Scheduling System etc.

3 System Functions and Structure

3.1 System Structure

According to the using logic, train headway calculation and simulation system for high-speed railway could be divided into three parts, data management, train traction effort calculation and train headway calculation. The structure of the system is shown in Fig. 1.

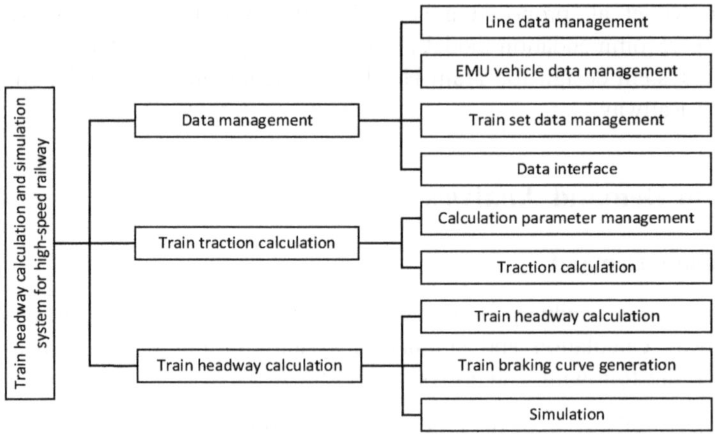

Fig. 1. System structure

3.2 System Functions

The main functions of the system could be designed as follows:

1. Data management
   (1) Rail line data management
   (2) Station data management
   (3) EMU data management
2. Train traction effort calculation
   (1) Train traction effort calculation pretreatment
   (2) Traction effort calculation pretreatment function includes: Rail line extraction, single track line automatic reverse, train speed limit serializes.
   (3) Automatic train traction effort calculation
3. Train headway calculation
   V-S data calculation according to train ATP, train safety protection distance and signal working time.

4 Key Issues for System Design

4.1 Automatic Train Traction Effort Calculation

The driving tactics is an extremely important part of automatic train traction effort calculation. The strategies used by the system includes time saving, energy saving, fixed time, and the following aspects should be considered for the working condition:

1. Automatic working condition selection of train driving tactics.
2. Automatic optimization of train driving tactics.

Workflow of automatic train traction effort calculation could be shown as follows (Fig. 2):

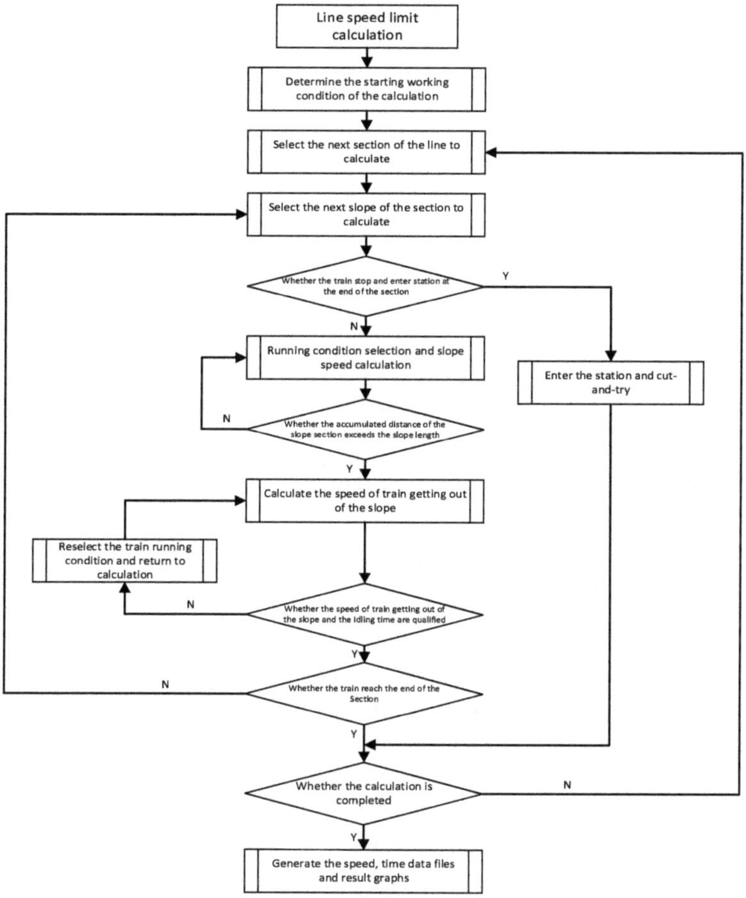

Fig. 2. Workflow of automatic train traction effort calculation

4.2 Train Headway Calculation

Train headway calculation is based on the train traction effort calculation and train simulation. Therefore, it's necessary to design the train simulation scheme and calculation method.

The simulation process could be divided into three stages: start and traction acceleration process, intermediate running process and braking process. According to above stages, the train simulation result could be obtained.

Usually, EMU uses a continuous braking mode curve for train operation control which could be calculated by reverse iteration method. The reverse iteration process could be shown as following (Fig. 3):

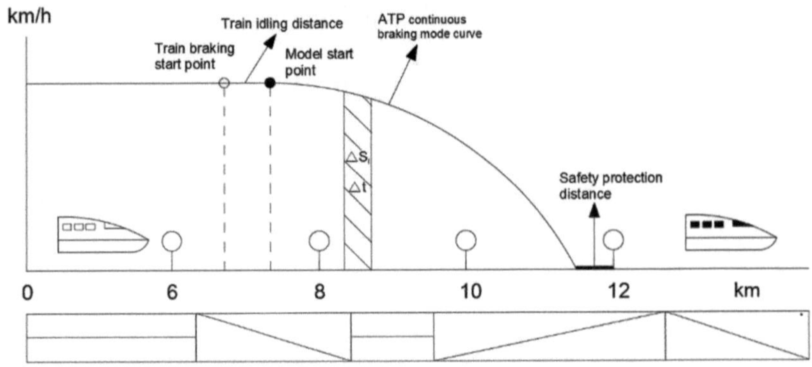

Fig. 3. Reverse iteration process of the continuous braking mode curve

The description of symbols in the calculation could be listed as Table 1.

Table 1. Description of symbols

Symbols	Description
S_1	Stop position
v_1	Target limit speed
v_0	Initial speed
S_0	Initial position
$v_{[]}$	Speed array
$s_{[]}$	Position array
$V^R_{[]}$	Corresponding point speed array
$S^R_{[]}$	Corresponding point position array

The target limit speed is recorded as $v_1 = 0$. Initialize the train acceleration, running speed, running time and other related data and make $v_{[i]} = v_1 = 0$, $s_{[i]} = 0$, $i = 0$.

Set the acceleration at the initial speed $v_{[i]}$ in time step as the acceleration value. The speed variation dv and location variation ds are calculated by kinematic formula.

The algorithm of calculating the braking mode curve by the above data is shown in Fig. 4.

4.3 Simulation Method

Train headway is determined by the train operating conditions, such as the EMU type, rail line, driving tactics, train equipment type and climatic conditions, so it's necessary to check the train headway in different conditions.

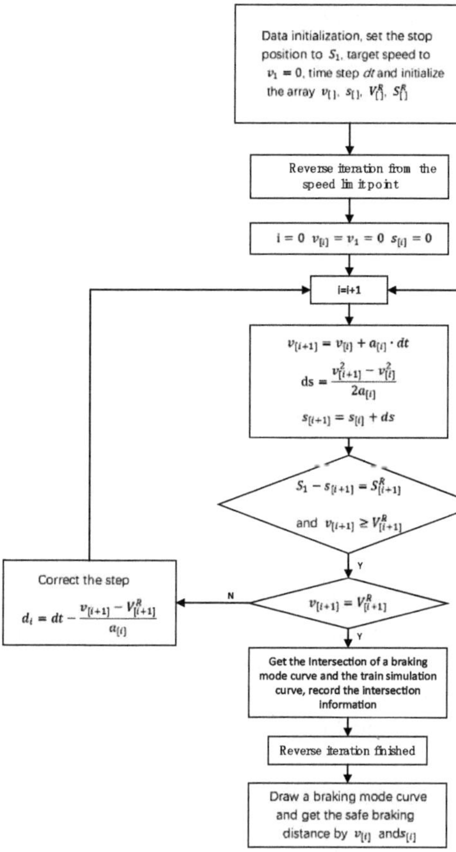

Fig. 4. Algorithm of braking mode curve calculating

For the current train control system used in China, users can check the train headway according to the following combinations:

(1) Train control system type: 200H, 200C, 300T, 300S, 3D and 300H.
(2) Driving tactics: maximum running speed, limit point retention and enter in speed control mode.
(3) Climatic conditions: wet and dry rail and wind speed.

For the series of combinations, they should be converted into system operating parameters by system, then the train running process simulation can be used to calculate the train headway under different conditions. Finally, generate the statistical charts to provide the basis for the analysis of train headway standard.

The statistical chart of train headway could be shown as following (Fig. 5):

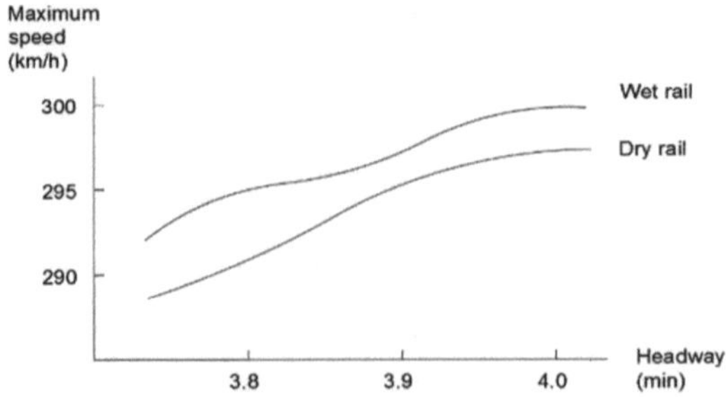

Fig. 5. Statistical chart of train headway

5 Conclusions

Based on train traction effort calculation theory and high-speed railway train operation control principle, a train headway calculation and simulation system was designed for high-speed railway by a certain train control program and train operating conditions. This system provides train tracing operation analysis under different equipment and operating conditions by simulation, and could provide a more scientific and effective method for determining the technical operation time standard of railway transportation system.

References

1. Gill, D.C., Goodman, C.J.: Computer-based optimisation techniques for mass transit signalling design. IEE Proc. Electric Power Appl. B **139**(3), 261–275 (1992)
2. Chang, C.S., Du, D.: Further improvement of optimisation method for mass transit signalling block-layout design using differential evolution. IEE Proc. Electric Power Appl. **146**(5), 559–569 (1999)
3. Wong, K.K., Ho, T.K.: Dynamic coast control of train movement with genetic algorithm. Int. J. Syst. Sci. **35**(13–14), 835–846 (2004)
4. IEEE Standard for Communications-Based Train Control (CBTC) Performance and Functional Requirements. IEEE Std. IEEE (2002)
5. IEEE Guide for the Calculation of Braking Distances for Rail Transit Vehicles (2009)
6. Dong, H., Ning, B., Cai, B., et al.: Automatic train control system development and simulation for high-speed railways. Circ. Syst. Mag. IEEE **10**(2), 6–18 (2010)

Train Scheduling for Heavy Haul Railway

Pei Wang[1,2,3], Jie Zhang[1,2,3(✉)], Baoheng Feng[1,2,3], and Wengao Peng[4]

[1] School of Transportation and Logistics,
Southwest Jiaotong University, Chengdu 610031, China
zhangjie@swjtu.cn

[2] National Railway Train Diagram Research and Training Center,
Southwest Jiaotong University, Chengdu 610031, China

[3] National and Local Joint Engineering Laboratory of Comprehensive Intelligent Transportation,
Southwest Jiaotong University, Chengdu 610031, China

[4] Department of Transportation, Guangzhou Railway Group, Guangzhou 510000, China

Abstract. Train Scheduling is very important for Heavy haul railway transportation organization. Based on the train scheduling procedure and rules of heavy haul train operation in China, this paper discusses some key problems for heavy haul train scheduling, which included train line plan for single track and double track line combined section, time window plan, train combined and decomposition problem, and train loading/unloading problem. Then the relative direction balanced departure method was put forward for train line plan at combined section, making the time window with the size of the interval capacity loss for time window plan, balanced train line layout for the combined and decomposition problem. For train loading and unloading problem, train plan should be adjusted to meet the working time at the arrived station.

Keywords: Heavy haul railway · Train scheduling

1 Introduction

Train diagram is very important for railway operation safety and efficiency, which is the foundation of railway transportation organization [1]. At present, heavy haul railway transportation develops rapidly, and the scheduling has great influence on the efficiency of train operation, therefore, the quality of train scheduling should be improved.

By review, research on the train diagram mainly focuses on passenger railway, high speed railway, and rail transit. However, few research was focused on heavy haul

This research was supported by the National Natural Science Foundation of China (Project No. 61403317, 61703351), National Key Research and Development Program of China (2016YFC0802208), Science and Technology Plan of Sichuan province (Project NO. 2017ZR0149, 2017RZ0007), and Science and Technology Plan of China Railway Corporation (Project No. 2016X006-D), and the Fundamental Research Funds for the Central Universities (2682017ZDPY04, 2682017CX022, 2682017CX018).

J.-S. Pan et al. (eds.), *Advances in Smart Vehicular Technology, Transportation, Communication and Applications*, Smart Innovation, Systems and Technologies 86,
https://doi.org/10.1007/978-3-319-70730-3_13

railways. Because of the character of heavy haul train operation, such as less train stops, train decomposition, which were different from the regular rail line, so the methods of train scheduling should be considered.

In this paper, with the characteristics of heavy haul transportation organization, the principles and procedure for train diagram of heavy haul railway are summarized. Then some key problems for heavy haul train scheduling was put forward. In addition, a composite method for train scheduling was analyzed to improve the quality of heavy haul railway train diagram.

2 The Process of Heavy Haul Train Scheduling

The Characteristics of train diagram of heavy haul railway could be summarized as no speed difference, less train stops, and intensive train combination and decomposition at some connection station. For the heavy haul railways, there are few passenger train, and the passenger train line is usually fixed and does not change with other conditions. Therefore, this paper only studies the process and the drawing method of the heavy haul freight train diagram.

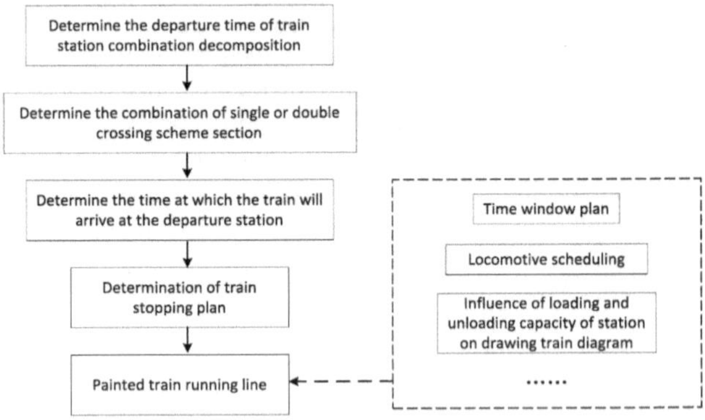

Fig. 1. Train scheduling process for heavy haul train diagram

2.1 Principles

(1) Meet all time standards.
(2) Ensure the transportation efficiency of customs.
(3) Reasonable economic locomotive and vehicle arrangement.
(4) Ensure the coordination of train running line at all stations and sections.
(5) Reasonable time window layout.
(6) Reasonable utilization of rail line passing capacity and station capacity.
(7) Reasonable crew schedule.

2.2 Procedure

For the process of heavy haul train scheduling, first determine the departure time of the train in the special station. For example, single-double track, loading and unloading station. Then, according to the station and other restrictions of the interval, draw the running line. The detail process of train scheduling of heavy haul railway could be described as Fig. 1.

3 Key Problems for Heavy Haul Railway Train Scheduling

According to the characteristics of heavy haul transportation, four problems are essential point for the process of heavy haul train scheduling, which includes train line plan of single and double line combined section, time window plan, the combined and decomposition problem and train loading/unloading problem.

3.1 Train Line Plan for Single-Track and Double-Track Line Combined Section

For the heavy haul train at single-double track section, if train line was not reasonable, which would lead to longer stop time at the station for waiting [2]. Therefore, train line should start from single-track section because there is less capacity for intensive arriving or departing than double-track section. In addition, if the same direction of the departure interval is shorter, and train line was not layout balanced, which would lead to waiting time greatly increased for the train from the single-track. Therefore, when scheduling,

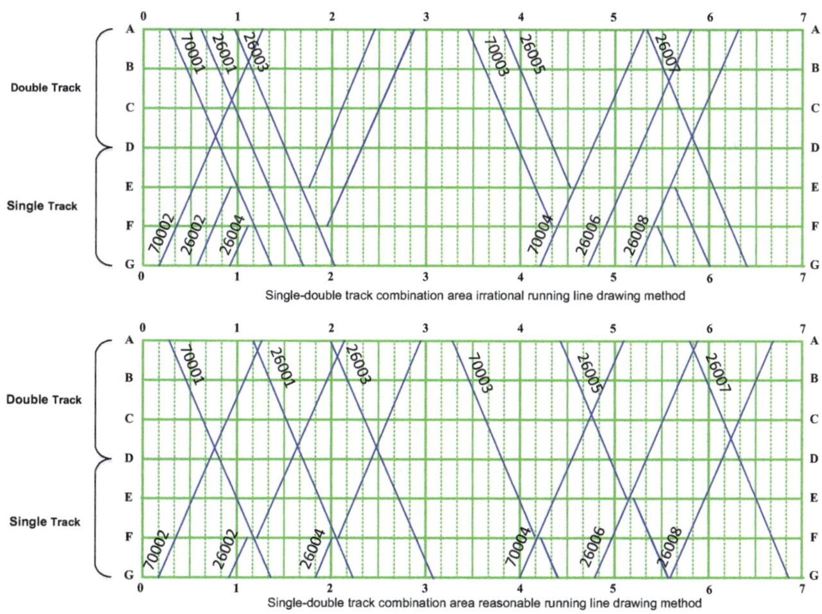

Fig. 2. Train diagram sketch for single-track and double-track combined area

it is necessary to start from the two opposite directions as balanced as possible, and appropriately extend the departure interval, thus reducing the relative time of the train in the single track section, as shown in Fig. 2.

In addition, parking and start-up of heavy haul train are more difficult than empty train, the empty train line should be arranged to avoid the heavy haul train line.

3.2 Time Window Plan

At present, there are three kinds of time windows used in the train diagram of heavy haul railway, such as rectangular time window, V-type time window and mixed time window. The time window of heavy haul train operation diagram mainly consists of three parts, section blocking time, construction auxiliary time, influence time [3]. The different schemes of time windows will affecting the capacity through section. The influence time can not be used for the extra loss of train operation. According to the capacity loss calculation of time window after size selection, it can evaluate the quality of the time window plan.

When opening the V-type time window and rectangular time window, the setting scheme are shown in Fig. 3.

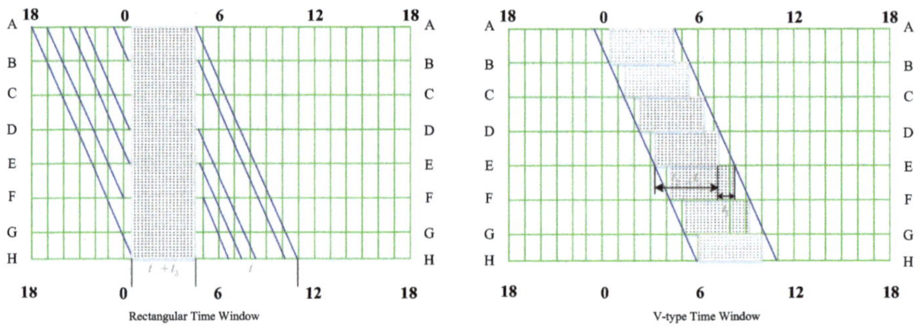

Fig. 3. Sketch map of rectangular and V-type time window scheme

For using a rectangular time window, the loss of its capacity can be determined by press type (Table 1):

$$N = (t_1 + t_2 + t_3)/I - M - 1 \tag{1}$$

Table 1. Description of symbols

Symbols	Description
N	The time window in the influence area loss of the number of goods trains
t_1	The time window effect of time, min
t_2	The time window closed operation time, min
t_3	The time window auxiliary operation time, min
I	Minimum train tracking interval time, min
M	The line number of the intermediate station section

For using the V-type time window, the loss of its capacity can be determined by press type:

$$N = (t_1 + t_2 + t_3)/I - 1 \tag{2}$$

When a rectangular time window is opened, the capacity loss is related to the number of stations and the number of departure track in the stations. When V-type time window is opened, the decrease capacity only depends on the region of the longest running time within the power supply.

For the mixed time window, according to the space and time, the combination can be divided into four types, the decrease capacity is calculated respectively for the rectangular time window and the V-type time window.

3.3 Train Combined and Decomposition Problem

The train station in the process of decomposition, if the large length train reach more intensive, and they are limited by the station departure device, then the short length train be started with tracking interval method after large length train decompose. On the one hand, it will bring great pressure to the station. On the other hand, it also increases the waiting time of the short length train at station after the large length train is decomposed, lead to increase the transport time of goods. As a result, large length train arrival can be arranged as evenly as possible. When the large length train first accessed are decomposed, the short length train which from the large length train are arranged immediately, and the other large length train are accessed after the short length train is departure, as shown in Fig. 4.

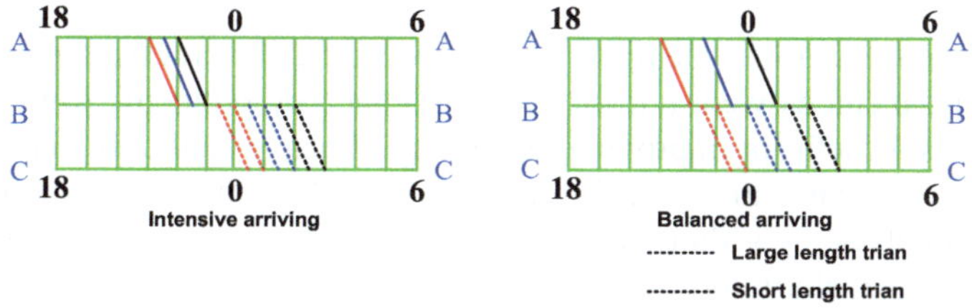

Fig. 4. Sketch map of train connection scheme of combined decomposition station

For the combination process, the principle of "first come, first group" should be adopted as far as possible. A large length train made by combining such principles. The departure will be more balanced and will not bring pressure to the station's departure equipment. At the same time, it can also make the train diagram more balance.

3.4 Train Loading and Unloading Problem

In order to avoid the pressure made by cargo operation, the train operation line should be drew under the cargo capacity of the station. And when there are more trains, the departure time of the subsequent train can be changed, it will make the train running line distribute more evenly in the whole running diagram. Or it can also allow subsequent trains to wait in front of qualified sections or stations.

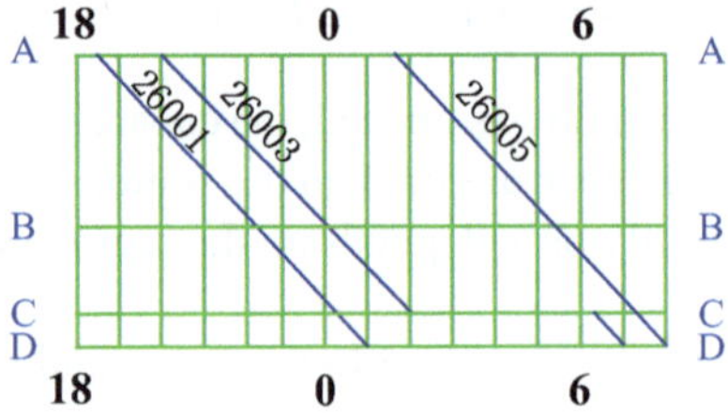

Fig. 5. Sketch map of considering loading and unloading capacity

As shown in Fig. 5, the D station is the station where the unloading work is needed, now assume that the unloading capacity is limited, only one unloading line is available, and the unloading time is 6 h. When the station is connected to the 26001 train to start unloading, the follow-up train 26003 and 26005 can not immediately pass the station. Enlarge the departure interval between the train and the 26001, so that the train line can be balanced. Only when the train station at A start tracking, and is not affected by the D station unloading capacity, can be 26003 arrangement conditions in front of the station, for example, stop at C station for wait, and when train unloading operation finished at D station, and then the 26003 access. Considering the loading capacity of the station, similar to the unloading situation, it will not be repeated here.

4 Conclusion

This paper have summarized the principle and procedure of heavy haul train scheduling, and then some methods were put forward for some key problems while scheduling, which would be benefit for improving the quality of heavy haul train diagram. But the methods in this paper is some tactics by common situation or practical experience, which was not by models or algorithms, and some constraint such as locomotive routing were not considered. For the future research, the intelligent methods for heavy haul train scheduling should be study in order to improve the organization level of heavy haul transportation.

References

1. Qiyuan, P., Ciguang, W.: Railway Traffic Organization. China Railway Publishing House, Beijing (2015)
2. Siji, H.: Railway Traffic Organization. China Railway Publishing House, Beijing (2009)
3. Caiwen, M.: Railway Heavy Transport Organization. China Railway Publishing House, Beijing (2015)

Research on Train Delay Diagnosis in Train Diagram Based on Big Data Technology

Yong Zhang[4], Wenqing Li[1,2,3], Shaoquan Ni[1,2,3](✉), and Xiaowei Liu[1,2,3]

[1] School of Transportation and Logistics, Southwest Jiaotong University, Chengdu 610031, China
464986388@qq.com

[2] National Railway Train Diagram Research and Training Center, Southwest Jiaotong University, Chengdu 610031, China

[3] National and Local Joint Engineering Laboratory of Comprehensive Intelligent Transportation, Southwest Jiaotong University, Chengdu 610031, China

[4] China Railway First Survey and Design Institue Group Co. Ltd., Xi'an 710043, China

Abstract. This paper studies the method of applying big data technology to the analysis of train delay in train diagram. The actual data of train operation is obtained based on field collection, which formed data warehouse by data cleaning, data integration and other preprocessing techniques, then we compare it with the basic train diagram. According to different types of train delays, this paper proposes methods to identify the areas self-delay or aftereffect-delay occurred, evaluate redundant capability of station and interval, analyze relationship between train delay and stop time in station or and running time in interval based on big data technology and computer graphic technology.

Keywords: Train delay · Train diagram · Big data technology · Computer graphic technology

1 Introduction

Train diagram is the basis of train operation organization. At present, from the realization situation of train operation, the passenger train diagram realization rate is high, while the realization rate of CRH passenger train diagram is higher, basically can reach more than 99%, while the other passenger train realization rate is at around 98%. Freight train realization rate is relatively low, generally not more than 80%. Which reasons that train diagram can't deliver is a lot, but the most common factor is that the operation ability

This research was supported by the National Natural Science Foundation of China (Project No. 61403317, 61703351), National Key Research and Development Program of China (2017YFB1200702, 2016YFC0802208), Science and Technology Plan of Sichuan province (Project NO. 2017ZR0149,2017RZ0007), and Science and Technology Plan of China Railway Corporation (Project No.: 2016X006-D), and the Fundamental Research Funds for the Central Universities (2682017ZDPY04, 2682017CX022, 2682017CX018).

J.-S. Pan et al. (eds.), *Advances in Smart Vehicular Technology, Transportation, Communication and Applications*, Smart Innovation, Systems and Technologies 86,
https://doi.org/10.1007/978-3-319-70730-3_14

and train transportation equipment does not match, among the factor the people can't exert the ability of the transportation equipment in the application of equipment. The emergence of these problems is relevant to the factor that the data of the correct diagnosis can't be obtained completely before the train diagram is compiled, and is also relevant to that the relevant parameters of the train diagram can't be revised correctly.

2 Research Contents

First, we deal with the investigation in railway train diagram execution department to understand the performance of train diagram, solicit opinions and suggestions in the implementation of train diagram and collect actual train diagram data. Secondly, we preprocess the collected data and put it into the data warehouse. Then, we analyze the operation law of train delay by comparing the basic train diagram with the actual results of the train operation with big data technology and computer graphic analysis technology. The structure of research contents is showed in Fig. 1.

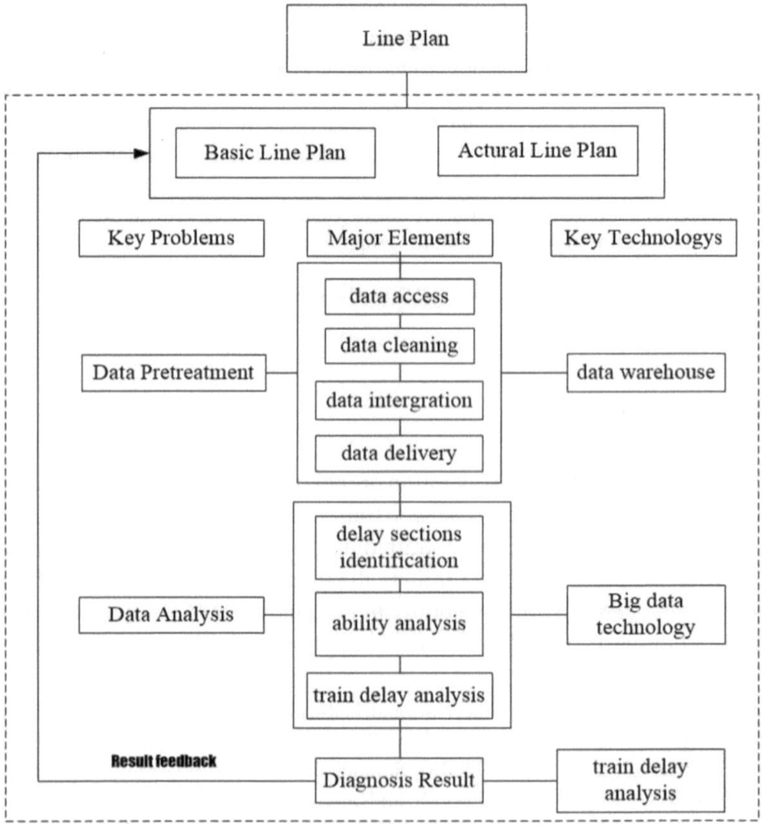

Fig. 1. Structure of research contents

3 Data Classification and Definition

Simple as it looks, train diagram actually contains a considerate number of data that it can be claimed that all kinds of data related to the operation of the railway can be found in it. The basic data collected can be divided into two major categories, numeric type and logical type. In order to describe the relationship between various basic data, corresponding correlation data is required.

3.1 Numeric Data

This type of data is stored in the database in numeric, char, varchar, and other formats. Depending on the type of operation, it can be divided as:

- Basic data of train operation: line name, station name, train number, running time in the section, additional time of starting and stopping, station interval, tracking interval of automatic block train, successive departure interval of semi-automatic section, station stop period of passenger train, etc.
- Device maintenance related data: maintenance time window, additional time due to construction slow, etc.;
- Technical operation related data: operation time of change of locomotive, locomotive crew transfer time, vehicle technical inspection time, water supply of passenger train, train connecting time at technical station, etc.

3.2 Logical Data

This kind of data is stored in the database in int, bool and other formats, which can be divided into:

- Train level: passenger train—EMU(C, G, D), Direct train, express train, Pu fast train, Puke train; cargo train—baggage train, direct train, through train, section train, transfer train, express train, etc.
- Line level: main level, branch line, etc.

3.3 Associated Data

This type of data is usually stored in the form of tables in the database, and depending on the different related objects, it can be divided as:

- Infrastructure associated data: adjacent relationship between station and line, etc.
- Train operation associated data: relationship between train number and station,relationship between train number and line, train number and its corresponding train stop information, meeting and avoiding information in the intermediate station, etc.

4 Analysis of Train Delay Based on Big Data

All the time, the diagnostic analysis of train diagram always adopts the traditional statistical and evaluation methods, and lacks effective technical means and equipment to deal with the mass data generated by train operation. Limited samples lead to difficult to obtain objective and accurate conclusions, and the research process is too rough and partial qualitative. With the help of big data technology, it can make up for the shortage of samples in traditional methods, reduce the influence of the subjective will of researchers, and improve the reliability of results.

4.1 Frequent Regional Identification of Train Self-delay

Self-delay refers to the delay in the operation of the train, which is not caused by interference from other trains, but by the direct loading on the train. There are two forms of self-delay, one is the delay of the train at the station showed in the Fig. 2, and the other is the delay on the way showed in the Fig. 3.

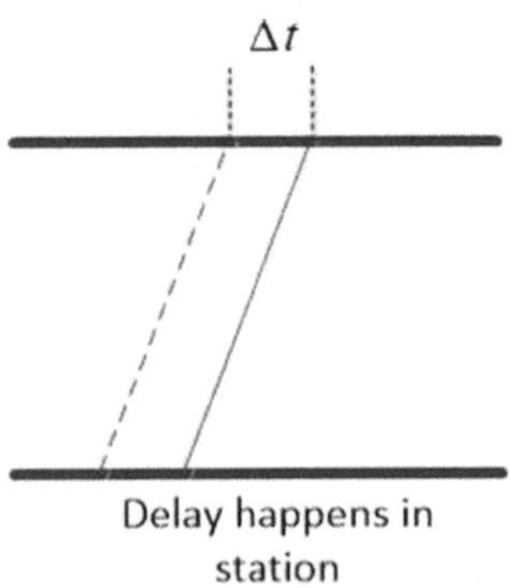

Fig. 2. Delay of the train at the station

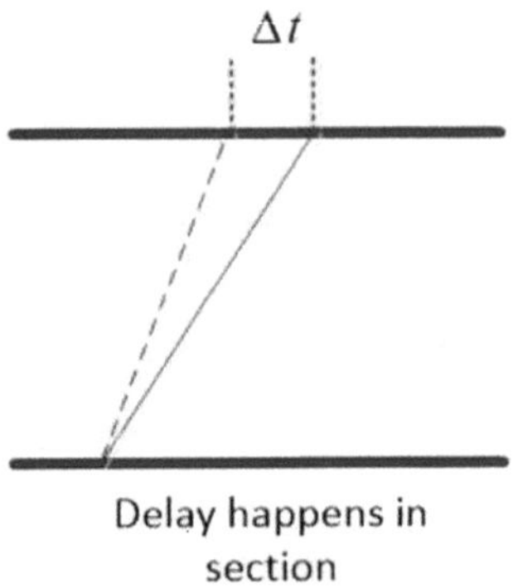

Fig. 3. Delay of the train at the section

The delay of train itself is often the cause of large area delay. Therefore, it is necessary to identify the train train's frequent delayed area to adjust the parameters of train diagram so as to reduce the delay. In order to visually understand the station delay, we extract all self-delay trains from the database, drawing a line chart by taking the station number as abscissa and the number of delayed trains at each station as ordinate, showed in the Fig. 4.

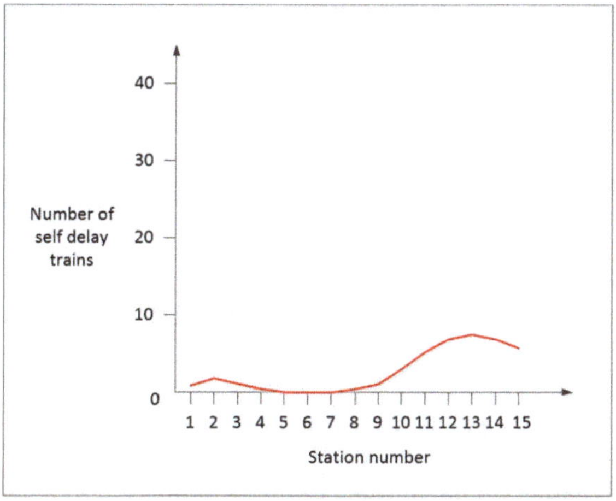

Fig. 4. Frequent regional identification of train self-delay

As can be seen from the Fig. 4, the station at the number 9–15 are high frequency area where the train self-delay happened. It is necessary to make a critical evaluation of its parameters, station scheduling, personnel level and equipment capacity.

4.2 Frequent Regional Identification of Train Enhanced Aftereffect-Delay

Aftereffect-delay refers to the disturbance of the train to the train due to the delay of the train ahead, which results in the delay of the train. There are three forms of aftereffect-delay, attenuated aftereffect-delay means the delay time of the train ahead is larger than that of the train which is showed in the Fig. 5, equivalent aftereffect-delay means the delay time of the train ahead is equal to that of the train which is showed in the Fig. 6, enhanced aftereffect-delay means the delay time of the train ahead is less than that of the train which is showed in the Fig. 7.

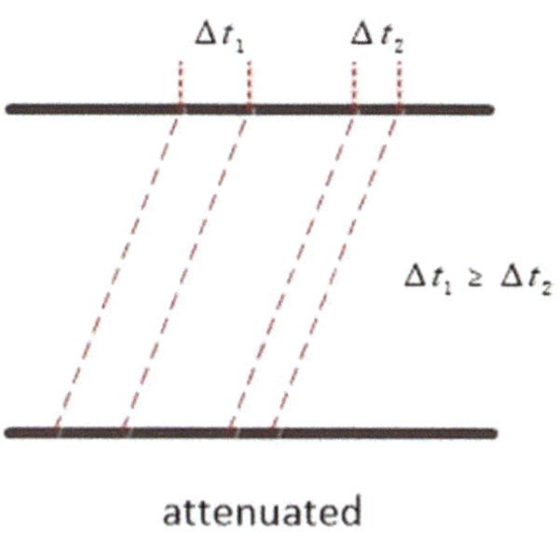

Fig. 5. Attenuated aftereffect-delay

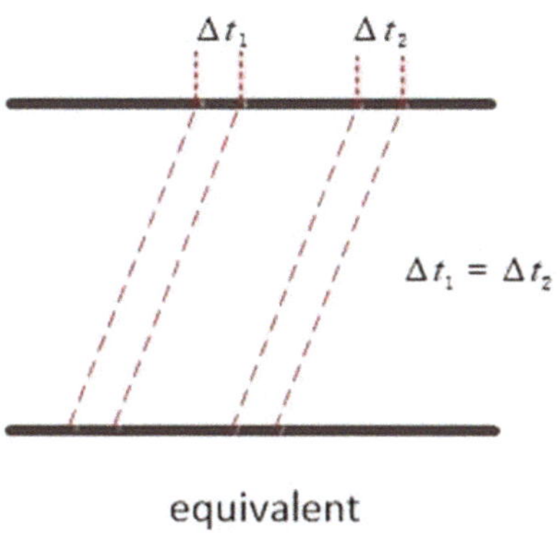

Fig. 6. Equivalent aftereffect-delay

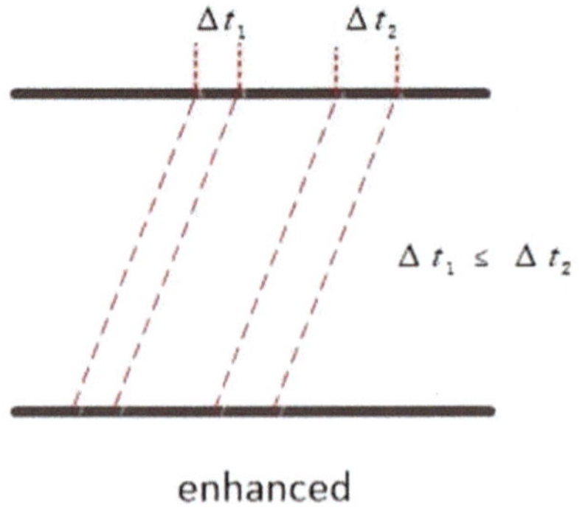

Fig. 7. Enhanced aftereffect-delay

The aftereffect enhanced delay is caused by external interference once the train is delayed again. It will cause two impacts on the railway transportation organization and cause greater harm. According to the station number, the train stop time data of all the aftereffect delays is extracted form the database, then we calculate the difference between the delay time of each train and the former delayed train according to the time series to select enhanced aftereffect-delay trains. The total number of delay trains at NO.i stations is M_i, of which the number of enhanced aftereffect-delay trains is M_i^Z, then the probability of aftereffect-delay occurs at NO.i stations is:

$$P_i^Z = \frac{M_i^Z}{M_i} \tag{1}$$

Drawing the line graph with station number as abscissa and P_i^Z as ordinate, showed in Fig. 8. As can be seen from the diagram, the enhanced aftereffect-delay

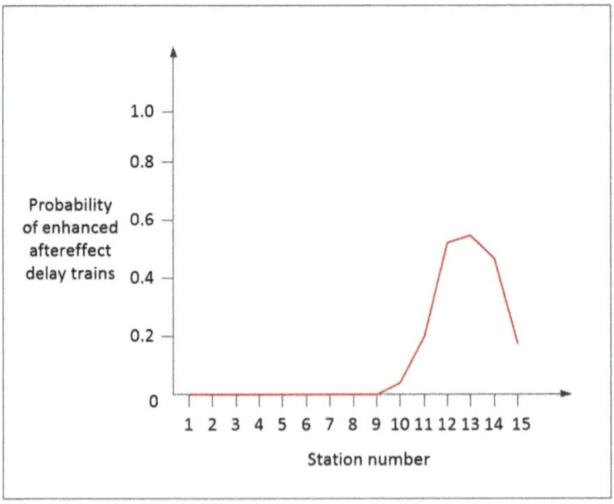

Fig. 8. Frequent regional identification of train enhanced aftereffect-delay

is more serious at the station NO.11–NO.15. It is necessary to make a critical evaluation of its capacity.

4.3 Analysis of Station Redundancy

By comparing the characteristics of the train at a specific site, the actual capacity of the station can be used to determine whether the capacity of the station is sufficient. Select a station which stop time is 3 min, from the database to read all the stops' actual stop time and draw the histogram showed in the Fig. 9. From the figure can be intuitively see the station's redundant time is large, in order to save station capacity, can be appropriate to shorten the provisions of the station train stop time.

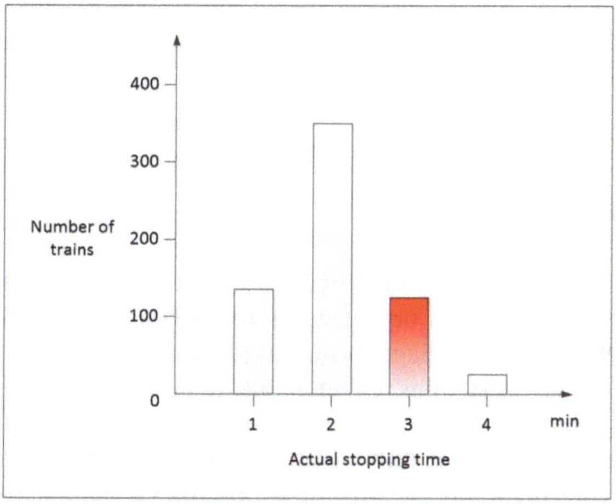

Fig. 9. Analysis of station redundancy

4.4 Analysis of Section Redundancy

Select a interval which the provisions of run-time is 30 min, and read the actual run-time of all trains through the interval from the database and draw a histogram, shown in the Fig. 10. From the figure can be intuitive to see the train in the interval occurred more late, about 60%. But the late departure time is not large, indicating that the interval is more tense, resulting in the reasons may be the front of the tension or the station interval caused by interference, it may be the interval map set the run-time set unreasonable, need to be treated as appropriate.

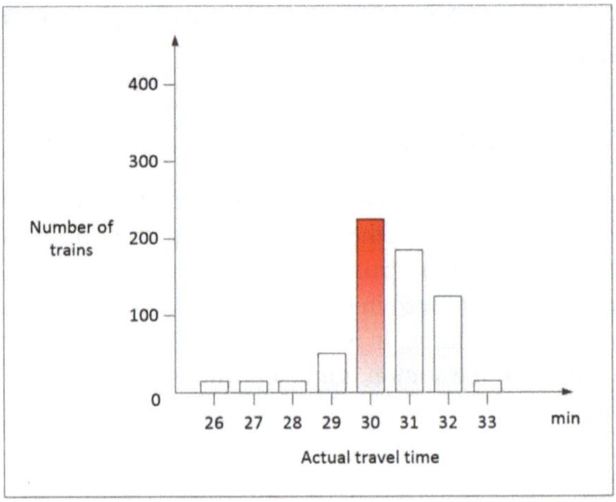

Fig. 10. Analysis of station redundancy

5 Conclusions

Application of big data technology can effectively use the massive data generated in the process of railway transportation organization, to analysis objectively the situation of the train delay in the actual train diagram such as identification of frequency delay areas, to assess the redundancy of stations and intervals, to optimize and adjust train diagram parameters. The application of big data technology is of great significance to optimize and promote the development of railway information, to build the railway marketing system, to improve the quality of railway transportation service and to enhance the core competitiveness of railway.

References

1. Qiyuan, P., Songnian, Z., Haifeng, Y.: A system for evaluation of train diagram elasticity. J. Southwest Jiaotong Univ. **33**(4), 367–371 (1998)
2. Yijian, Y., Yuqiang, H.: Evaluation of stability of railway working diagram based on max-plus algebra. J. China Railway Soc. **31**(4), 14–19
3. Junhua, C.: Research on establishment and evaluation of the stability-based train woring diagram of passenger dedicated lines. Beijing Jiaotong university Ph.D. thesis
4. Malachy, C.: Optimizing scheduled times, allowing for behavioural response. Transp. Res. Part B Methodological **33**(7), 329–342 (1999)
5. Goverde Rob, M.P.: Railway timetable stability analysis using max-plus system theory. Transp. Res. Part B Methodological **41**(2), 179–201 (2007)

On Linear Precoding of Unitary Space-Time Modulation for Spatial-Temporal Correlated Flat Fading Channel

Caihong Yu[1(✉)], Kejun Jia[2], and Zheng Yang[3]

[1] Fujian Provincial Key Laboratory of Digital Equipment, Fujian University of Technology, Fuzhou 350000, Fujian, China
yucaihong@fjut.edu.cn

[2] School of Computer and Communication, Lanzhou University of Technology, Lanzou 730050, Gansu, China
kjjia@lut.cn

[3] Fujian Provincial Engineering Technology, Research Center of Photoelectric Sensing Application, Fujian Normal University, Fuzhou 350000, Fujian, China
zyfjnu@163.com

Abstract. In this paper, we present a precoding scheme of unitary space-time modulation (USTM) system considering that the fading channel is spatial-temporal correlated, which is a typical wireless communication scenario for high speed railway. The design of the precoder uses transmit correlation matrix of the channel, aiming at minimizing the mean-square error of channel estimation, where the receiver utilizes all of channel correlation information to the non-coherent detection method. Simulation results show that the proposed scheme can significantly outperform the non-coherent receiver without precoding.

Keywords: USTM · Precoding · Spatial-temporal correlated · Channel estimation

1 Introduction

With the rapid development of the high-speed railway (HSR) communications, the traditional multi-antenna transmission technology, such as space-time coding, is facing great challenges as the high mobility makes the channel estimation difficult. As a non-coherent transmission scheme, unitary space-time modulation (USTM) [1] is expected to be an alternative for HSR communications, where the receiver is not necessary to know the exact channel coefficients. For high speed railway (HSR), however, the poor-scattering wireless channel may cause high spatial-temporal correlation, which greatly impact performance of USTM system. Reference [2] analyzed that spatial correlation reduces the coding gain of USTM system and degrades system performance. In fact, even if the spatial correlation information is known at receiver, it is no use to improve the system performance [3]. Reference [4] designed a precoder for USTM that can exploit the spatial correlation, and [5] proposed a MMSE precoder for USTM, both assuming that the channel is quasi-static in one space-time block.

J.-S. Pan et al. (eds.), *Advances in Smart Vehicular Technology, Transportation, Communication and Applications*, Smart Innovation, Systems and Technologies 86,
https://doi.org/10.1007/978-3-319-70730-3_15

Nevertheless, little attention has been pay to the precoding problem in the case where the channel is time-varying successively.

In this study, we present a linear MMSE precoder for USTM in spatially-correlated, successively-time-varying, flat fading channel with channel spatial correlation available at the transmitter and all channel spatial-temporal correlation available at the receiver. The simulation results show that our proposed scheme can efficiently improve the error performance compared with the non-precoding case.

Notations used in this paper are as follows: (*) denotes Khatri-Rao product; $(\cdot)^{\dagger}$, $(\cdot)^{\mathrm{T}}$ and $(\cdot)^{\mathrm{H}}$ denotes pseudo-inverse, transpose and conjugate transpose of a matrix, respectively; vec(·) denotes the vectorization of a matrix; tr(·) and det(·) denotes the trace and the determinant of a matrix, respectively; $\mathbf{I}_n$ is identity matrix with $n \times n$; diag $\{a_1, \ldots, a_n\}$ is a diagonal matrix with $n \times n$. The zero-mean, unit-variance, circularly symmetric, complex Gaussian distribution is denoted by CN (0, 1).

2 System Model

Consider a communication link comprising M transmit antennas and N receive antennas that operates in a Rayleigh flat-fading environment. Each receiver antenna responds to each transmitter antenna though a statistically correlated fading coefficient that is successively time-varying for T symbol periods. At time t, we transmit the signal $\{s_{tm}, t = 1, \ldots, T, m = 1, \ldots, M\}$ on M transmit antennas, and we receive the noisy signal $\{y_{tn}, t = 1, \ldots, T, n = 1, \ldots, N\}$ on N receive antennas

$$y_{tn} = \sqrt{\rho} \sum_{m=1}^{M} s_{tm} h_{tmn} + w_{tn} \tag{1}$$

Here h_{tmn} is the complex-valued fading coefficient at time t between the mth transmit antenna and the nth receive antenna. The fading coefficients are sample-to-sample variant and spatially-correlated. The additive noise at time t and receive antenna n is denoted as w_{tn}, and is independent and identically distributed CN (0, 1). The quantities in the signal model (1) are normalized so that ρ represents the expected signal-to-noise ratio (SNR) at each receive antenna. With a matrix notation, we can rewrite (1) compactly to obtain

$$\mathbf{Y} = \sqrt{\rho}\left(\mathbf{I}_T * \mathbf{S}^{\mathrm{H}}\right)^{\mathrm{H}} \underline{\mathbf{H}} + \mathbf{W} \tag{2}$$

where

$$\mathbf{S} = \begin{bmatrix} s_{11} & \cdots & s_{1M} \\ \vdots & \ddots & \vdots \\ s_{T1} & \cdots & s_{TM} \end{bmatrix}, \quad \mathbf{Y} = \begin{bmatrix} y_{11} & \cdots & y_{1N} \\ \vdots & \ddots & \vdots \\ y_{T1} & \cdots & y_{TN} \end{bmatrix},$$

$$\mathbf{H}_t = \begin{bmatrix} h_{t11} & \cdots & h_{t1N} \\ \vdots & \ddots & \vdots \\ h_{tM1} & \cdots & h_{tMN} \end{bmatrix}, \quad \mathbf{W} = \begin{bmatrix} w_{11} & \cdots & w_{1N} \\ \vdots & \ddots & \vdots \\ w_{T1} & \cdots & w_{TN} \end{bmatrix},$$

and $\underline{\mathbf{H}} = \left[\mathbf{H}_1^T \ldots \mathbf{H}_T^T\right]^T$.

Suppose that we use unitary space–time modulation with precoding, i.e., $\mathbf{S} = \sqrt{T/M}\Phi_l\mathbf{P}(l = 1,\ldots,L)$, where unitary space-time codes $\mathbf{\Phi}_l$ satisfy $\Phi_l^{\mathrm{H}}\Phi_l = \mathbf{I}_M(l = 1,\ldots,L)$ and the precoding matrix $\mathbf{P}$ satisfies tr($\mathbf{PP}^{\mathrm{H}}$) = M, then $\mathbf{\Phi}_l\mathbf{P}$ is launched into MIMO channel.

Let $\underline{\mathbf{h}}$ = vec($\underline{\mathbf{H}}$). In order to analyze conveniently, we assume that the channel correlation matrix is based on Kronecker structure [3], that is

$$\mathbf{R}_{\underline{\mathbf{h}}} = \mathbb{E}\left[\underline{\mathbf{h}}\underline{\mathbf{h}}^{\mathrm{H}}\right] = \mathbf{R}_N \otimes \left(\mathbf{R}_T \otimes \mathbf{R}_M\right) \tag{3}$$

where $\mathbf{R}_M$, $\mathbf{R}_N$ and $\mathbf{R}_T$ denote the transmit, receive and time correlation matrix, respectively. Thus, the fading channel can be considered as spatial-temporal correlated.

3 Joint Design of Precoder and Receiver

3.1 Precoder Design

Considering that the transmit correlation of the channel matrix may be utilized at transmitter, we rewritten the channel matrix $\underline{\mathbf{H}}$ in (1) as

$$\underline{\mathbf{H}} = \mathbf{I}_T \otimes \left(\mathbf{R}_M^{1/2}\mathbf{H}_w\right) + \underline{\mathbf{E}} \tag{4}$$

where $\mathbf{H}_w$ is a $M{\times}N$ independent channel matrix which is constant for T symbol periods; $\underline{\mathbf{E}}$ is a zero-mean Gaussian matrix concerned with receive correlation and time correlation of fading channel. Thus (1) is

$$\begin{aligned}\mathbf{Y} &= \sqrt{\frac{\rho T}{M}}\mathbf{\Phi}_l\mathbf{P}\mathbf{R}_M^{1/2}\mathbf{H}_w + \sqrt{\frac{\rho T}{M}}\left(\mathbf{I}_T * (\mathbf{\Phi}_l\mathbf{P})^{\mathrm{H}}\right)^{\mathrm{H}}\underline{\mathbf{E}} + \mathbf{W} \\ &= \mathbf{X}\left(\mathbf{R}_M^{1/2}\mathbf{H}_w + \mathbf{E}\right) + \mathbf{W}\end{aligned} \tag{5}$$

where

$$\mathbf{X} = \sqrt{\frac{\rho T}{M}}\mathbf{\Phi}_l\mathbf{P} \tag{6}$$

and

$$\mathbf{E} = (\mathbf{\Phi}_l\mathbf{P})^{\dagger}\left(I_T * (\mathbf{\Phi}_l\mathbf{P})^{\mathrm{H}}\right)^{\mathrm{H}}\underline{\mathbf{E}} \tag{7}$$

Obviously, Eq. (7) contains both the time correlation and receive correlation information of the channel. Let $\mathbf{y}(n)$, $\mathbf{h}(n)$, $\mathbf{w}(n)$, $\mathbf{h}_w(n)$ and $\mathbf{e}(n)$ be the nth column of $\mathbf{Y}$, $\mathbf{R}_M^{1/2}\mathbf{H}_w + \mathbf{E}$, $\mathbf{W}$, $\mathbf{H}_w$ and $\mathbf{E}$, respectively. Then we have

$$\mathbf{y}(n) = \mathbf{X}\mathbf{h}(n) + \mathbf{w}(n) \tag{8}$$

where

$$\mathbf{h}(n) = \mathbf{R}_M^{1/2}\mathbf{h}_w(n) + \mathbf{e}(n) \tag{9}$$

Thus far, Eqs. (8) and (9) now can be seen as state-space equations describing a two-state linear time-varying system. A minimum variance unbiased estimate of channel gains can be obtained via a Kalman filter [5]. Suppose that two states of Kalman filter is the kth state and the $(k + 1)$th state respectively. According to Eq. (6.6.20) in [6], the one-step ahead forecast mean-square error (MSE) matrix of channel estimates is given by

$$\begin{aligned}\mathbf{\Sigma}_{k+1|k} &= \mathbf{R}_M^{1/2}\mathbf{\Sigma}_k\left(\mathbf{R}_M^{1/2}\right)^{\mathrm{T}} + \mathbf{Q}\\ &= \mathbf{R}_M^{1/2}\left(\mathbf{\Sigma}_{k|k-1}^{-1} + \mathbf{X}^{\mathrm{H}}\mathbf{X}\right)^{-1}\left(\mathbf{R}_M^{1/2}\right)^{\mathrm{T}} + \mathbf{Q}\\ &= \mathbf{R}_M^{1/2}\left(\mathbf{\Sigma}_{k|k-1}^{-1} + \frac{\rho T}{M}\mathbf{P}^{\mathrm{H}}\mathbf{P}\right)^{-1}\left(\mathbf{R}_M^{1/2}\right)^{\mathrm{T}} + \mathbf{Q}\end{aligned} \tag{10}$$

where $\mathbf{Q}$ is the covariance matrix of $\mathbf{e}(n)$, $\mathbf{\Sigma}_k$ is the mean-square error matrix of the state estimates. As SNR $\to \infty$, we have

$$\left(\mathbf{\Sigma}_{k|k-1}^{-1} + \frac{\rho T}{M}\mathbf{P}^{\mathrm{H}}\mathbf{P}\right)^{-1} \to \frac{M}{\rho T}\left(\mathbf{P}^{\mathrm{H}}\mathbf{P}\right)^{-1} \tag{11}$$

Thus, $\mathbf{\Sigma}_{k+1|k} \to (M/\rho T)\mathbf{R}_M^{1/2}\left(\mathbf{P}^{\mathrm{H}}\mathbf{P}\right)^{-1}\left(\mathbf{R}_M^{1/2}\right)^{\mathrm{T}} + \mathbf{Q}$ at high SNR. Chose the appropriate precoding matrix $\mathbf{P}$ to minimize the one-step ahead forecast MSE matrix, we model precoding design problem as

$$\begin{aligned}&\min_{\mathbf{P}} f(\mathbf{P}) = tr\left\{\mathbf{R}_M\left(\mathbf{P}^{\mathrm{H}}\mathbf{P}\right)^{-1}\right\}\\ &s.t. \qquad tr\left\{\mathbf{P}^{\mathrm{H}}\mathbf{P}\right\} = M\end{aligned} \tag{12}$$

By the singular value decomposition (SVD) of the matrix P and the eigenvalue decomposition (ED) of the matrix $\mathbf{R}_M$, i.e., $\mathbf{P} = \mathbf{V}_{\mathbf{P}}\mathbf{D}_{\mathbf{P}}\mathbf{U}_{\mathbf{P}}^{\mathrm{H}}$ and $\mathbf{R}_M = \mathbf{V}_{\mathbf{R}}\mathbf{\Lambda}_{\mathbf{R}}\mathbf{V}_{\mathbf{R}}^{\mathrm{H}}$, where $\mathbf{V_P}$, $\mathbf{U_P}$, $\mathbf{V_R}$ is unitary matrix, $\mathbf{D_P} = \mathrm{diag}\{p_1, \ldots, p_M\}$ and $\mathbf{\Lambda_R} = \mathrm{diag}\{r_1, \ldots, r_M\}$. The cost function in (12) is reduced to

$$f(\mathbf{P}) = tr\left\{\left(\mathbf{U_P^H V_R}\right)\mathbf{\Lambda_R}\left(\mathbf{U_P^H V_R}\right)^{\mathrm{H}}\mathbf{D_P^{-2}}\right\} \tag{13}$$

Then we orderly design $\mathbf{V_P}$, $\mathbf{U_P}$ and $\mathbf{D_P}$.

(1) Since the matrix $\mathbf{V_P}$ does not affect the cost function, which means it can be any unitary matrix, we let $\mathbf{V_P} = \mathbf{I}_M$.
(2) According to Lemma 11 in [7], it satisfies $\mathbf{U_P^H V_R} = \mathbf{I}_M$ in order to minimizing the trace function in (13), then we set $\mathbf{U_P} = \mathbf{V_R}$.
(3) The cost function is further reduced to

$$f(\mathbf{P}) = tr\left\{\mathbf{\Lambda_R D_P^{-2}}\right\} \tag{14}$$

Using Lagrangian multiplier algorithm, it can be easy to derive each diagonal value of the matrix $\mathbf{D_P}$

$$p_i = \left(r_i \frac{M}{\sum_{j=1}^{M} r_j^{1/2}}\right)^{1/2}, i = 1, \cdots, M \tag{15}$$

(4) The resulting precoding matrix $\mathbf{P}$ is given by

$$\mathbf{P} = \mathbf{D_P V_R^H} \tag{16}$$

3.2 Receiver Design

At receiver, rewriting both sides of (2) by vectorization operation, we obtain

$$\mathbf{y} = \sqrt{\frac{\rho T}{M}}\left(\mathbf{I}_T \otimes \left(\mathbf{I}_T * (\mathbf{\Phi}_l \mathbf{P})^{\mathrm{H}}\right)^{\mathrm{H}}\right)\underline{\mathbf{h}} + \mathbf{w} \tag{17}$$

where $\mathbf{y} = \mathrm{vec}(\mathbf{Y})$, $\underline{\mathbf{h}} = \mathrm{vec}(\underline{\mathbf{H}})$ and $\mathbf{w} = \mathrm{vec}(\mathbf{W})$. Under the condition of known $\mathbf{\Phi}_l$, the received signal vector $\mathbf{y}$ follows complex Gaussian distribution with zero mean and covariance matrix

$$\mathbf{R}_{\mathbf{y}_l} = \mathbf{I}_{TN} + \frac{\rho T}{M} R_N \otimes \left(\mathbf{\Phi}_l \mathbf{P} \mathbf{R}_M (\mathbf{\Phi}_l \mathbf{P})^{\mathrm{H}} \odot \mathbf{R}_T\right) \tag{18}$$

A non-coherent maximum likelihood detection [1] is used and then we get the optimal non-coherent receiver

$$\begin{aligned} &\arg\max_{\boldsymbol{\Phi}_l}\{p(\mathbf{y}|\boldsymbol{\Phi}_l)\} \\ &= \arg\max_{\boldsymbol{\Phi}_l}\left\{\frac{\exp\left(-\mathbf{y}^{\mathrm{H}}\mathbf{R}_{\mathbf{y}_l}^{-1}\mathbf{y}\right)}{\det\left(\mathbf{R}_{\mathbf{y}_l}\right)}\right\} \\ &= \arg\min_{\boldsymbol{\Phi}_l}\left\{\mathbf{y}^{\mathrm{H}}\mathbf{R}_{\mathbf{y}_l}^{-1}\mathbf{y}\right\} \end{aligned} \tag{19}$$

4 Simulation Result

The Monte Carlo simulation results of the USTM system are provided here. For simplicity, let the number of transmit antennas be $M = 2$, the number of receive antennas be $N = 1$. Let the period of a space-time code be $T = 5$. Let the number of unitary space-time code set be $L = 2$. The space correlation matrix of the channel is modeled as $\mathbf{R}_M = [1\ r;\ r\ 1]$ and $\mathbf{R}_N = 1$. We consider that the channel time correlation follows Jakes' model [8], i.e., $[\mathbf{R}_T]_{ij} = J_0(2\pi f_d T_s|i–j|)$, where $f_d T_s$ denotes the normalized Doppler frequency shift.

Set $r = 0.9$ and $f_d T_s = 0.05$. Figure 1 shows the bit-error rates (BER) of USTM system with and without precoding, marked as "SCTV channel, proposed" and "SCTV channel, optimal receiver", respectively. We see that the system with a precoder outperforms the system with no precoder at all SNRs. Besides, our proposed scheme even outperforms the scheme without precoding under spatially-independent and time-

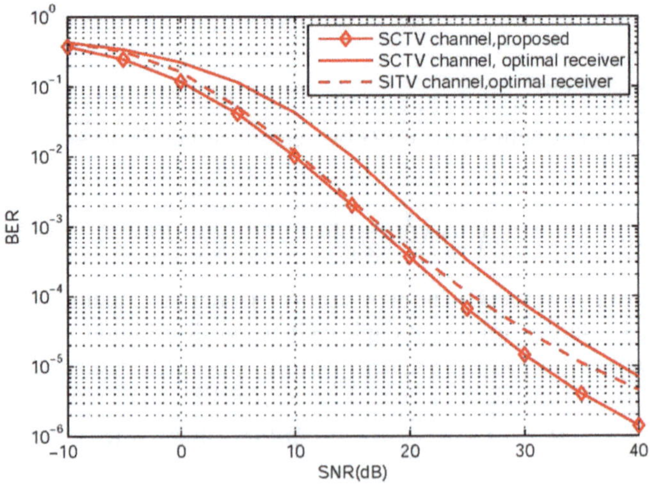

Fig. 1. BER performance of the USTM system with and without precoding under different channel condition.

varying channel marked as "SITV channel, optimal receiver", which means our proposed scheme obtain additive performance gain because of the use of transmit correlation information of the channel at transmitter.

5 Conclusion

In this paper, we have designed a precoding scheme for USTM system under the spatial-temporal correlated flat fading channel. The exact channel coefficient can be unknown at both transmitter and receiver. Compared with USTM without precoding, simulation results demonstrate a significant improvement of error performance of USTM system with precoding.

Acknowledgment. This work was supported by Fujian University of Technology research fund project and the NSFC projects (No. 61701118 and No. 61461026).

References

1. Hochwald, B.M., Marzetta, T.L.: Unitary space-time modulation for multiple-antenna communications in Rayleigh flat fading. IEEE Trans. Inf. Theor. **46**(2), 543–564 (2000)
2. Guo, Y., You, X.: Impact of channel correlation on performance of unitary space-time codes. Chin. J. Electron. **20**(3), 545–549 (2011)
3. Yu, C., Li, H.: Research on velocity-information-based Unitary space-time modulation in train-wayside communication under high-speed railway scenario. J. China Railway Soc. **37**(9), 69–74 (2015)
4. Guo, Y., Zhu, S., Liang, Z.: Linear precoding of unitary space-time code for GLRT decoder. IEICE Trans. Fundam. Electron. Commun. Computerences **91**(A.2), 695–699 (2008)
5. Nguyen, V.K., Xiang, Y.: MMSE precoder for unitary space-time codes in correlated time-varying channels. IEEE Signal Process. Lett. **12**(8), 569–572 (2005)
6. Zhao, S., Zhao, J.: Detection and Estimation Theory. Tsinghua University Press, Beijing (2005)
7. Palomar, D.P., Cioffi, J.M., Lagunas, M.A.: Joint Tx-Rx beamforming design for multicarrier MIMO channels: a unified framework for convex optimization. IEEE Trans. Sig. Process. **51**(9), 2381–2401 (2003)
8. Jakes, W.C.: Microwave Mobile Communications. IEEE Press, Piscataway (1993)

Innovative Electrical Technology

The Integration of DFX Principles with TRIZ for Product Design – A Case Study of Electric Scooter

Tien-Lun Liu(✉), Yi-Chen Li, and Ji-Ze Xiao

Department of Industrial Engineering and Management, St. John's University, 499, Section 4, Tam King Road, Tamsui District, New Taipei City 25135, Taiwan, R.O.C.
tliu@mail.sju.edu.tw

Abstract. In order to improve product development, enterprises need to strengthen product lifecycle management (PLM) and collaborative design capability. For the description of product design we have proposed a four-level structural classification scheme, which includes Design Intentions, Design Requirements, Design Parameters and TRIZ Engineering Parameters. This study is based on the classification scheme, and introduces the concept of Design for X (DFX) and then discusses the design principles from four different perspectives including manufacturing aspect, customer aspect, maintenance aspect and environmental aspect. We deliberate the meaning of these aspects and further establish their relevance with the TRIZ 48 engineering parameters. An electric scooter design is presented to illustrate the application.

Keywords: Design for X · TRIZ · Product design · Electric scooter

1 Introduction

In current product development, enterprises often need to cooperate as a project team, or even need to work with other enterprises to accelerate the development process. This paper explores the problems of industrial innovation, and tries to combine the TRIZ method to integrate the related operations in product design. TRIZ is mostly used at the conceptual design stage to generate design concepts and solve problems. Therefore, the main purposes of this study are as follows:

1. To assist collaborative design and provide better understanding for team members to communicate each other more efficiently during product development process.
2. To combine the characteristics of TRIZ into the product design and then help resolve design problems through TRIZ methods when needed.

J.-S. Pan et al. (eds.), *Advances in Smart Vehicular Technology, Transportation, Communication and Applications*, Smart Innovation, Systems and Technologies 86,
https://doi.org/10.1007/978-3-319-70730-3_16

2 Literature Review

2.1 Design Collaboration

Fan (2014) depicted that how to assure the designate processes are easy to follow-up and clearly to understand is the key to success. If have to build the collaborative development working environment, better to do it at the earliest phase. It's a hard requirement to reach every aspect of agreement from different departments to functions. While agreements define the quality of a project, thus all project success elements including guidelines, priorities, communications, or endorsements must be unambiguously identified at project kick-off and agreed upon related parties. Clearly defining the ways to execute jobs help teams from confusing or arguments during issue managements. This research is considered on the base of the collaborative platform for the product development due to the different backgrounds of the team members. The design requests from the various departments in the process of product development are integrated, and the product design information is then compiled and analyzed to achieve the goal of the project.

2.2 Design for EXcellence

In the spirit of collaborative development, there is a very important concept to product design - early involvement in the product life cycle, which results in a "Design for X (DFX)" methodology. Since the 1970s, many DFX ideas have been proposed by experts and scholar. Kuo et al. (2001) pointed out that DFX covers Design for Assembly (DFA), Design for Manufacturing (DFM), Design for Disassembly (DFD), Design for Environment, DFE) and Design for Quality (DFQ) and so on. The main purpose of DFX is to hopefully fulfill different design considerations in the product life cycle, thereby enhancing the product capabilities. It may also refer to "DFX" as "Design for eXcellence" for representing the achievement of a variety of design goals. This study takes DFX as the initial design intention and hopes to integrate the design requests from various departments.

2.3 TRIZ

The TRIZ theory is a systematic innovation approach, which was mainly established by Russian scientist Altshuller (2000) in 1946. TRIZ is abbreviated from Russian and interpreted in English as "Theory of Inventive Problem Solving". TRIZ theory was concluded methodically via analyzing more than twenty hundred thousand patents, and inferring the resolving techniques involved in these patents. There are several problem defining methods and solving tools in TRIZ, including System Operator, Ideality, Resource, Contradiction Matrix, 40 Inventive Principles, Separation Principles, Substance-Field Analysis, Patterns of Evolutions, ARIZ, etc.

The concept of Contradictions is one of the fundamental philosophies in TRIZ. Contradictions existing in the problem imply potential opportunities to innovation.

Contradictions are classified into two different categories. One is technical contradiction and the other is physical contradiction. A technical contradiction is defined as two different engineering parameters conflict each other in the problem; for instance, a car needs to be fast but energy-saving. A physical contradiction means an engineering parameter conflicts by itself, such as a car needs to be longer for better space but also needs to be shorter for easy parking. Contradiction Matrix is formulated to deal with technical contradictions, and the Separation principles are developed to solve physical contradictions.

3 Four-Level Structural Classification

In our previous research (Liu et al. 2016), authors discussed product design requests from project team members and made connection to TRIZ engineering parameters according to their characteristics and attributes of the design. In such a way the TRIZ contradiction analysis can be applied once encountering conflicts. The classification hierarchy was developed into four levels, which are Design Intention, Design Requirement, Design Parameter and Engineering Parameter, as explained below:

- First level - Design Intentions (DIs): The concept of DFX is applied as the first level to comprehend each collaborative team member's aim on his/her design demands. The typical DFXs used in this level are Design for Manufacture (DFM), Design for Assembly (DFA), Design for Customer (DFC) and Design for Environment (DFE). However the DFXs can be expanded if necessary.
- Second level - Design Requirements (DRs): These are explicit requests from the team members to state their purposes on DFXs. The design requirements should be concrete and meaningful.
- Third level - Design Parameters (DPs): These are the product specifications which are typically related to the parts or components. DPs are usually quantifiable.
- Fourth level - Engineering Parameters (EPs): The definitions of EPs are the same as discussed in TRIZ theory. However the 48 engineering parameters proposed by Mann (2003) are applied instead of 39 in classical TRIZ as shown in Table 1.

Founded on the four-level structural hierarchy a more comprehensive scheme for the multi-disciplinary design requests is constructed. A basic conceptual example is illustrated in Fig. 1. By arranging the design requests in such a way, the cross-functional team members can understand what and why the product design is deliberated from different viewpoints. Particularly linking to TRIZ engineering parameters, the design conflicts may be examined in more details and possibly resolved through the contradiction matrix and inventive principles. Therefore the efficiency of collaborative product development could be improved to achieve the effectiveness of product design quality.

Table 1. The 48 engineering parameters in TRIZ (Matrix 2003)

1	Weight of moving object	25	Loss of substance
2	Weight of non-moving object	26	Loss of time
3	Length of moving object	27	Loss of energy
4	Length of stationary object	28	Loss of information
5	Area of moving object	29	Noise
6	Area of stationary object	30	Harmful emissions
7	Volume of moving object	31	Other harmful effects generated by system
8	Volume of stationary object	32	Adaptability/versatility
9	Shape	33	Compatibility/connectivity
10	Amount of substance	34	Trainability/operability/controllability
11	Amount of information	35	Reliability/robustness
12	Duration of action of moving object	36	Repairability
13	Duration of action of stationary object	37	Security
14	Speed	38	Safety/vulnerability
15	Force/torque	39	Aesthetics/appearance
16	Energy used by moving object	40	Other harmful effects acting on system
17	Energy used by stationary object	41	Manufacturability
18	Power	42	Manufacturing precision/consistency
19	Stress/Pressure	43	Automation
20	Strength	44	Productivity
21	Stability	45	System complexity
22	Temperature	46	Control complexity
23	Illumination intensity	47	Ability to detect/measure
24	Function efficiency	48	Measurement precision

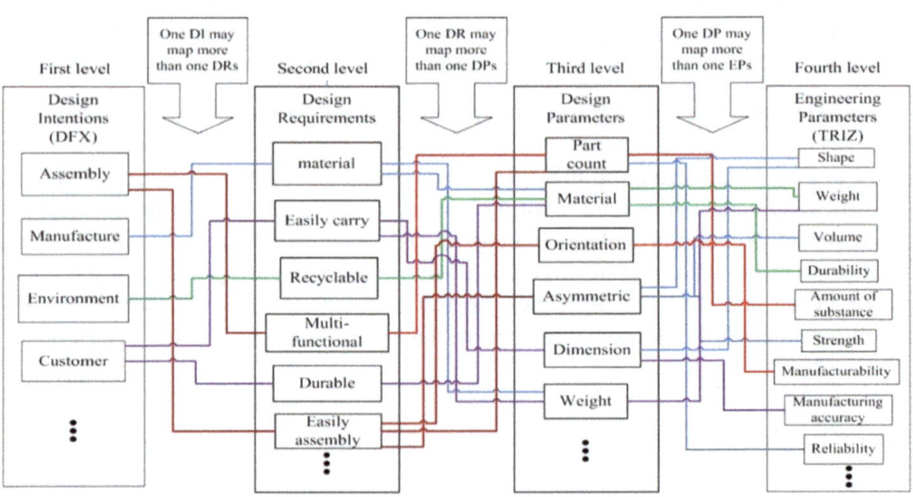

Fig. 1. The four-level structural classification

4 Case Study

An electric scooter design is applied to demonstrate the proposed methodology as we consider the DFXs for four different design purposes - manufacturing, customer, maintenance and environmental protection. In order to understand the applicability of the four-tier architecture, this section will assume two design situations of the electric scooter. The first situation is considered as the design requirements without setting any premise of the design and the second situation is discussed for the design integration on the scooter frame.

4.1 The Situation Without Premise

The assumption of this situation is that the departments do not have any design premise at the beginning. As the product project proceeds, the team members put forward their design requests based on their backgrounds. For examples from the manufacturing point of view, it is desirable that the parts of the scooter frame can be reduced in assembly, and the degree of adhesion between the hulls is improved. In the maintenance point of view, the motor needs to be easily repaired and the battery can be easily picked up when replaced. As for a customer's viewpoints, a good battery capacity and long travelling distance are preferred as well as handling the battery can be lightweight and convenient. In addition, the storage space should be larger. From the environmental point of view, the scooter should be energy-saving and the battery has to be recycled. To sum up the design requests from different aspects, the corresponding four-level data structure is shown in Table 2.

Table 2. Four-level data structure of design requests from different aspects

DFX	Design requirement	Design parameter	No. engineering parameter
Manufacturing	Reduce the number of parts	Frame	10. Amount of substance 41. Manufacturability 45. System complexity
		Outer shell	
	Reduce part variation	Outer shell	41. Manufacturability 42. Manufacturing precision
Maintenance	Good accessibility	Battery	16. Energy used by moving object 36. Repairability
	Easy detection of diagnostic accuracy	Motor	46. Control complexity 47. Ability to detect/measure 48. Measurement precision

(*continued*)

Table 2. (*continued*)

DFX	Design requirement	Design parameter	No. engineering parameter
Customer	Suitable size and space to use	Storage space	9. Shape 34. Operability/controllability 40. Other harmful effects acting on system
	Reduce the physical burden	Battery	16. Energy used by moving object 24. Function efficiency
	Increase battery capacity	Battery	7. Volume of moving object 10. Amount of substance 40. Other harmful effects acting on system
	Long travelling distance	Battery Motor	12. Duration of action of moving object 24. Function efficiency 27. Loss of energy
Environment	Reduce energy consumption	Motor	27. Loss of energy
	Designed to be recyclable or reusable	Battery	25. Loss of substance 31. Other harmful effects generated by system

After examining the four-tier structure in Table 2, we can find that the immediately improved design parameters and request conflicts. The battery and the motor are significant targets to be focused. Furthermore, possible conflicts can be identified. For instance, the customer wants a bigger battery capacity, but the battery capacity is usually proportional to the volume and weight of the battery such that will impact the storage space of the scooter. Therefore after checking the Contradiction Matrix in TRIZ, we can find several inventive principles as shown in Table 3 for improvement directions. Note that there are total 40 inventive principles in TRIZ, the details can be found in Ref. ?.

Table 3. Contradition analysis for battery

Improving engineering parameters	Worsening engineering parameters	Suggested inventive principles number
10. Amount of substance	12. Duration of action of moving object	35, 40, 34, 10
	31. Other harmful effects generated by system	1, 35, 30, 31
	40. Other harmful effects acting on system	35, 9, 21, 22

The suggested inventive principles of this scenario are numbered 1, 9, 10, 21, 22, 30, 31, 34, 35, 40, and the most frequently appeared principle is 35 - *parameter change/changing properties*, which may be sought as the first attempt for solution.

4.2 Frame Integration Design

The scooter frame and the outer shell are assembled together, and then the frame will be manufactured in one forming. This way can reduce the number of parts and enhance the reliability of the frame as well as reduce the gap between the parts to achieve better appearance for customers. Nevertheless from the manufacturing aspect, the number of parts is reduced, the technical requirement of the frame manufacturing becomes higher. Parts reduction may help maintenance, but the integrated component needs be replaced entirely once damaged. As from the environmental aspect, the material usage is simpler, however the waste may increase if replaced as a whole. In addition the more complex manufacturing process may cause more pollution. In this situation, the corresponding four-level data structure is shown in Table 4.

Table 4. Four-level data structure of frame integration design

DFX	Design requirement	Design parameter	No. engineering parameter
Manufacturing	Reduce the parts count	Frame	10. Amount of substance
		Outer shell	45. System complexity
	Easy manufacture	Frame	41. Manufacturability 44. Productivity
Maintenance	Simplify the product and maintenance	Frame	10. Amount of substance 36. Repairability 45. System complexity
Customer	Good quality and appearance	Frame	35. Reliability/robustness
		Outer shell	35. Reliability/robustness 39. Aesthetics/appearance
Environment	Raw material type is single	Frame	10. Amount of substance
		Outer shell	33. Compatibility/connectivity 40. Other harmful effects acting on system
	Simplify manufacturing process	Frame	25. Loss of substance 27. Loss of energy

Frame integration is mainly to decrease body parts of the scooter, but the manufacturing process may become difficult and harmful to the environment, resulting in the conflict between DFM and DFE. Similarly by querying the Contradiction Matrix in TRIZ, we could find the corresponding inventive principles to look for possible solution as shown in Table 5.

Table 5. Contradition analysis for frame integration design

Improving engineering parameters	Worsening engineering parameters	Suggested inventive principles number
10. Amount of substance	41. Manufacturability	10, 2, 35, 1
	44. Productivity	1, 13, 3, 35, 36
	27. Loss of energy	35, 7, 18, 19
45. System complexity	41. Manufacturability	3, 12, 1, 28, 26
	44. Productivity	16, 12, 29, 8
	27. Loss of energy	35, 28, 13, 10

The suggested inventive principles of this scenario are numbered 1, 2, 3, 7, 8, 10, 12, 13, 16, 18, 19, 28, 35, 36, and the more frequently appeared principles are 35 (*Parameter change/changing properties*), 1 (*Segmentation*) and 12 (*Equipotentiality*). These inventive principles may be applied to solve the conflict problem.

5 Conclusion

This model may be applied to a cloud-based product development platform to aggregate the design requests from collaborative team members. Through the proposed hierarchical product design contents, members of the product development team can effectively express their intentions of the design, and communicate each other better to reduce the gap between different professional fields. Combined with TRIZ engineering parameters, the conflicts inbetween may be resolved by contradiction analysis in TRIZ. The product quality and time to market can then be improved effectively.

References

Altshuller, G.: The Innovation Algorithm: TRIZ, Systematic Innovation and Technical Creativity. Technical Innovation Center, Inc. (2000)

Fan, C.S.: A study on a product development and collaborative design process improvement of a transnational company—take the D company as an example. Master Thesis, National Taiwan University (2014)

Kuo, T.C., Huang, S.H., Zhang, H.C.: Design for manufacture and design for X: concept, applications, and perspectives. Comput. Ind. Eng. **41**(3), 241–260 (2001)

Liu, T.L., Li, Y.C., Li, M.L.: A framework of cloudbased collaborative platform to integrate product design requests and contradiction analysis. In: Proceedings of 2016 International Conference on Advanced Materials for Science and Engineering (ICAMSE), pp. 475–478 (2016)

Mann, D.L., Dewul, S.: Updating the TRIZ contradiction matrix. In: 5th Annual International Conference of Altshuller Institute for TRIZ Studies (2003)

Adaptive Nonsmooth Attitude Tracking Control of Quadrotor UAV with Dynamic Uncertainties

Dongwei He[1,2](✉), Pei Gao[3], Lisang Liu[1,2], Xuecheng Jiang[4], and Jishi Zheng[1,2]

[1] School of Information Science and Engineering, Fujian University of Technology, Fuzhou 350118, China
yzak_juel@hotmail.com

[2] Fujian Provincial Key Laboratory of Digital Equipment, Fujian University of Technology, Fuzhou 350118, China

[3] Department of Information Management Engineering, Fujian Business University, Fuzhou 350012, China

[4] Department of Electronics Engineering, Minjiang University, Fuzhou 350118, China

Abstract. This paper addresses a new adaptive attitude controller for quadrotor UAV in the presence of parameters uncertainty and bounded external disturbances. First, an attitude dynamic model is built for the quadrotor UAV, considering parameters uncertainty and bounded external disturbances. Moreover, an adaptive nonsmooth attitude controller is developed by integrating nonsmooth control and adaptive control techniques, based on backstepping approach. And the stability of the closed-loop system is analyzed based on the Lyapunov method. Finally, the effectiveness and advantages of the proposed method are illustrated via some simulation results and comparisons.

Keywords: Quadrotor · Attitude tracking control · Adaptive control · Nonsmooth control

1 Introduction

Quadrotor UAV (Unmanned Aerial Vehicle) has been developed for various military and civil missions, due to its simple structure, VTOL (Vertical Taskoff and Landing) and great maneuverability [1]. Quadrotor UAV is a typically coupled and underactuated nonlinear system. Its control system usually consist of attitude part and position part, whose position part guarantee the Quadrotor UAV to track the desired positions by changing the attitudes and total thrust, and the attitude part guarantee the quadrotor UAV to track the desired attitudes determined that given by position controller. It means that the position control performance is affected by the attitude tracking performance, besides several maneuver tasks require high attitude tracking performance. Unfortunately, it is not easy to achieve high attitude tracking performance, due to the coupled character and parameter uncertainties. Moreover, external disturbances also bring more challenge. Many techniques have been proposed to solve the above

J.-S. Pan et al. (eds.), *Advances in Smart Vehicular Technology, Transportation, Communication and Applications*, Smart Innovation, Systems and Technologies 86,
https://doi.org/10.1007/978-3-319-70730-3_17

problem. Tradition method such as PID, LQ control [2], robust feedback linearization [3], robust control [4] etc. are adopted to improve the attitude tracking performance for quadrotor UAV to some extent, but they can't achieve high performance because of ignoring the coupled character and parameter uncertainties during controller design. So many novel controller methods are proposed. The work [5] proposed a robust disturbance rejection control method based on robust disturbance observer (DOB), robust DOB was designed to compensate the uncertainty. In [6], An robust control method was addressed, adaptation laws were designed to learn and compensate the modeling error and external disturbance. In [7], A proportional-integral control method was proposed based on additive-state-decomposition (ASD) dynamic inversion design. By ASD and a new definition of output, the considered uncertain system is transformed into a first-order system, in which all original uncertainties were lumped into a new disturbance at the new defined output. And a dynamic inversion control was applied to reject the lumped disturbance.

In this brief, an adaptive nonsmooth attitude tracking controller is developed for quadrotor with dynamic uncertainties, integrating the nonsmooth control and adaptive control technology. Compared with the aforementioned method, the proposed method has fast transient, high-precision performances and simpler controller structures.

2 Preliminaries

2.1 Model Dynamics of Quadrotor UAV Attitude System

An attitude system model of the rigid-body quadrotor UAV with dynamic uncertainties can be written as (see [1–4]):

$$
\begin{aligned}
\ddot{\phi} &= \dot{\theta}\dot{\psi}(I_y - I_z)/I_x + blU_\phi/I_x + \Delta_\phi \\
\ddot{\theta} &= \dot{\phi}\dot{\psi}(I_z - I_x)/I_y + blU_\theta/I_y + \Delta_\theta \\
\dot{\psi} &= \dot{\phi}\dot{\theta}(I_x - I_y)/I_z + dU_\psi/I_z + \Delta_\psi
\end{aligned} \tag{1}
$$

Where $\phi \in \mathrm{R}$, $\theta \in \mathrm{R}$, $\psi \in \mathrm{R}$ denote the Euler angles (roll, pitch, yaw) respectively, $I_x \in \mathrm{R}^+$, $I_y \in \mathrm{R}^+$, $I_z \in \mathrm{R}^+$ are the moments of inertia of quadrotor, b is the lift coefficient, d is the drag coefficient, $l \in \mathrm{R}^+$ is the distance from the epicenter of the quadrotor UAV to the rotor axis, $\Delta_\varphi \in \mathrm{R}$, $\Delta_\theta \in \mathrm{R}$, $\Delta_\psi \in \mathrm{R}$ denote the uncertainties caused by bounded external disturbance. And U_ϕ, U_θ, U_ψ are given as follows:

$$
\begin{aligned}
U_\phi &= \omega_2^2 - \omega_4^2 \\
U_\theta &= -\omega_1^2 + \omega_3^2 \\
U_\psi &= \omega_1^2 - \omega_2^2 + \omega_3^2 - \omega_4^2
\end{aligned} \tag{2}
$$

Where $\omega_i \in \mathrm{R}, i = 1, 2, 3, 4$ denote the angular speed of the rotor i. For the design procedure of adaptive nonsmooth attitude tracking controller, the dynamic model will be transformed as follow.

Define $k_1 = bl$, $k_2 = d$, $p_{x1} = I_x/k_1$, $p_{y1} = I_y/k_1$, $p_{z1} = I_z/k_1$ $p_{x2} = I_x/k_2$, $p_{y2} = I_y/k_2$, $p_{z2} = I_z/k_2$, we can write Eq. (1) in the following form

$$\begin{aligned} U_\varphi &= p_{x1}\ddot{\varphi} - \dot{\theta}\dot{\psi}p_{y1} + \dot{\theta}\dot{\psi}p_{z1} + T_\varphi \\ U_\theta &= p_{y1}\ddot{\theta} - \dot{\varphi}\dot{\psi}p_{z1} + \dot{\theta}\dot{\psi}p_{x1} + T_\theta \\ U_\psi &= p_{z2}\dot{\psi} - \dot{\varphi}\dot{\theta}p_{x2} + \dot{\varphi}\dot{\theta}p_{y2} + T_\psi \end{aligned} \tag{3}$$

With: $T_\varphi = p_{x1}\Delta_\varphi$, $T_\theta = p_{y1}\Delta_\theta$, $T_\psi = p_{z2}\Delta_\psi$.

Then Eq. (3) can be represented as:

$$\boldsymbol{U} = \boldsymbol{M}\ddot{\boldsymbol{q}} + \boldsymbol{C}(\dot{\boldsymbol{q}}) + \boldsymbol{T} \tag{4}$$

where, $\boldsymbol{q} = [\,\varphi\ \theta\ \psi\,]^{\mathrm{T}}$, $\dot{\boldsymbol{q}} = [\,\dot{\varphi}\ \dot{\theta}\ \dot{\psi}\,]^{\mathrm{T}}$, $\ddot{\boldsymbol{q}} = [\,\ddot{\varphi}\ \ddot{\theta}\ \ddot{\psi}\,]^{\mathrm{T}}$, $\boldsymbol{C}(\dot{\boldsymbol{q}}) = [\,-\dot{\theta}\dot{\psi}p_{y1} + \dot{\theta}\dot{\psi}p_{z1} - \varphi\dot{\psi}p_{z1} + \dot{\theta}\dot{\psi}p_{x1} - \dot{\varphi}\dot{\theta}p_{x2} + \dot{\varphi}\dot{\theta}p_{y2}]^{\mathrm{T}}$, $\boldsymbol{U} = [\,U_\varphi\ U_\theta\ U_\psi\,]^{\mathrm{T}}$, $\boldsymbol{T} = [\,T_\varphi\ T_\theta\ T_\psi\,]^{\mathrm{T}}$, $\boldsymbol{M} - \mathrm{diag}(\,p_{x1}\ p_{y1}\ p_{z2}\,)$.

If we notate: $\boldsymbol{p} = [\,p_{x1}\ \ p_{y1}\ \ p_{z1}\ \ p_{x2}\ \ p_{y2}\ \ p_{z2}\ \ T_\varphi\ \ T_\theta\ \ T_\psi\,]^{\mathrm{T}}$ and $\boldsymbol{\Phi} \triangleq \boldsymbol{\Phi}(\,\dot{\boldsymbol{q}}\quad \ddot{\boldsymbol{q}}\,) = \begin{bmatrix} \ddot{\varphi} & -\dot{\theta}\dot{\psi} & -\dot{\theta}\dot{\psi} & & & & 1 & & \\ \dot{\theta}\dot{\psi} & \ddot{\theta} & -\dot{\varphi}\dot{\psi} & & & & & 1 & \\ & & & -\dot{\varphi}\dot{\theta} & \dot{\varphi}\dot{\theta} & \dot{\psi} & & & 1 \end{bmatrix}$.

Equation (3) could be derived as:

$$\boldsymbol{U} = \boldsymbol{\Phi}\boldsymbol{p} \tag{5}$$

Then the dynamic model of the quadrotor UAV with uncertainties could be represented as:

$$\boldsymbol{U} = \boldsymbol{\Phi}(\boldsymbol{p}_{\mathrm{n}} + \boldsymbol{p}_{\mathrm{u}}) \tag{6}$$

Where $\boldsymbol{p}_{\mathrm{n}}$ is the parameter vector of nominal system, $\boldsymbol{p}_{\mathrm{u}}$ is the unknown parameter vector of the quadrotor UAV dynamic model.

2.2 New Version of Barbalate's Lemma

Theorem 1. If $x(t)$ is a uniformly continuous function and if there exist a lower bounded scalar function V and a continuous positive definite and radially unbounded scalar function M such that $-\dot{V} \geq M(x(t))$, then $x(t) \to 0$ as $t \to \infty$ [8].

3 Adaptive Nonsmooth Attitude Tracking Control of Quadrotor UAV

Assumption 1. The reference attitude trajectory $\boldsymbol{q}_{\mathrm{r}} \in C^2$, and the sample rate of the controller is fast enough, so $\dot{\boldsymbol{p}} \approx 0$.

Backstepping approach will be adopted to design the adaptive nonsmooth attitude tracking controller (ANSATC) as follow.

Step 1. Let $\boldsymbol{x}_1 \triangleq \boldsymbol{q}$ and $\boldsymbol{x}_2 \triangleq \dot{\boldsymbol{q}}$, lead to $\dot{\boldsymbol{x}}_1 = \boldsymbol{x}_2$. Define $z_1 = \boldsymbol{x}_1 - \boldsymbol{x}_{1\mathrm{r}}$ and $z_2 = \boldsymbol{x}_2 - \boldsymbol{\beta}$, where $\boldsymbol{x}_{1\mathrm{r}} = \boldsymbol{q}_\mathrm{r}$ and $\boldsymbol{\beta}$ is the auxiliary part which will be designed later.

Consider the candidate Lyapunov function as follow

$$V_1 = z_1^\mathrm{T} z_1 / 2 \tag{7}$$

Differentiating V_1 versus time along the trajectories of system (1) yields

$$\dot{V}_1 = z_1^\mathrm{T} \dot{z}_1 \tag{8}$$

With the definitions above, it could be derived as

$$\dot{V}_1 = z_1^\mathrm{T}(\boldsymbol{x}_2 - \boldsymbol{x}_{2\mathrm{r}}) = z_1^\mathrm{T} z_2 + z_1^\mathrm{T}(\boldsymbol{\beta} - \boldsymbol{x}_{2\mathrm{r}}) \tag{9}$$

$\boldsymbol{\beta}$ is designed as

$$\boldsymbol{\beta} = \boldsymbol{x}_{2\mathrm{r}} - \boldsymbol{K}_1 \mathrm{sig}(z_1)^\alpha \tag{10}$$

Where, $\boldsymbol{K}_1 \in \mathrm{R}_+^{3\times 3}$ is a diagonal positive-definite gain matrix, $\alpha \in \mathrm{R}^+, 0 < \alpha < 1$. And the notation $\mathrm{sig}(z_1)^\alpha$ is defined as [9]

$$\mathrm{sig}(z_1)^\alpha \triangleq \left[\, |z_{1\varphi}|^\alpha \mathrm{sign}(z_{1\varphi}) \quad |z_{1\theta}|^\alpha \mathrm{sign}(z_{1\theta}) \quad |z_{1\psi}|^\alpha \mathrm{sign}(z_{1\psi}) \,\right]^\mathrm{T} \tag{11}$$

From (10) and (11), $\dot{V}_1$ could be deduced as

$$\dot{V}_1 = z_1^\mathrm{T} z_2 - \boldsymbol{K}_1 z_1^\mathrm{T} \mathrm{sig}(z_1)^\alpha \tag{12}$$

It is easy to deduce that $\dot{V}_1 = \boldsymbol{K}_1 z_1^\mathrm{T} \mathrm{sig}(z_1)^\alpha \leq 0$ as $z_2 = 0$.

Step 2. We define parameter matrixes of the nominal system as $\boldsymbol{M}_\mathrm{n}$, $\boldsymbol{C}_\mathrm{n}$ and $\boldsymbol{T}_\mathrm{n}$, and consider an desired system (parameters are known, without external disturbances) as the nominal system, from (4) and (6) it is easily obtained that

$$\boldsymbol{U} = \boldsymbol{\Phi} \boldsymbol{p}_\mathrm{u} + \boldsymbol{M}_\mathrm{n} \ddot{\boldsymbol{q}} + \boldsymbol{C}_\mathrm{n}(\dot{\boldsymbol{q}}) \tag{13}$$

Where $\boldsymbol{T}_\mathrm{n} = 0$, $\boldsymbol{M}_\mathrm{n}$ is a diagonal positive-definite matric. Then consider a new candidate Lyapunov function as follow

$$V = V_1 + \left(z_2^\mathrm{T} \boldsymbol{M}_n z_2 + \bar{\boldsymbol{p}}^\mathrm{T} \boldsymbol{\Gamma} \bar{\boldsymbol{p}}\right)/2 \tag{14}$$

Where $\bar{\boldsymbol{p}} = \boldsymbol{p} - \hat{\boldsymbol{p}}$, $\hat{\boldsymbol{p}}$ is the estimation result obtained by adaption laws, which will be designed next. $\boldsymbol{\Gamma} \in \mathrm{R}_+^{3\times 3}$ is a diagonal positive-definite gain matrices.

Differentiating V with respect to time and from (12) yields

$$\dot{V} = z_1^{\mathrm{T}} z_2 - \boldsymbol{K}_1 z_1^{\mathrm{T}} \mathrm{sig}(z_1)^{\alpha} + z_2^{\mathrm{T}} \boldsymbol{M}_n \dot{z}_2 + \bar{\boldsymbol{p}}^{\mathrm{T}} \boldsymbol{\Gamma} \dot{\bar{\boldsymbol{p}}} \tag{15}$$

Substituting Eq. (13) into (15), we obtain

$$\begin{aligned} \dot{V} &= z_1^{\mathrm{T}} z_2 - \boldsymbol{K}_1 z_1^{\mathrm{T}} \mathrm{sig}(z_1)^{\alpha} + z_2^{\mathrm{T}} \boldsymbol{M}_n \dot{z}_2 + \bar{\boldsymbol{p}}_{\mathbf{u}}^{\mathrm{T}} \boldsymbol{\Gamma} \dot{\bar{\boldsymbol{p}}}_{\mathrm{u}} \\ &= z_1^{\mathrm{T}} z_2 - \boldsymbol{K}_1 z_1^{\mathrm{T}} \mathrm{sig}(z_1)^{\alpha} + z_2^{\mathrm{T}} \left[-\boldsymbol{C}_{\mathrm{n}}(\dot{\boldsymbol{q}}) - \boldsymbol{\Phi} \boldsymbol{p}_{\mathrm{u}} - \boldsymbol{M}_{\mathrm{n}} \dot{\boldsymbol{\beta}} + \boldsymbol{U} \right] + \bar{\boldsymbol{p}}_{\mathbf{u}}^{\mathrm{T}} \boldsymbol{\Gamma} \dot{\bar{\boldsymbol{p}}}_{\mathrm{u}} \end{aligned} \tag{16}$$

Design $\boldsymbol{U}$ as follow

$$\boldsymbol{U} = \boldsymbol{U}_{\mathrm{nsc}} - z_1 + \boldsymbol{M}_{\mathrm{n}} \dot{\boldsymbol{\beta}} + \boldsymbol{C}_{\mathrm{n}}(\dot{\boldsymbol{q}}) + \boldsymbol{\Phi} \hat{\boldsymbol{p}}_{\mathrm{u}} \tag{17}$$

Where $\boldsymbol{U}_{\mathrm{nsc}} = -\boldsymbol{K}_2 \mathrm{sig}(z_2)^{\alpha}$, $\boldsymbol{K}_2 \in \mathrm{R}_{+}^{3\times 3}$ is a diagonal positive-definite gain matrices.

From (16), (17) can be derived as

$$\dot{V} = -\boldsymbol{K}_1 z_1 \mathrm{sig}(z_1)^{\alpha} - \boldsymbol{K}_2 \mathrm{sig}(z_2)^{\alpha} + \left(\dot{\bar{\boldsymbol{p}}}_{\mathbf{u}}^{\mathrm{T}} \boldsymbol{\Gamma} - z_2^{\mathrm{T}} \boldsymbol{\Phi}\right) \bar{\boldsymbol{p}}_{\mathrm{u}} \tag{18}$$

We let $\dot{\bar{\boldsymbol{p}}}_{\mathbf{u}}^{\mathrm{T}} \boldsymbol{\Gamma} - z_2^{\mathrm{T}} \boldsymbol{\Phi} = 0$, from Assumption 1 we can deduce the adaptation laws as follow [10]

$$\dot{\hat{\boldsymbol{p}}}_{\mathbf{u}}^{\mathrm{T}} = -z_2^{\mathrm{T}} \boldsymbol{\Phi} \boldsymbol{\Gamma}^{-1} \tag{19}$$

And we get $\dot{V} \le 0$. For the system analysis, we define scalar function $N\left(z_1^{\mathrm{T}}, z_2^{\mathrm{T}}\right)$ as

$$N\left(z_1^{\mathrm{T}}, z_2^{\mathrm{T}}\right) = \boldsymbol{K}_1 z_1 \mathrm{sig}(z_1)^{\alpha} + \boldsymbol{K}_2 \mathrm{sig}(z_2)^{\alpha} \tag{20}$$

From (14) and (20), it is easily to deduce that the infimum of V ($\inf\{V\}$) exists, and supremum of V satisfies $\sup\{V\} \le V(0)$, so z_1, z_2, $\bar{\boldsymbol{p}}_{\mathrm{u}}$ are bounded.

From (10), we can obtain

$$\dot{z}_1 = z_2 - \boldsymbol{K}_1 \mathrm{sig}(z_1)^{\alpha} \tag{21}$$

From (10), (13) and (17), we can obtain

$$\boldsymbol{M}_{\mathrm{n}} \dot{z}_2 = -z_1 - K_2 sig(z_2) + \boldsymbol{\Phi} \bar{\boldsymbol{p}}_{\mathrm{u}} \tag{22}$$

From (21) and (22), we can easily deduce that z_1 and z_2 are bounded, so z_1, z_2 are uniformly continuous function and $N\left(z_1^{\mathrm{T}}, z_2^{\mathrm{T}}\right)$ is radially unbounded. By Theorem 1 we can obtain that then $z_1 \to 0$, $z_2 \to 0$ as $t \to \infty$, which means the real attitude will tracking reference attitude trajectory without error as $t \to \infty$.

Remark 1. When $\alpha = 1$, the adaptive nonsmooth controller in (17) will become a type of normal adaptive controller, and it will become a PD controller when $\alpha = 1$ and $\hat{\boldsymbol{p}}_{\mathrm{u}} = 0$.

4 Simulation Results

The parameters of the quadrotor VAU are $m = 2\,\text{kg}$, $l = 0.2\,\text{m}$, $I_x = 1.25\,\text{kg}\cdot\text{m}^2$, $I_y = 1.25\,\text{kg}\cdot\text{m}^2$, $I_z = 2.5\,\text{kg}\cdot\text{m}^2$, $b = 2.98 \times 10^{-6}\,\text{N}\cdot\text{s}^2/\text{rad}^2$, $d = 1.14 \times 10^{-7}\,\text{N}\cdot\text{s}^2/\text{rad}^2$. And the sampling rate of controller is 10 kHz. For simplicity, the altitude control of the quadrotor UAV is ignored, which means the quadrotor UAV is holding the altitude eternally. The reference attitude trajectory could be represented as $\varphi_{\text{r}} = (\pi/6)\sin(2\pi)(\text{rad})$,$\theta_{\text{r}} = (\pi/6)\sin(2\pi)(\text{rad})$,$\psi_{\text{r}} = 0$.

And $\Delta_\phi = 10\%mgl/I_x$, $\Delta_\theta = 10\%mgl/I_y$, with the assumption that the parameters are unknown to controller and the nominal parameters are given as $k_{1\text{n}} = 0.8k_1$, $k_{2\text{n}} = 0.6k_2$, $I_{\text{xn}} = 0.7I_{\text{x}}$, $I_{\text{yn}} = 0.7I_{\text{y}}$, $I_{\text{zn}} = 0.7I_{\text{z}}$. Four experiments are designed with $\boldsymbol{K}_1 = 5 \times \text{diag}(1 \quad 1 \quad 1)$, $\boldsymbol{K}_2 = 4 \times 10^6 \times \text{diag}(1 \quad 1 \quad 1)$, and the different part of the experiments are shown as Table 1, which represent normal PD controller (Method 1), normal adaptive controller (Method 2), nonsmooth controller (Method 3), and the proposed method (Method 4) respectively.

The simulation results are illustrated in Fig. 1. Figure 1(a) presents the tracking results of ϕ, it is clearly that the proposed method (Method 4) has fastest transient and most precise tracking performances among the four experiments. The similar tracking results of θ are shown in Fig. 1(b). And the tracking results of ψ are shown in Fig. 1(c). Figure 2 shows the control input of the four experiments, which are all continuous without chattering.

Table 1. Table of experiment parameters.

Experiment	Parameters of controller	
1	$\alpha = 1$	$\Gamma^{-1} = 0 \times \text{diag}(1 \quad 1 \quad 1 \quad 1 \quad 1 \quad 1 \quad 1 \quad 1 \quad 1)$
2	$\alpha = 1$	$\Gamma^{-1} = 5 \times 10^6 \times \text{diag}(1 \quad 1 \quad 1 \quad 1 \quad 1 \quad 1 \quad 1 \quad 1 \quad 1)$
3	$\alpha = 0.8$	$\Gamma^{-1} = 0 \times \text{diag}(1 \quad 1 \quad 1 \quad 1 \quad 1 \quad 1 \quad 1 \quad 1 \quad 1)$
4	$\alpha = 0.8$	$\Gamma^{-1} = 5 \times 10^6 \times \text{diag}(1 \quad 1 \quad 1 \quad 1 \quad 1 \quad 1 \quad 1 \quad 1)$

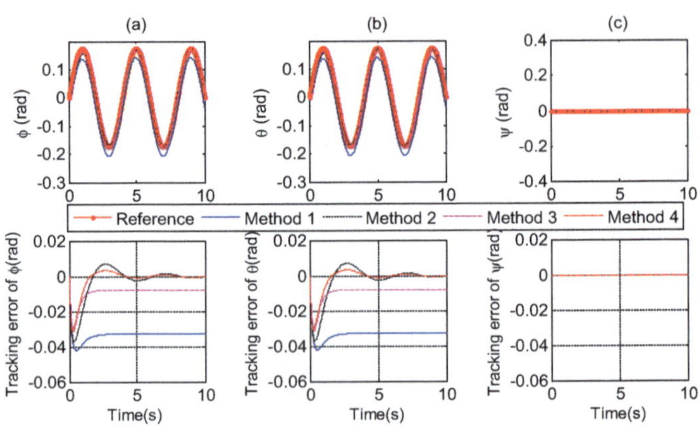

Fig. 1. Results of simulations

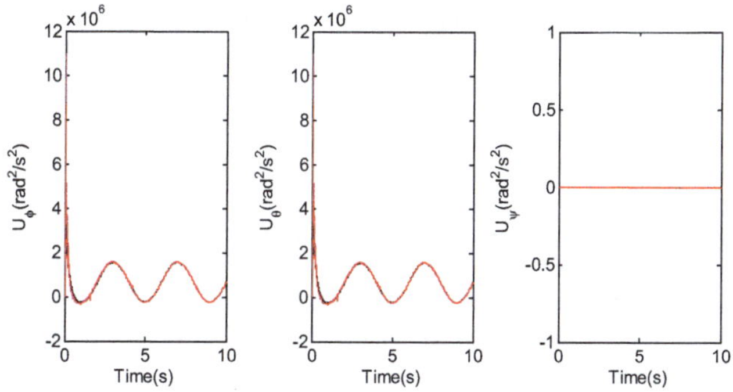

Fig. 2. Comparison of control input

Further, to evaluate the relationship between tracking performance and control parameter α, four experiments are carried out with $\alpha = 1$, $\alpha = 0.9$, $\alpha = 0.8$ and $\alpha = 0.7$. Figure 3 illustrates the tracking result of ϕ, we can easily find that tracking performance are the best when $\alpha = 0.7$, which means attitude tracking performance can be improved by reducing α. But α could not be arbitrary small, because an arbitrary small may lead to chatting in real system.

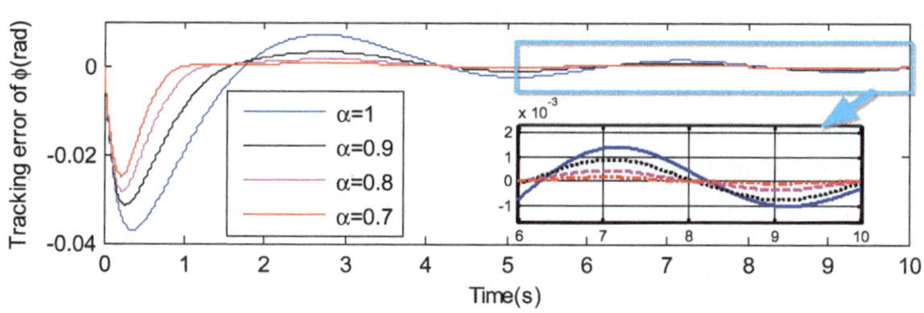

Fig. 3. Tracking result comparison of ϕ with different α

5 Conclusion

In this paper, an adaptive nonsmooth attitude tracking control of quadrotor UAV based on backstepping approach is presented. A dynamic model was built considering the dynamic uncertainties, an ANSATC was designed in detail, and the system stability is analyzed. Finally the effectiveness of the ANSATC was validated via serial simulations, which showed satisfactory transient and precise tracking performance while rejecting dynamic uncertainties.

Acknowledgements. The authors would like to thank the anonymous reviewers for their valuable comments. This work was supported in part by Initial Scientific Research Fund in Fujian University of Technology under Grant GY-Z13101, in part by the Natural Science Foundation of

Fujian Province under Grant 2017H11171120, and in part by the Surveying, Mapping and Geoinformation Technology Project of Fujian Province under Grant 2017JK05.

References

1. Bouabdallah, S., Murrieri, P., Siegwart, R.: Design and control of an indoor micro quadrotor. In: IEEE International Conference on Robotics and Automation, pp. 4393–4398 (2004)
2. Bouabdallah, S., Noth, A., Siegwart, R.: PID vs. LQ control techniques applied to an indoor micro quadrotor. In: Proceedings of the 2004 IEEE/RSJ International Conference on Intelligent Robots and Systems, pp. 2451–2456 (2004)
3. Mokhtari, A., Benallegue, A., Daachi, B.: Robust feedback linearization and H^{∞} controller for a quadrotor unmanned aerial vehicle. In: 2005 IEEE/RSJ International Conference on Intelligent Robots and Systems, pp. 1198–1203 (2005)
4. Lee, T., Leok, M., Mcclamroch, N.H.: Nonlinear robust tracking control of a quadrotor UAV on SE(3). In: Proceedings of the American Control Conference, vol. 15, no. 2, pp. 4649–4654 (2012)
5. Wang, L., Su, J.: Robust disturbance rejection control for attitude tracking of an aircraft. IEEE Trans. Control Syst. Technol. **23**(6), 1 (2015)
6. Islam, S., Liu, P.X., Saddik, A.E.: Robust control of four-rotor unmanned aerial vehicle with disturbance uncertainty. IEEE Trans. Ind. Electron. **62**(3), 1563–1571 (2015)
7. Quan, Q., Du, G.X., Cai, K.Y.: Proportional-integral stabilizing control of a class of MIMO systems subject to nonparametric uncertainties by additive-state-decomposition dynamic inversion design. IEEE/ASME Trans. Mechatron. **21**(2), 1092–1101 (2016)
8. Hou, M., Duan, G., Guo, M.: New versions of Barbalat's lemma with applications. J. Control Theor. Appl. **8**(4), 545–547 (2010)
9. Su, Y., Zheng, C.: Global finite-time inverse tracking control of robot manipulators. Robot. Comput. Integr. Manuf. **27**(3), 550–557 (2011)
10. Slotine, J.E., Li, W., Daizhan, C.: Applied Nonlinear Control. China Machine Press, Beijing (2006)

Design of the Classroom Intelligent Light Control System Based on ARM9

Xiu-Zhen Zhang(✉) and Li-Sang Liu

College of Information Science and Engineering,
Fujian University of Technology, Fuzhou 350118, China
yxz@fjut.edu.cn, liu_ls@163.com

Abstract. With aim to save energy that seriously wasted in classroom, this paper propose an intelligent light control system based onARM9 as core processor CPU by dividing the classroom lighting into several square area. The TSL230 programmable optical frequency converters were adopted to detect the current light intensity of the classroom while the pyroelectric electric infrared sensors were adopted to detect the person information in the room. The data was sent to ARM9 to make a reasonable switch lights decisions by comparing the data with the setting threshold of illuminance and that of people. The system was proved with the merits of simple operation, low cost, high reliability and energy saving.

Keywords: Classroom light · Embedded system · Pyroelectric infrared sensor · Energy saving

1 Introduction

Higher education has been paid great attention and the education population has also increased significantly. Thus the number of classrooms in the campus continues to increase. However, the lights in the classroom are always turned on even if there was few people in it due to the lack of awareness of energy conservation [1]. At night, it is common to see the whole lights are turned on in many classrooms while only a few students studying in the classroom. What is worse, the lights are still on after people leaving [2]. As a result, this kind of energy waste brings the water and electricity costs of the high school a heavy burden. A little waste of one classroom though, a great waste to the school day by day [3]. It is estimated that a school has 240 classrooms and each classroom save electricity about 3° a day, then more than 10 million RMB of electricity fee can be saved a year. As a result, a more intelligent, more effective control of the lights system, which will not affect the students' vision and learning, is urgently need. The system, with the purpose of energy conservation, can automatically turn on and off according to the actual degree of illumination in the daytime [4]. The intelligent light control system not only can achieve energy conservation, but also promise a comfortable lighting environment. Meanwhile, the main source of electricity in China is coal, the lighting energy saving can also reduce harmful gas emissions and the air pollution.

J.-S. Pan et al. (eds.), *Advances in Smart Vehicular Technology, Transportation, Communication and Applications*, Smart Innovation, Systems and Technologies 86,
https://doi.org/10.1007/978-3-319-70730-3_18

2 Principle of Classroom Lighting System

The intelligent classroom lighting control system determines the lights on and off according to the distribution of person and the illumination. The standard classroom is divided into five square areas named A, B, C, D, E area, shown in Fig. 1. The optical frequency conversion sensor and infrared pyroelectric sensors are installed in the ceiling of each area to detect the indoor light intensity and personnel flow, then the collected data was sent to the core processor to determine whether the light should be on or not.

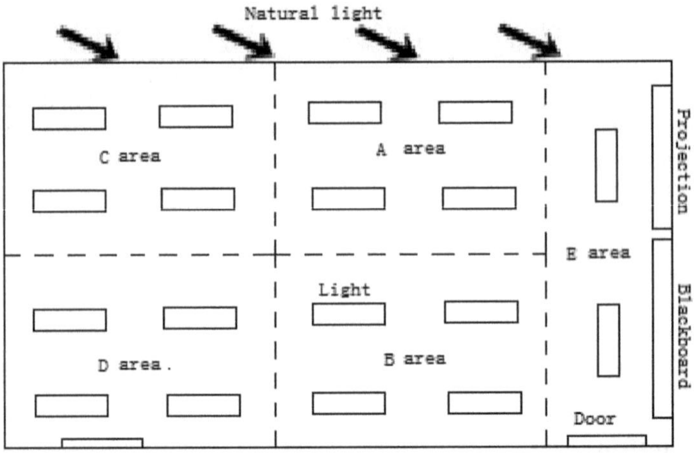

Fig. 1. Classroom light layout area

The international standard classroom lighting is controlled at 300Lx–500Lx, where Lx is the illumination unit, which refers to the luminous flux received per unit area on the illuminated surface [5]. If the natural illumination is not satisfied, the light illumination will be more than 300Lx, then the system will be mistakenly meet the requirements of light and turn off the lights. In order to avoid the occurrence of this phenomenon, an upper limit of 500Lx illumination is proposed taking the overlap of the natural light and light generation into account. The lights will be turn off when the natural light and light superimposed illumination is greater than 500Lx and turn on when that is less than 300Lx.

When the daylight illumination is not enough or at night, the lights should be turn on. Assuming the total number of a classroom is Y, then each area can commodate 1/4 of Y, let Y/4 = X. Take A Area as an example, when someone is detected in the A area, then turn on the A area lamps and other areas of lighting remain turn off. There are four lights in each area and only two of them will be turn on if person number is less than 1/4 of X, no matter where the person is sitting in this area. The scheme can save the energy efficiently.

The concrete rules of the scheme are as follows:

When P < X/4, turn on two lamps; When X/4 < P <= X/2, turn on 3 lamps; When P > X/2, turn on four lights; Where, P is the number of person.

On the contrary, the lights keep off if the illumination is sufficient.

3 Hardware Design

The S3C2410 chip with embedded ARM920T core was adopted as the main controller of the system, ARM is the abbreviation of Advanced RISC Machine, has the advantages of small size, low power consumption, low cost, high performance, data processing capacity and so on [6]. S3C2410 can support external expansion SDRAM and NAND FLASH, addressing space is 128 MB/block (8 blocks in total, that is, 1 GB), the chip has 4 KB SRAM, integrated MMU and rich on-chip peripherals, integrated independent 16 KB instruction cache and 16 KB data cache, support μC/OS-II, Window CE and embedded operating system, especially suits for mobile devices and small general-purpose embedded applications [7].

The overall structure of the system is shown in Fig. 2. The external components mainly include the human body signal acquisition module, the optical signal acquisition module, the clock module, the display module, the storage module and the I/O interface. The infrared pyroelectric sensor and optical frequency converter are used for real-time monitoring and detection of the classroom.

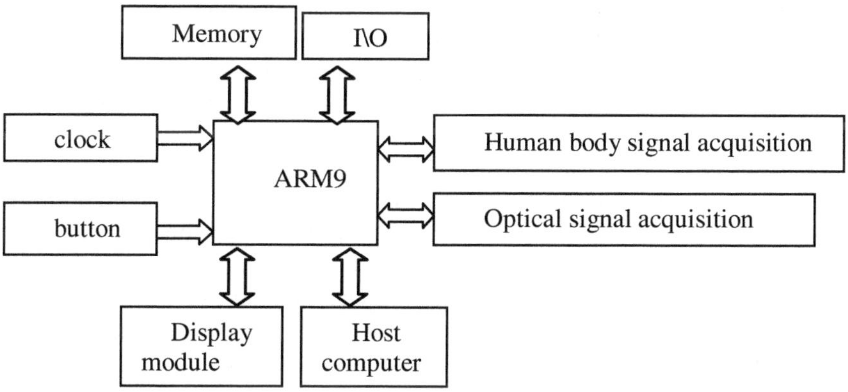

Fig. 2. Diagram of system hardware architecture block

3.1 Lighting Module Interface Design

The programmable optical frequency converter TSL230 is adopted to judge whether the classroom lighting is bright enough or not.., It is a high-performance, low-cost smart sensor with programmable sensitivity and full-scale output frequency [8]. As shown in Fig. 3, the logic inputs S0 and S1 are corresponding to the sensitivity, S2 and S3 are corresponding to the frequency, OUT is the output pulse. That is, the optical frequency is converted to frequency via TSL230, then transfer the frequency to the CPU, where the frequency will be converted to illumination and compared with the threshold. In order to ensure the accuracy of the collected signal data, the sensors should be placed away from direct sunlight and lighting and collect the scattered light of its own area.

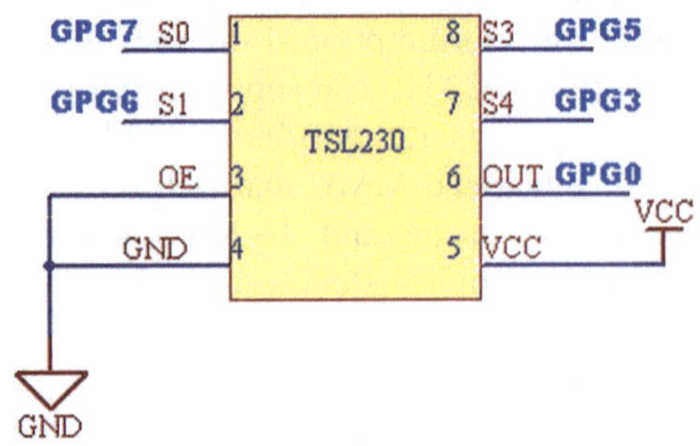

Fig. 3. Diagram of lighting module interface

3.2 Person Detection Module

The person can be detected by the infrared sensor according to the human body temperature. The paper adopts CS9803GP infrared pyroelectric sensor for personnel information collection. The pyroelectric infrared detector has two key components: one is pyroelectric infrared sensor (PIR), which can be turn wavelength between 8 and 12 μm into electrical signals and inhibit the natural white light and other interference signals; the other is Fresnel lens, which can refract the pyroelectric infrared signal on the PIR and divide the detection area into a number of bright and dark areas [9]. The working principle is: the infrared ray emitted by the human body about 10 μm is enhanced by the Fresnel filter to the infrared sensor source. Then the sensing element will lose the charge balance when the infrared radiation temperature of the human body is received, and the charge is released, Finally, a detection signal will be produced, that is, the infrared signal is converted to electrical signal output.

3.3 RTC Module

The function of adjustment and display of the date and time is necessary for the time is one of the lighting control decision factors. It can also be used as a clock. RTC is inside of the CPU and the external circuit supplies power for the RTC, wiring diagram is shown in Fig. 4.

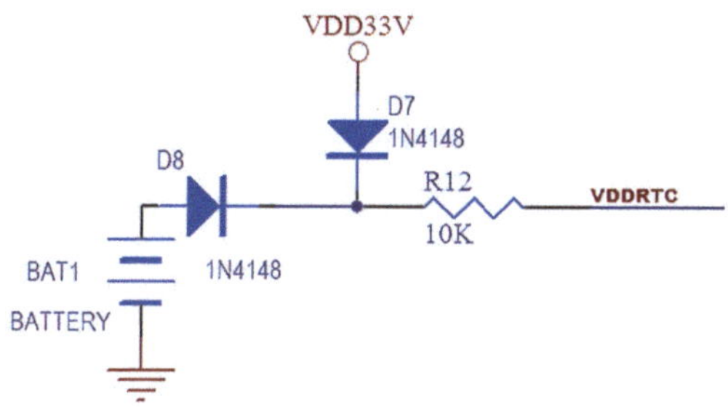

Fig. 4. Diagram of RTC power supply circuit

3.4 LCD Module and Key Module

The LCD display and the key module are designed to facilitate the system function module selection, parameter setting, system initialization, etc. TSXM, TSXP, TSYM, TSYP are four ports of the four-wire touch screen and can be directly connected to the touch screen; six buttons of the key module and 8 external interrupt ports EINT8, EINT11, EINT13, EINT14, EINT15, EINT19 are used to achieve man-machine interface. Due to limitation of the number of words, the system displays in multi-screen, using the button to select the screens or modify the parameter value.

Users can set the parameters like the thresholds of person numbers, illumination threshold and working time. The password also can be set for system security.

4 Software Programming

The program is operated in μ c/os II operating system, which is a priority-based pre-emptive real-time kernel. In μ c/os II, there are multi-task and each task has its own dedicated stack, the capacity of which can be arbitrary [10].

4.1 Main Program Flow Chart

There are several tasks: ① automatic manual switch; ② record person distribution; ③ manual mode task; ④ automatic mode task.

System work flow chart is shown in Fig. 5. When in the automatic mode task, firstly judge whether the lights are on or off, then turn on the lights if the illumination is less than 300lx and the lights are off, keep the lights off if the illumination is greater than 500lx. Turn off the lights if the illumination is greater than 500lx and the lights are on, otherwise keep the lights on.

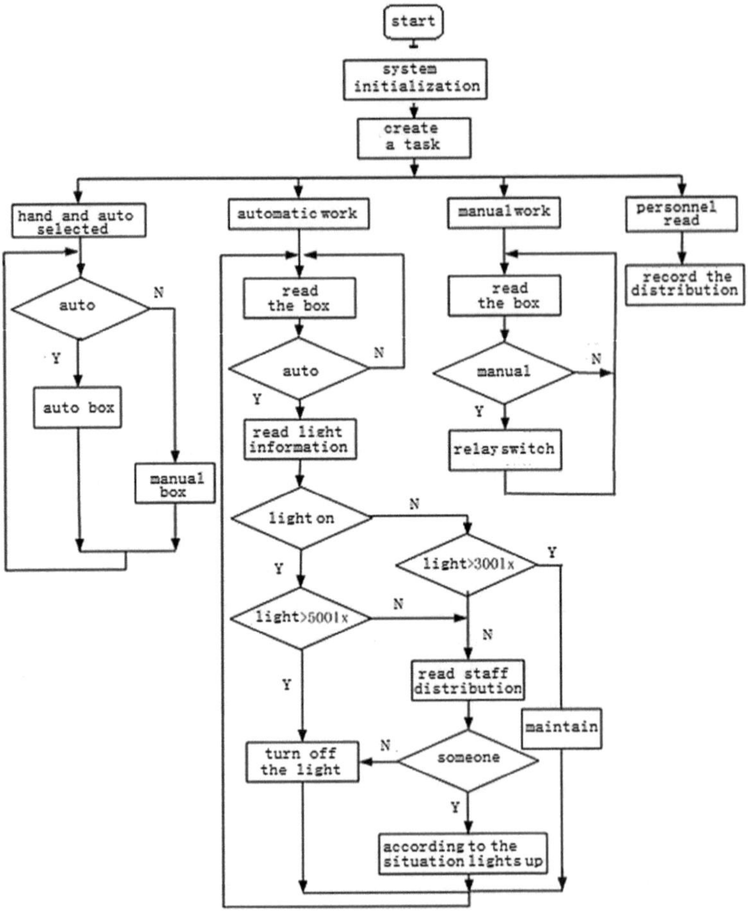

Fig. 5. Chart of main system flow

The main task is:

```
#define TASK_STK_SIZE    512
OS_STK      Task1Stk[TASK_STK_SIZE];
OS_STK      Task2Stk[TASK_STK_SIZE];
OS_STK      Task3Stk[TASK_STK_SIZE];
OS_STK      Task4Stk[TASK_STK_SIZE];
void Task1(void *data);
void Task2(void *data);
void Task3(void *data);
void Task4(void *data);
void main (void)
{   OSInit();

DevInit (); // device initialization
OSTaskCreate(Task1, (void *)0, &Task1Stk[TASK_STK_SIZE - 1], 0);
   OSTaskCreate(Task2, (void *)0, &Task2Stk[TASK_STK_SIZE - 1], 1);
   OSTaskCreate(Task3, (void *)0, &Task3Stk[TASK_STK_SIZE - 1], 2);
   OSTaskCreate(Task4, (void *)0, &Task4Stk[TASK_STK_SIZE - 1], 3);
   OSStart();
}
// task① hand automatically switch tasks
Void Task1 (void * pdata)
{Pdata = pdata;
While (1)
{/ * Task specific code ... * /
```

```
OSTimeDlyHMSM (0, 0, 0, 100); // Set the timer
}
}
// task②record the distribution of classroom staff
Void Task2 (void * pdata)
{Pdata = pdata;
While (1)
{/ * Task specific code ... * /
OSTimeDlyHMSM (0, 0, 0, 500); // Set the timer
}
}
// task③manual mode task
Void Task3 (void * pdata)
{Pdata = pdata;
While (1)
{/ * Task specific code ... * /
OSTimeDlyHMSM (0, 0, 0, 800); // Set the timer
}
}
// task④automatic mode task
Void Task4 (void * pdata)
{Pdata = pdata;
While (1)
{/ * Task specific code ... * /
OSTimeDlyHMSM (0, 0, 1, 0); // Set the timer
}
}
```

4.2 Other Control Modules

1. System initialization

System initialization includes the configuration initialization of keys, sensors, LCD display and real-time clock.

2. Real-time clock module

The real-time clock (RTC) unit can work with a backup battery when the system is power off. The RTC transfers 8-bit data to the CPU by STRB/LDRB ARM. The data contains seconds, minutes, hours, days, months, and years. The external crystal is 32.768 kHz and the alarm function is provided.

3. Display driver module

The driver module is mainly to configure the ports of LCD, including the display state and the starting line. LCD screen need to be clear before use to avoid incorrect information display. While data displaying, the coordinates, the text and graphics of the data should be pre-written to the LCD.

4. Key module

In the determinant keyboard mode, ARM scans the keyboard repeatedly to catch the users' command order, and it will no longer respond to other requests while executing or processing the command until the ARM returns to the scanning mode.

5 Conclusion

Based on ARM technology, the proposed classroom lighting intelligent control system, automatically control the light by the ambient light and the person numbers, is more reasonable and effective to reduce the cost of power. The timing parameter is set to make the control system adapt to the school's schedule. This intelligent light control system can be applied to all types of schools, universities and colleges by simply transforming the original lighting system.

References

1. Lin, B.: Design and analysis of intelligent lighting control system with FPGA as the core control. J. Hubei Univ. (Nat. Sci.) **35**(3), 308–314 (2013)
2. Ma, J., Hui, C.: Intelligent light control system based on CAN/LIN bus. Process Autom. Instrum. **30**(9), 40–45 (2009)
3. Ling, Z., Zou, D., Xu, M.: Design and realization of intelligent lighting system for classroom. J. Jilin Univ. (Inf. Sci. Ed.) **27**(4), 435–440 (2009)
4. Zhang, Y., Zhou, X., Zhu, S.: Based on single chip computer classroom saving energy control system design research. Manuf. Autom. **34**(2), 103–105 (2012)
5. Wang, J., Wei, Y.: Design of classroom lighting energy-saving control system based on AT89S51 and RE200B. J. Pingdingshan Univ. **28**(2), 36–38 (2013)
6. Zhang, P.: Developing network video monitor system based on embedded S3C2410. Microcomput. Inf. **24**(12), 174–178 (2008)
7. Chen, Q., Li, J.: Investigation on ARM9-based embedded fingerprint identification system. J. Chongqing Univ. **27**(9), 22–25 (2004)
8. Ding, R., Li, G.: Programmable color light - to - frequency converter TCS230 and its application. Glob. Electron. Compon. **7**(1), 59–62 (2005)
9. Shi, L.: Measuring techniques based on pyroelectric infrared sensors. Electron. Des. Eng. **11**(21), 162–165 (2013)
10. Zhang, Y.: ARM Theory & C Program Design. Xi'an University of Electronic Science and Technology Press (2009)

Electromagnetic Compatibility Analysis of PIND Equipment for Rocket Engine Attached Pipes

Guotao Wang[1,2(✉)], Qun Ding[1], Leizhen Gao[2], Qiang Wang[2], and Liang Guo[2]

[1] School of Electronic Engineering, Heilongjiang University, Harbin, China
star5892@163.com
[2] Reliability Institute for Electric Apparatus and Electronic, Harbin Institute of Technology, Harbin, China

Abstract. The remainders in attached pipelines of rocket engine are great harm to the reliable operation of engine. Signal noise and misjudgment caused by electromagnetic interference reduce the detection accuracy seriously. This paper analyzes the common-mode voltage of the rectifier bridge side and the inverter side of the detection equipment. According to the cut-off frequency, the fundamental voltage drop and the fundamental current and other parameters, the common-mode voltage filter is designed. The crosstalk coupling process between cable bundles is analyzed in this paper. Based on the finite element method, the distribution parameters of the cable bundle are analyzed and the structure of the cable bundle is optimized. With these methods, the impact of electromagnetic interference on the accuracy of remainders detection is reduced effectively.

Keywords: Attached pipelines of rocket engine · Remainder · PIND · Electromagnetic compatibility

1 Introduction

Remainder refers to all the material which exist in the product and has nothing to do with the requirement. It may be entered by external or internally generated [1]. Rocket engine as a power plant for aerospace systems, its reliability is a key factor in for the launch mission [2]. The structure of rocket engine is complex, and the assembly process is numerous, so remainders are not easy to be perceived. Free remainder will cause propel system exceptions, mechanical failure, control system damage and other

This research was supported by National Natural Science Foundation of China (51607059, 61271347, 51077022); Natural Science Foundation of Heilongjiang Province (QC2017059); Postdoctoral Fund in Heilongjiang Province (LBH-Z16169); Science and Technology Innovative Research Team in Higher Educational Institutions of Heilongjiang Province (No. 2012TD007); Heilongjiang University Youth Science Fund Project (QL201505).

J.-S. Pan et al. (eds.), *Advances in Smart Vehicular Technology, Transportation, Communication and Applications*, Smart Innovation, Systems and Technologies 86,
https://doi.org/10.1007/978-3-319-70730-3_19

issues, even lead to casualties. Therefore, the detection of remainders in attached pipelines of rocket is a necessary measure to exclude rocket engine work failure.

2 Research Background

Based on Particle Impact Noise Detection (PIND), Harbin Institute of Technology designed automatic detection system for remainders in attached pipelines of rocket engine. The schematic diagram is shown in Fig. 1.

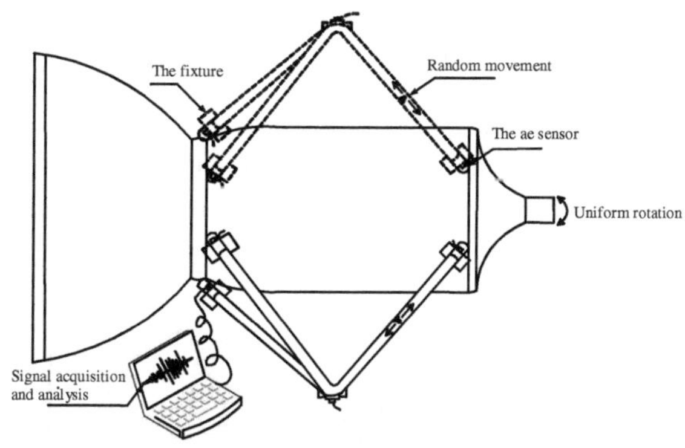

Fig. 1. The detection system of remainders in attached pipelines of rocket engine

2.1 Existing Problems

In order to provide sufficient and adjustable excitation capacity, the frequency conversion is employed. A lot of data show that the electromagnetic interference makes the background noise increased by 30%–55%. In addition, the electromagnetic interference increases detective complexity of remainder signal, and result in misjudgment.

The main problems of electromagnetic compatibility in this system are:

- As a result of PWM technology, the output pulse of inverter contains a large number of high-order harmonic components.
- The pulse output voltage generated by the high-speed power switching devices causes serious electromagnetic interference to the sensor.
- The power lines of motor and the signal lines of acoustic emission sensor pass through the center shaft, and crosstalk occurs between them.

2.2 Research Status at Home and Abroad

In the electromagnetic compatibility of the variable frequency drive system, B. Basavaraja proposed the improved second-order RLC low-pass filter [5]. A. Consoliand designed a four-phase inverter to eliminate the common-mode voltage, and proposed "auxiliary phase" and second-order LC filter [6]. Sun Li have studied the electromagnetic interference generated in the integration of electrical systems [7].

The methods of studying the electromagnetic compatibility of shielded cable mainly include the field method and the equivalent road method. Taylor put forward a variety of field-line coupling model, and domestic research is mainly based on transmission line theory [8]. Sun Beiyun proposed the calculation of the distributed capacitance of the Multi-conductor cable [9]. Yi Bin study the coupling response of voltage and current in the shielded cable with ANSYS [10].

At present, there are few studies on the electromagnetic interference in the detection of remainders in attached pipelines of rocket engine, so it needs further analysis.

3 Electromagnetic Compatibility Analysis of Variable Frequency Drive System

The variable frequency drive system is composed of frequency converter and asynchronous motor, the IGBT in the inverter circuit produce a higher common-mode voltage [11], which is the main source of electromagnetic interference. The simulation schematic diagram of variable frequency drive system is shown in Fig. 2.

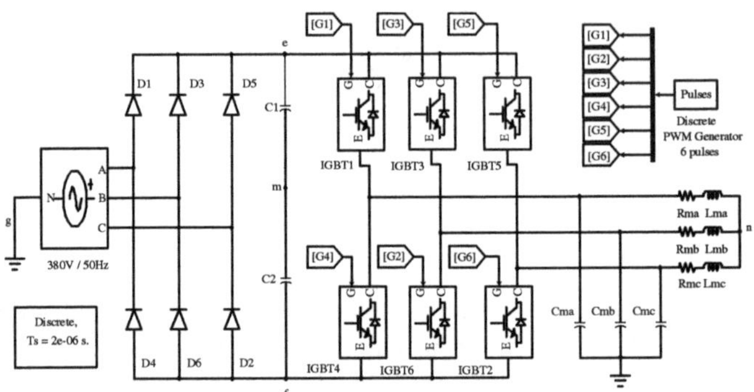

Fig. 2. Simulation schematic diagram of variable frequency drive system

3.1 Common-Mode Voltage of the Rectifier Bridge

From Fig. 2, the function is as follows:

$$U_{\mathrm{mg}} = \frac{U_{\mathrm{eg}} + U_{\mathrm{fg}}}{2} \tag{1}$$

$$U_{\mathrm{mg}} = \frac{3\sqrt{2}}{8\pi} U_{\mathrm{AB}} \sin(3wt) + \frac{3\sqrt{2}}{80\pi} U_{\mathrm{AB}} \sin(9wt) + \ldots \tag{2}$$

The Fourier series expansion formula in (2) [12]. Where w is the fundamental frequency of AC input voltage, and U_{AB} is the AC input line voltage.

The model of variable frequency drive system is established by Simulink, the waveform and spectrum of U_{mg} are shown in Fig. 3. There is 6k − 1 (k = 1, 2, 3,…)

harmonics, where there is no harmonics at the fundamental frequency, the amplitude of the third harmonic is the largest and the amplitude of the harmonics decreases with increasing frequency.

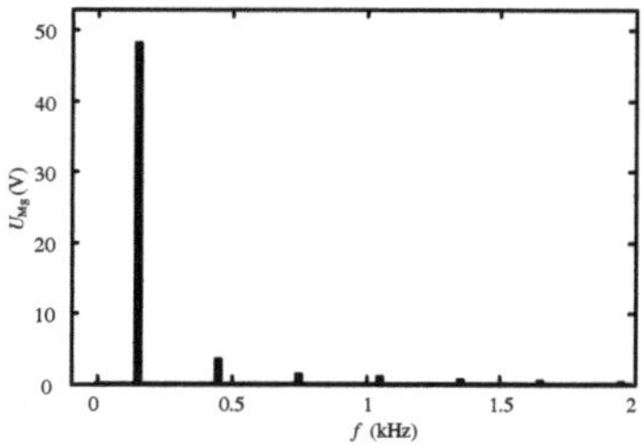

Fig. 3. Original waveform spectrum

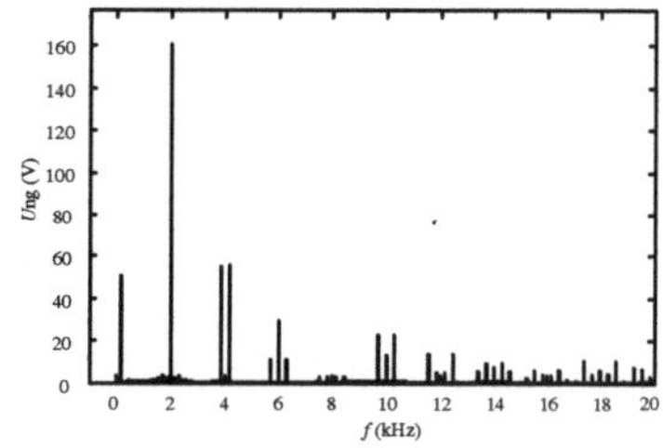

Fig. 4. New spectrum

3.2 Common-Mode Voltage of the Inverter

Figure 4 shows the simulation waveform of the common-mode voltage. The Fourier expression of the common-mode voltage $U_{\rm ng}$ is as follows [12]:

when $n = 1, 3, 5, \ldots, k = 6j, j = 1, 2, 3, \ldots$

$$U_{\rm ng} = \sum_{n=1}^{\infty} (-1)^{(n-1)/2} \frac{E_{\rm DC}}{2} (\frac{4}{n\pi}) \left\{ J_0(\frac{an\pi}{2}) \cos(nw_s t) + 3 \sum_{n=1}^{\infty} J_k(\frac{an\pi}{2}) [\cos(nw_s + kw_1)t + \cos(nw_s - kw_1)t] \right\} \quad (3)$$

when $n = 2, 4, 6, \ldots, k = 6j - 3, j = 1,2,3,\ldots$

$$U_{\rm ng} = 3 \sum_{n=1}^{\infty} (-1)^{n/2} \frac{E_{\rm DC}}{2} (\frac{4}{n\pi}) J_k(\frac{an\pi}{2}) [\sin(nw_s + kw_1)t + \sin(nw_s - kw_1)t] \quad (4)$$

Where w_s is the carrier angular frequency, w_1 is the modulation wave angular frequency, J_k is the k-order Bessel function, and E_d is the DC bus voltage.

From (3), (4) and Fig. 6, if the modulated wave is the sine wave, w_1 of the modulation wave does not exist in the common-mode voltage. The harmonics will exist only when the carrier frequency is odd times of w_s, and the harmonic amplitude of the double carrier is largest. The inter-harmonics exists at the frequency $nw_s \pm kw_1$.

3.3 Design of Common-Mode Voltage Filter for Three-Phase Inverter

As shown in Fig. 5, this paper selects the second order RLC passive filter to suppress the differential mode voltage and the common mode voltage. The model parameters of the induction motor as follows: $R_{\rm ma}$ is 22Ω, $L_{\rm ma}$ is 44 mH, $C_{\rm ma}$ is 6nF.

The common-mode voltage of the inverter:

$$U_{ng} = \frac{V_1 + V_2 + V_3}{3} = \frac{V_{1m} + V_{2m} + V_{3m}}{3} + V_{mg} \tag{5}$$

From circuit analysis:

$$U_{ng} = \frac{1}{3}(R_f i_o + \frac{1}{C_f}\int i_o dt) + U_{fm} + U_{mn} \tag{6}$$

According to (6), U_{ng} is proportional to R_f, inversely proportional to C_f. The filter parameters can be determined synthetically by cutoff frequency, fundamental voltage drop, fundamental current [13].

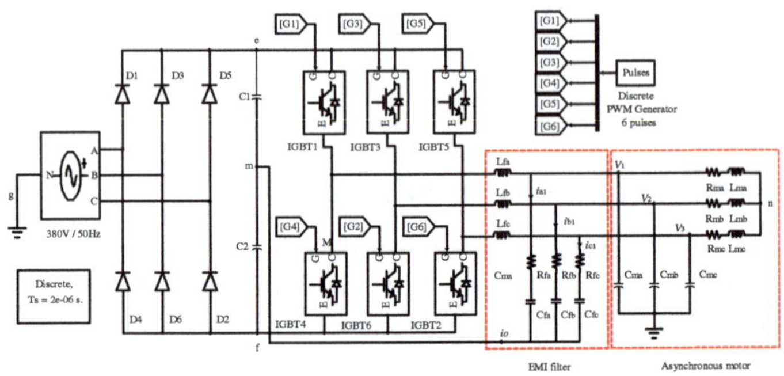

Fig. 5. Simulation schematic of variable frequency drive system with filter

- **Cut-off frequency.** The harmonic amplitude of the common-mode voltage is largest at the carrier frequency. And, the cutoff frequency should less than one tenth of carrier frequency. If the operating frequency of the filter is too low, it will bring the capacitor current and phase shift. Therefore, the cutoff frequency is usually ten times higher than fundamental frequency. Considering the conditions and verified by experiments, the cut-off frequency is determined to be 400 Hz.
- **Filter inductance.** Usually the fundamental voltage drop on the inductor should be less than 3% ~ 5%, the inductance range is

$$L_f = (3\% \sim 5\%)L_{ma} = 1.32\,\text{mH} \sim 2.2\,\text{mH}, \text{select } L_f = 1.6\,\text{mH}.$$

- **Filter capacitor.** In the no-load conditions, the fundamental current of the capacitance should be less than 10% of the inverter output current, the inverter rated current is 30A, rated power E_d = 17.5 kW. Then $I_c = aE_d \pi f C/\sqrt{2} = 3\text{A}$, and the solution is 1.54 μF.
- **Damping resistance.** Too large value of resistance will affect the dead time of switch. In addition, the capacitor capacitance is less than 20% of the rated current of

the inverter when circuit modulates. Considering the conditions and after simulating and experimenting, select $R_f = 2\Omega$.

The simulation results are shown in Fig. 6 to Fig. 8. From Figs. 4 and 6, the high-frequency common-mode voltage of 2 kHz or more is well suppressed after filtered. Figures 7 and 8 show that it also has a good effect on differential mode voltage.

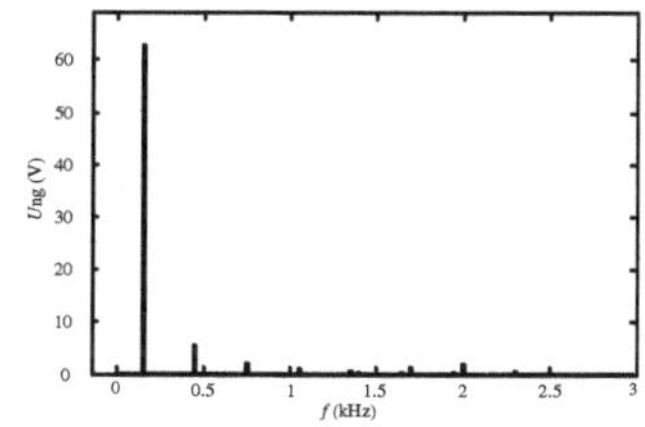

Fig. 6. Common voltage spectrum of filtered inverter

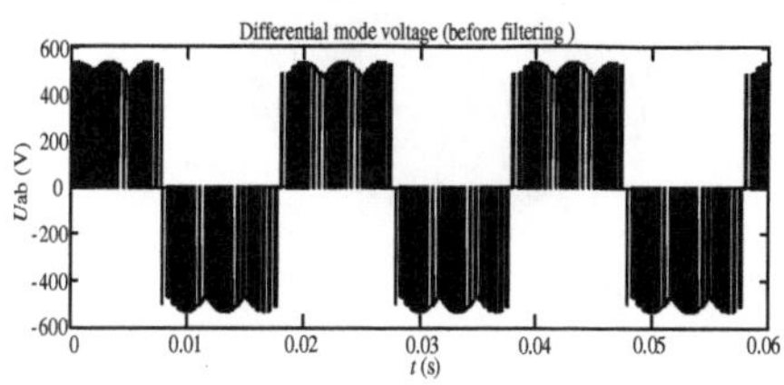

Fig. 7. mode voltage before filtering

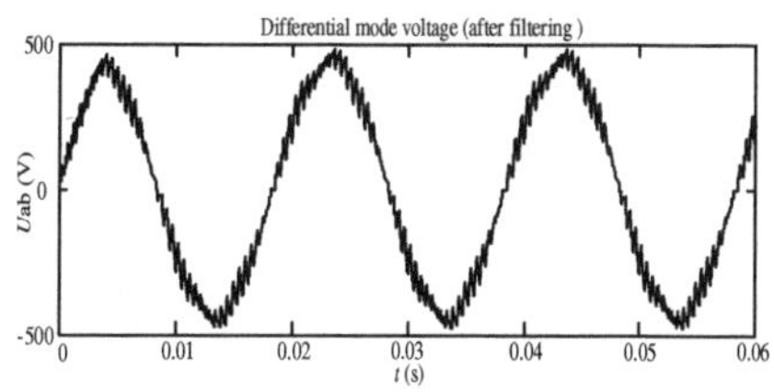

Fig. 8. mode voltage after filtering

4 Electromagnetic Compatibility Analysis of Sensor Cable

4.1 Coupling Theory of Capacitance and Inductance

The electromagnetic energy causes crosstalk when the distributed capacitance and the distributed inductance are coupled. The capacitive coupling between the cables is:

$$U_2 = \frac{Z_i}{Z_i + \frac{1}{\mathrm{j}wc}} U_i \tag{7}$$

(7) displays that the electromagnetic interference voltage U_2 is related to the value of coupling capacitance C and input impedance Z_i, and decreasing the values of w, C and Z_i can reduce the electromagnetic interference caused by the capacitive coupling.

The inductance coupling between the cables is:

$$U_{\mathrm{N}} = \frac{\mathrm{d}}{\mathrm{d}t} M I_1 = M \frac{\mathrm{d}}{\mathrm{d}t} I_1 = \mathrm{j} w M I_1 \tag{8}$$

(8) shows U_{N} is proportional to $\mathrm{d}I_1/\mathrm{d}t$ and M, therefore reducing M and decreasing the frequency w can suppress the inductance coupling voltage [10].

4.2 Finite Element Analysis of the Sensor Cable

Based on the finite element principle, this paper calculates the distribution parameters between cables.

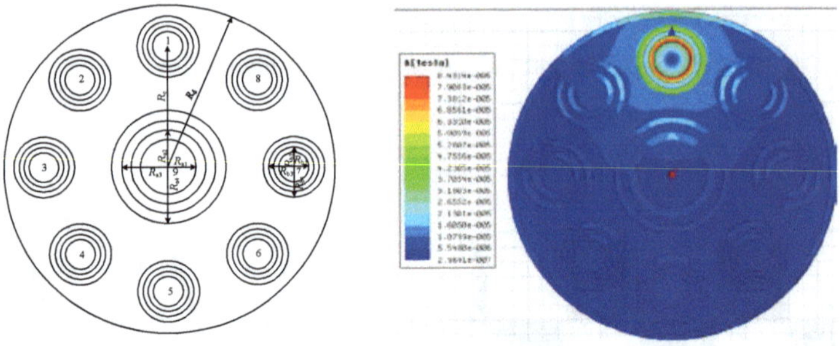

Fig. 9. Cable simulation diagram

Figure 9 shows the 9-conductor cable model that is established by the Maxwell. Each wire is composed of core layer, inner insulating layer, shielding layer and outer insulating layer from inside to outside. According to the actual cable, the relative permittivity of the inner and outer insulating layers and the intermediate filling layer is set to 2.5, 2.6 and 2.1. The relative permeability of all the media is 1, the structural parameters are as follows: $R_{\mathrm{a1}} = 3$ mm, $R_{\mathrm{a2}} = 4$ mm, $R_{\mathrm{a3}} = 5$ mm, $R_{\mathrm{a4}} = 6$ mm, $R_{\mathrm{b1}} = 1$ mm, $R_{\mathrm{b2}} = 1.5$ mm, $R_{\mathrm{b3}} = 2$ mm, $R_{\mathrm{b4}} = 2.5$ mm, $R_{\mathrm{c}} = 9.5$ mm, $R_{\mathrm{d}} = 13$ mm.

By calculating the distributed capacitance and inductance between the cables, It can be seen from Tables 1 and 2, the distributed capacitance between the sensor cores is 0 when the shield is applied, and it exists only between the shielding layers and the sensor cables. Without shielding, there is a pF-level capacitance. If the shielding layer is applied, the self- capacitance is calculated in the reference system, so the self-capacitance value is larger than that without shielding layer.

Table 1. Capacitance simulation results with shielding layer (pF)

C	1	2	3	9	10	16	17	18
1	137.22	0	0	0	0	0	0	0
2	0	137.22	0	0	0	0	0	0
3	0	0	137.22	0	0	0	0	0
:	:	:	:	:	:	:	:	:
9	0	0	0	139.39	−139.39	0	0	0
10	0	0	0	−139.39	310.3	−14.598	−14.598	−14.598
11	−137.22	0	0	0	−14.598	−0.2083	−0.6506	−12.152
12	0	−137.22	0	0	−14.598	−0.1301	−0.2084	−0.6507
13	0	0	−137.22	0	−14.598	−0.2084	−0.1301	−0.2085
14	0	0	0	0	−14.598	−0.6507	−0.2085	−0.1301
15	0	0	0	0	−14.598	−12.15	−0.6507	−0.2085
16	0	0	0	0	−14.598	180.62	−12.15	−0.6501
17	0	0	0	0	−14.598	−12.15	180.62	−12.15
18	0	0	0	0	−14.598	−0.6501	−12.15	180.62

Table 2. Capacitance simulation results without shielding layer (pF)

C	1	2	3	4	5	6	7	8	9
1	23.92	−6.600	−0.838	−0.294	−0.238	−0.294	−0.838	−6.600	−5.972
2	−6.600	23.92	−0.838	−0.838	−0.838	−0.294	−0.294	−0.838	−5.972
3	−0.838	−6.600	23.92	−6.600	−0.838	−0.294	−0.238	−0.294	−5.972
4	−0.294	−0.838	−6.600	23.92	−6.600	−6.600	−0.294	−0.238	−5.972
5	−0.238	−0.294	−0.838	−6.600	23.92	6.600	−0.838	−0.294	−5.972
6	−0.294	−0.238	−0.294	−0.838	−6.600	23.92	−6.600	−0.838	−5.972
7	−0.838	−0.294	−0.238	−0.294	−0.838	−6.600	23.92	−6.600	−5.972
8	−6.600	−0.838	−0.294	−0.238	−0.294	−0.838	−6.600	23.92	−5.972
9	−5.972	−5.972	−5.972	−5.972	−5.972	−5.972	−5.972	−5.972	48.521

5 Conclusion

This paper analyzes the principle of generating electromagnetic interference in the variable frequency transmission system and multicore pin cable in the automatic detection system of remainders in attached pipelines of rocket engine, and simulates the common mode voltage of the inverter and the distribution parameters of the cable. The main conclusions are as follows. The mechanism and characteristics of the common mode voltage of the rectifier and the inverter are analyzed. RLC passive filter is designed to suppress the common mode voltage. The principle of crosstalk between sensor cables is analyzed, and the simulation model of 9-conductor cable is established.

References

1. QJ 2850-1996: Prevention and control for foreign object debris (FOD) of space products. Beijing: Commission of Science, Technology and Industry for National Defense (1996)
2. Lu, X.: Inexhaustible power that recommend the "Long March" rocket - the development of "Long March" rocket engine in aerospace propulsion technology research institute. Aerosp. Indus. Manage. **6**, 18–22 (2007)
3. Zhen, X., et al.: Current analysis of remainder's control in rocket testing proces. J. Sichuan Ordnance **36**(6), 74–76 (2015)
4. Qin, Y., et al.: Method of man-machine-environment control over foreign object debris in pipelines used in rocket engine test. J. Rocket Propul. **41**(6), 80–85 (2015)
5. Basavaraja, B., Siva Sarma, D.V.S.S.: Modeling and simulation of dv/dt filters for AC drives with fast switching transients. In: Power India Conference IEEE, pp. 10–14 (2006)
6. Consoli, A., et al.: An innovative EMI reduction design technique in power converters. IEEE Trans. Electromagn. Compat. **38**(4), 567–575 (1996)
7. Sun, L., Nie, J.: Analysis of electromagnetic compatibility in integrated motor system. Servo Control **6**, 33–35 (2010)
8. Wang, C., Zhu, C.: The present research and development trend of electromagnetic fields coupling to shielding cables. Electr. Wire Cable **3**, 1–5 (2011)
9. Sun, B., et al.: Method of moment for calculating capacitance of shielded multiconductor cable. High Power Laser Part. Beams **12**(6), 749–752 (2000)
10. Yi, B., Wang, Z.: Parameters calculation of shield cable and crosstalk between shielding layer and core wires. High Voltage Eng. **34**(4), 804–808 (2008)
11. Dai, S., et al.: Design of common-mode and differential-mode voltage dv/dt filter at inverter output terminals. Chin. J. Power Sources **39**(9), 1985–1988 (2015)
12. Jiang, Y., et al.: Research on common-mode voltage generated by a PWM inverter and its cancellation technology. Proc. CSEE **25**(9), 47–53 (2005)
13. Yan, B., Chen, X.: The engineering design of VVVF output PLC sine-wave filter. Electr. Mach. Control **6**(3), 256–260 (2002)

Application of Double-Hidden Layer BP Neural Network in Transformer Fault Alarm

Xin Su[1,2,3](✉), Xin-hua Jiang[1], Shun-miao Zhang[2], and Yao He[2]

[1] Fujian University of Technology, Fuzhou 350108, Fujian, China
xjlw66@163.com, xhjiang@fjut.edu.cn
[2] Beidou Navigation and Smart Traffic Innovation Center of Fujian Province Fuzhou, Fuzhou 350108, Fujian, China
zshunmiao@163.com, 240678575@qq.com
[3] Fujian Key Laboratory for Automotive Electronics and Electric Drive, Fujian University of Technology, Fuzhou 350108, Fujian, China

Abstract. The traditional power transformer fault judgment is through the oil chromatography online monitoring data set a threshold to determine whether the transformer is faulty, so there is a problem of the correct rate of fault alarm. This paper adopts the data analysis method, based on the Fuzhou transformer oil chromatography and its repair order data, selects effective characterization of influence the characteristics of transformer running state as evaluation indexes. A double-hidden layer BP neural network models were constructed and applied into transformer fault alarm, and evaluation accuracy is higher than the common fault alarm model.

Keywords: Power transformer · Fault diagnosis · Double-hidden layer BP neural network

1 Introduction

Since the types and contents of dissolved gas in oil are different when an oil immersed power transformer operates in the different state, dissolved gas in oil analysis (DGA) has long been an effective means of judging the state of a transformer. With the development of artificial intelligence technology, domestic and foreign scholars proposed a variety of classification algorithms based on DGA, such as probabilistic neural network [1], limit learning machine [2], random forest [3], support vector machine [4] and so on. However, the above methods belong to the single-layer neural network learning method, the learning ability is limited, the model accuracy is unstable, and when the accuracy of the alarm reaches a certain height, it is difficult to improve greatly.

At present, the evaluation system of transformer state concentrated on the certain indicators exceeds the standard warning by the artificial experience sets the corresponding threshold. On the one hand, because of the uncertainty of

J.-S. Pan et al. (eds.), *Advances in Smart Vehicular Technology, Transportation, Communication and Applications*, Smart Innovation, Systems and Technologies 86,
https://doi.org/10.1007/978-3-319-70730-3_20

human experience, the power transformer has false alarm problems, and the on-site identification of the fault situation requires a lot of manpower and material resources; on the other hand, excessive warning only reflects some parts of the transformer and fails to combine use of multi-source information to comprehensive evaluation the power transformer. So that the transformer operation and status information scattered in a number of relatively independent information platform, which is not conducive to the multidimensional of information and the analysis and display of multi data. Therefore, a new type of alarm method is needed in order to improve the correct rate of transformer fault alarm.

From the view of combining genetic algorithm [5], the training speed is accelerated and the precision of the training is improved by introducing the mutation operator, the inertia weight factor and the Gaussian weighted global extreme in the genetic algorithm. From the view of clustering algorithm [6], which is suitable for the overheating fault diagnosis of the transformer and the performance is better than the three ratio method, but it is not good when dealing with mixed electric fault. From the research of fuzzy control [7], which avoids the traditional algorithm using the three ratio method in the border to determine the shortcomings of inaccurate.

BP (Back-Propagation) neural network was proposed by scientists Rumelhart and McClelland in 1986 which is a kind of multilayer feedforward neural network trained by error back propagation algorithm. In 2006, the University of Toronto professor, the field of machine learning Geoffrey Hinton [8] and his students put forward a view that the multi-hidden layer of artificial neural network with excellent feature learning ability. The characteristics of the study have a more fundamental characterization of the data, which is conducive to visualization or classification. Literature [9] proposed the transformer discharge pattern recognition based on convolution neural network. According to the segmentation of the discharge signal visual attention mechanism, and take the gray image and bilinear interpolation normalization as the convolution of the input of the neural network the correct rate of which is about 94% in transformer fault diagnosis. Literature [10] put forward based on the deep learning neural network (DLNN), build a number of hidden layer neural network, through the characteristics of transformation or data feature extraction to discover intrinsic properties, classification accuracy is superior to the traditional support vector machine (SVM) method.

In order to overcome the influence of the fault alarm results of the uncertainty of human experience, in this paper, a new method of fault alarm for the power transformer is constructed based on double-hidden layer BP neural network. First, the Fuzhou transformer oil chromatography and its repair order data were cleaned, fusion matched and generated a typical training data set for training and testing. Then by adjusting the number of hidden layer neurons and the number of hidden layers and the hidden layer activation function to select the appropriate construction of double-hidden layer BP neural network simulation model. Finally, the correctness and effectiveness of the method proposed in this article is verified through calculation examples and statistical result. Compared with the conventional method that applied to the fault alarm of power transformer, the method proposed in this article has a higher rate of positive judgment.

2 Implementation of Double-Layers BP Neural Network

2.1 Double-Hidden Layer BP Neural Network Principle and Implementation Steps

The BP neural network composed of nonlinear transformation units has good nonlinear mapping ability, which is widely used in power transformer fault prediction, alarm, classification and other fields. According to the number of hidden layer BP network can be divided into a single hidden layer and hidden multi-layer network. The results show that compared with the single hidden layer, the double-hidden layer BP neural network has superior generalization ability and high prediction accuracy. The number of hidden layers is considered from the network accuracy and training time. When the mapping relation is simple, choose a single hidden layer in order to improve the network speed. When the mapping relation is complex, choose a double-hidden layer in order to acquire higher the network accuracy. In this paper, the double hidden layer BP neural network is chosen, because the fault data format of the power transformer is various, the structure is complex, and it needs higher prediction accuracy. Specific BP neural network implementation steps are as follows:

(1) Randomly initialize weights.
(2) Implement forward propagation.
(3) Implement code to compute cost function.
(4) Implement back propagation to compute partial derivatives.
(5) Use gradient descent or advanced optimization method with back propagation to try to minimize cost function.

2.2 Double-Hidden Layer BP Neural Network Pseudo Code:

a means neurons, superscript l means number of layers, j represents a layer of j neurons.

Activation function:

$$g(x) = \frac{1}{1 + e^{-\chi}} \tag{1}$$

Cost function:

$$J(\Theta) = \frac{1}{2m}\sum_{i=1}^{m}(Y(\Theta)^{(i)} - y^{(i)})^2 \tag{2}$$

$\triangle_{ij}^{(l)}$ is the ith neuron in the weight matrix of the l layer influences the weight of the next layer of the jth neuron.

Algorithm 1. Double-layer BP neural network

Input: Training set A; learning rate α; Number of training sets m; Network Structure N; Number of layers L; $g(x)$: activation function; $J(\Theta)$: cost function;
Output: DBP's parameters ω, b;

initial network parameters ω, b randomly;
Set $\triangle_{ij}^{l}$=0(for all l,i,j);//use to compute the derivative of $J(\Theta)$
for each $i \in [1, m]$ **do**
  set $a^{(1)} = x^{(i)}$
  for each $l \in [1, L]$ **do**//perform forward propagation to compute $a^{(l)}$
    $a^{(l+1)} = g(a^{(l)}\Theta^{(l)})(adda_0^{(l+1)})$
  end for
  $\sigma^{(4)} = a^{(4)} - y^{(i)}$; //compute all previous layer error vector
  $\sigma^{(3)} = (\theta^{(3)})^T\theta^4. * g'(\theta^{(2)}a^{(2)})$;
  $\sigma^{(2)} = (\theta^{(2)})^T\theta^3. * g'(\theta^{(1)}a^{(1)})$;
  for each $i \in [1, L]$ **do** // compute weight matrix
    $\triangle_{ij}^{(l)} = \triangle_{ij}^{(l)} + a_j^{(l)}\delta_i^{(l+1)}$
  end for
  if $j = 0$ **then**
    $D_{ij}^{(l)} = \frac{1}{m}\triangle_{ij}^{(l)}$;
  else
    $D_{ij}^{(l)} = \frac{1}{m}\triangle_{ij}^{(l)} + \alpha\Theta_{ij}^{(l)}$;
  end if
end for

3 Double-Hidden Layer BP Neural Network Model Performance Debugging

The double-hidden layer BP neural network model is composed of the input layer, hidden layer, and output layer. The output layer classifier uses SVM, which is suitable for the two class problem, which can give the results of each classification in probability and combine with the neural network will get better classification performance. Applying Matlab explore and optimize the experimental conditions, such as the layer number, the number of layer neurons and the activation function of the layer, which markedly influenced the performance of the BP neural network.

3.1 Performance Analysis of Different Layers

As shown in Fig. 1, the neuron network has better performance when it is on the first floor and the two layers and the standard deviation is the lowest at the two levels, and the highest accuracy mean meets the engineering requirement. When the hidden layer increases, the mean precision does not improve significantly, and when the hidden layer is greater than four layers, the training time will increase exponentially. Therefore, it is suitable to choose double hidden layer BP neural network.

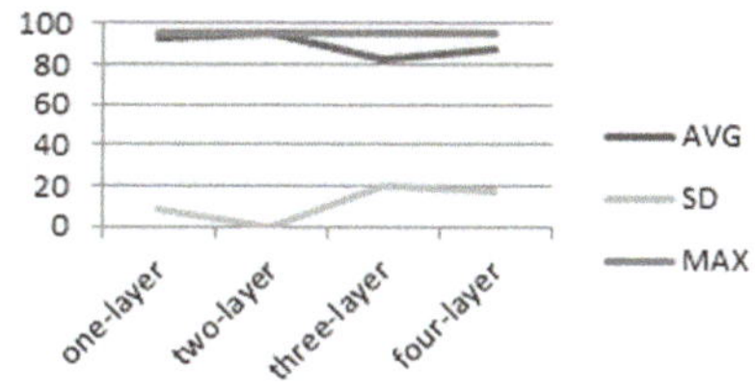

Fig. 1. Performance analysis of different layers

3.2 Performance Analysis of Different Neuron Numbers

As shown in Fig. 2, the neuron network has better performance when the number of neurons is 10 or 20. As the number of neurons increases, the training time will be increased. Therefore, it is appropriate to select the neural network structure with a smaller number of neurons to meet the requirements of accuracy.

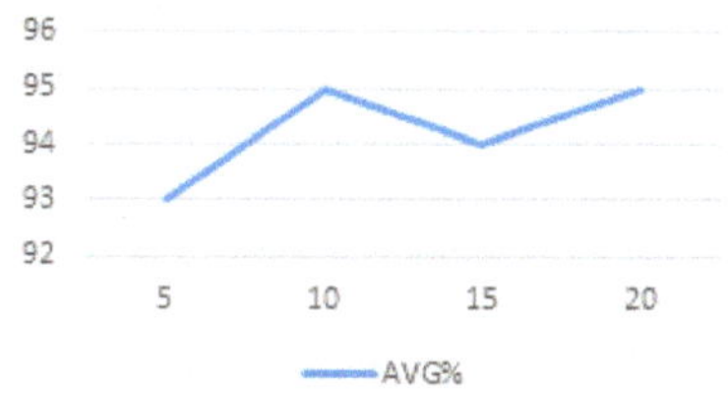

Fig. 2. Performance analysis of different neuron numbers

3.3 Performance Analysis of Different Activation Functions

As shown in Fig. 3, the neuron network has better performance when the activation function is logsig, purelin and tansig and purelin, the precision is higher and the standard deviation is lower. When the activation function is purelin and tansig, the precision is lower and the standard deviation is higher. Therefore, it is appropriate to choose the activation function of the combination of logsig and purelin.

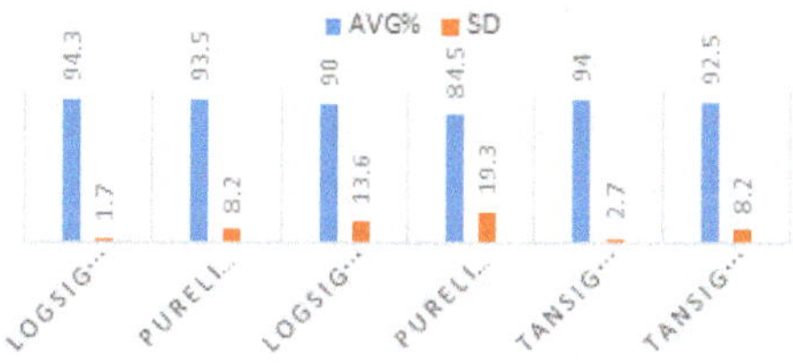

Fig. 3. Performance analysis of different activation functions

Finally, the selected BP neural network structure is double-layer, the number of neurons is 10, and the activation function is logsig and purelin. The constructed double-hidden layer BP neural network is shown in Fig. 4.

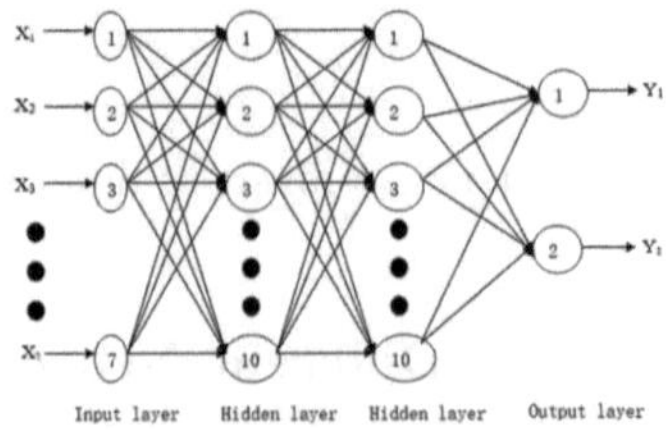

Fig. 4. Structure of double-layer BP neural network

In conclusion, the selected BP neural network structure is double-layer, the number of neurons is 10, and the activation function is logsig and purelin. The constructed double-layer BP neural network is shown in Fig. 4.

4 Experimental Results and Analysis

Selection of 2011 to 2016 Fuzhou 8 areas of the main transformer oil chromatography online monitoring data and the State Grid of Fuzhou transformer failure work order data. The work order data consists of "time of failure", "fault location", "alarm situation", "processing situation" and other components. Chromatographic online monitoring data consists of "transformer type" and "transformer position", "monitoring data acquisition time" "a device", "two equipment", "oil dissolved gas value" and other components. Applying Matlab match online monitoring data of Fuzhou City with the transformer fault data according to definite fault time matching transformer.

4.1 Data Cleaning

Cleaning data mainly follow the following principles:

(1) Selection: select Fuzhou oil chromatogram data and supporting the work list, a total of more than 200.
(2) Removal: oil chromatographic data due to the detection device did not upload the missing data.
(3) Interpolation value: Select the nearest day data to compliment when the alarm on the day of the oil chromatographic data is missing.
(4) Classification: different types of transformers, oil chromatography data types are different.

According to the engineering field of on-line monitoring data, taking into account the double-layer BP neural network has the characteristics of strong sample conversion and feature extraction ability, select $H_2, CH_4, C_2H_6, C_2H_4, C_2H_2, Co$ of the 6 characteristic variables of gas. In order to reduce the difference of gas content value and reduce the calculation error, formula (3) is used to standardize the gas content value.

x_{new} is the content of gas after standardization; x is the original value of gas; x_{mean} is the mean value of the gas content in the training set or test set x; x_{std} is the standard difference of the gas content in x.

$$x_{new} = \frac{x - x_{mean}}{x_{std}} \tag{3}$$

On the basis of actual engineering requirements, the input variables are $H_2, CH_4, C_2H_6, C_2H_4, C_2H_2, Co$ which are the 6 characteristic variables of gas. The output variable is the correct alarm and error alarm bar shown in the State Grid Fuzhou transformer fault list.

4.2 Experimental Verification

In order to verify the alarm capability of the algorithm, the interception of the national grid from 2011 to 2016 Fuzhou oil chromatography and corresponding work orders of which 192 fault data as a data set. Randomly selected 100 as a training sample, the remaining 92 as a test sample and repeat the test 300 times. Comparing the performance of the double hidden layer BP neural network with support vector machine algorithm, random forest algorithm, limit learning machine algorithm and probabilistic neural network algorithm .The performance of the five algorithms is shown in Table 1.

Table 1. Performance Comparison.

Method	Experimental mean		
	Precision	Recall	F1-measure
Random forest	0.9324	0.7388	0.7301
Support vector machine	0.8821	0.7385	0.7323
Probabilistic neural network	0.9021	0.7222	0.7311
Limit learning machine	0.8654	0.7716	0.7699
Double-layer BP neural network	**0.9589**	**0.8139**	**0.8146**

As can be seen from Table 1, the alert accuracy rate of the double hidden layer BP neural network algorithm is the highest among the 5 algorithms. The classification accuracy of random forest algorithm and the probabilistic neural network is close to that of double-hidden layer BP neural network. Among them, random forest and probabilistic neural network have higher diagnostic accuracy, but there is an important misjudgment which let the correct alarm judge into a false alarm. However, the double-hidden layer BP neural network misjudgment is to correct the false alarm to correct alarm, which is an admissible error in engineering practice. Therefore, the performance of the double-hidden layer BP neural network is better than the other 4 algorithms, either from the engineering practice or from the diagnosis performance.

5 Conclusion

Put forward a method of transformer fault alarm algorithm based on double-hidden layer neural network. The algorithm can analyze and judge the information of power grid fault and get the correct alarm information by setting up a double-hidden layer BP neural network structure. Compared with the conventional algorithms such as random forest algorithm, SVM algorithm, probabilistic neural network algorithm, and extreme learning machine algorithm, the method proposed in this article has a higher rate of positive judgment.

Acknowledgments. This work is supported by Fujian Science and Technology Department (No. 2017H0001), National Natural Science Foundation of China (No. 61304199), Fujian Science and Technology Department (No. 2014H0008), Fujian Transportation Department (No. 2015Y0008), Fujian Education Department (No. JK2014033, JA14209, JA1532), and Fujian University of Technology (No. GYZ13125, 61304199, GY-Z160064). Many thanks to the anonymous reviewers, whose insightful comments made this a better paper.

References

1. Malik, H., Khatri, A., Dohare, R.: Probabilistic neural network based incipient fault identification using DGA dataset. Procedia Comput. Sci. **58**, 665–672 (2015)
2. Li, S., Wu, G., Gao, B., Hao, C., Xin, D., Yin, X.: Interpretation of DGA for transformer fault diagnosis with complementary SaE-ELM and arctangent transform. IEEE Trans. Dielectr. Electr. Insul. **23**(1), 586–595 (2016)
3. Wang, D.W., Sun, Z.W.: Parallelization analysis of dissolved gases in transformer oil based on random forest algorithm. Appl. Mech. Mater. **519–520**(519–520), 98–101 (2014)
4. Sahri, Z.B., Yusof, R.B.: Support vector machine-based fault diagnosis of power transformer using k nearest-neighbor imputed DGA dataset. J. Comput. Commun. **02**(9), 22–31 (2014)
5. Cheng, S.F., Cheng, X.H., Lu, Y.: Application of wavelet neural network with improved particle swarm optimization algorithm in power transformer fault diagnosis. Power Syst. Prot. Control **42**(19), 37–42 (2014)
6. Islam, M.M., Lee, G., Hettiwatte, S.N.: A nearest neighbour clustering approach for incipient fault diagnosis of power transformers. Electr. Eng. **99**, 1–11 (2016)
7. Khan, S.A., Equbal, M.D., Islam, T.: A comprehensive comparative study of DGA based transformer fault diagnosis using fuzzy logic and ANFIS models. IEEE Trans. Dielectr. Electr. Insul. **22**(1), 590–596 (2015)

8. Hinton, G.E., Osindero, S., Teh, Y.W.: A fast learning algorithm for deep belief nets. Neural Comput. **18**(7), 1527–1554 (2014)
9. Zhao, S., Li, B., Wang, Y.: Partial discharge pattern recognition of transformer based on electric signal and ultrasonic comprehensive analysis. In: International Conference on Systems and Informatics, pp. 637–641. IEEE Press, Yantai (2012)
10. Xin, S., Zhu, Y., Ning, X., Wang, L., Gang, S., Chen, G.: Transformer fault diagnosis based on deep auto-encoder network. Electr. Power Autom. Equip. **36**, 122–126 (2016)

An Artificial Indoor Farm that Subverts Traditional Farming Patterns

Lin De Yao, Sheng-Hui Meng(✉), and Lin Shu Hua

National Demonstration Center for Experimental Electronic Information and Electrical Technology Education, Fujian University of Technology, Fuzhou 350118, Fujian, China
596166346@qq.com, menghui@fjut.edu.cn, 1226423394@qq.com

Abstract. An artificial indoor farm has been constructed to subvert traditional farming methods, mainly including a main control 32-bit microcontroller module; solar panel; battery; Finel lens; air temperature, humidity, wind, and other kinds of sensors; infrared lamp; water pump; spray; and other components. Room has been allotted for crop growth through the simulation of the natural processes of sunrise, sunset, water supply, and ventilation, to create a desired crop growth environment that includes indoor home plowing. This indoor farm is combined with Bluetooth, Apps, Wi-Fi, Internet, and other platforms, to achieve not only the restoration of arable land resources, but also a combined on-site and remotely controlled indoor farm.

Keywords: Solar energy indoor farm · Subversion · Farming mode

1 Introduction

With the rapid development of industrialization and urbanization, a large number of farmland and farmland have been developed. The reduction in cultivated land has resulted in a reduction in the supply of agricultural products, and the ability of grain production to support the population is declining. To improve the supply of agricultural products, we suggest that the supply of agricultural products should be reduced by:

1. changing the traditional agricultural cultivation practice, which was conducted exclusively outdoors, and making use of indoor space for cultivation and effective restoration of cultivated land resources.
2. simulating the natural ecology environment for the growth of plants.
3. achieving the realization of indoor farming, with and without automatic control, using the 32-bit single chip microcomputer as the core of control, with a variety of sensors integrated into control transmission technology.

J.-S. Pan et al. (eds.), *Advances in Smart Vehicular Technology, Transportation, Communication and Applications*, Smart Innovation, Systems and Technologies 86,
https://doi.org/10.1007/978-3-319-70730-3_21

2 Design Plan

2.1 Design Ideas

This design includes concepts from the following three patent applications: the "Land required for creating artificial indoor imitation natural sun light device", patent application number: 201710225452.3; "A technology recycling air intelligence land system", invention patent application number: 201510707967.8"; and a third frequency class C power amplifier, patent number: ZL 201210041362.6. The core content of the indoor-outdoor environment creates a generic natural system, required for the growth of plants, using features such as indoor artificial sunlight, ventilation, drainage, and other devices. This system integrates several technologies, forming an intelligent indoor farming system with the use of a personal computer or a mobile phone browser via the Internet or wireless data transmission technology, and implements a new farming model, combining indoor farming and remote management.

2.2 Design Diagram

The design scheme changes the traditional planting method, as shown in the block diagram in Fig. 1, mainly comprising a solar panel; battery; uninterruptible power supply

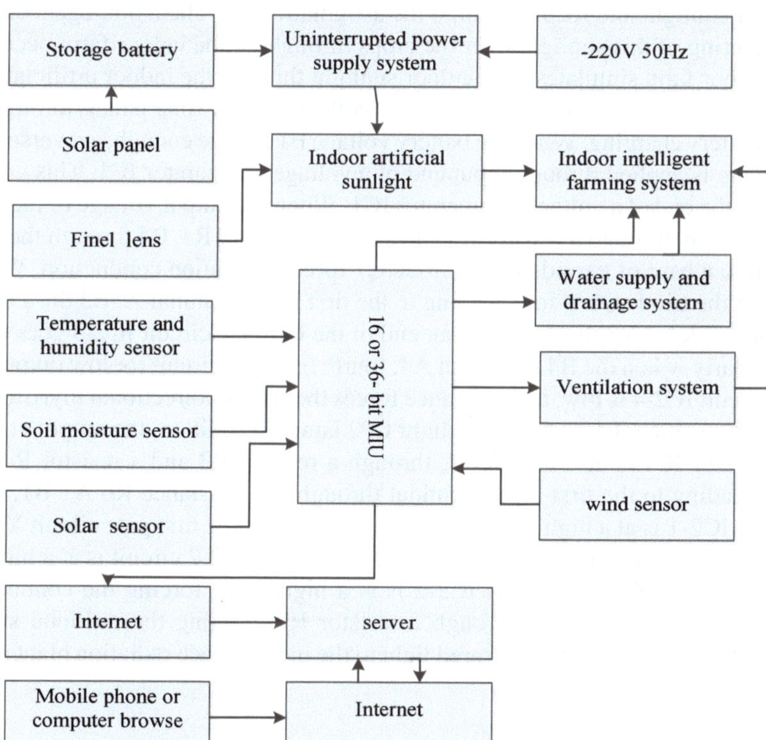

Fig. 1. Block diagram of artificial indoor farm system

system; Finel lens; indoor artificial sunlight: indoor farming system; drainage system; ventilation system; and a far and near range control system composed of air temperature, soil humidity, illumination, and wind sensors. The system uses a 32-bit MCU as the core processor for [4] signal acquisition, based on the law of plant growth, the preparation of scientific algorithms, and the realization of the light intensity, air humidity, ambient temperature, crop nutrients, and soil humidity. Real-time control combined with sensor technology simulates the natural environment for crop growth, so as to realize the automatic control of the growth of crops.

2.3 Introduction of Working Process of Subsystems

This model demonstrates the subversion of the traditional farming model, to create a natural environment by simulating the external environment for the growth of crops. It also demonstrates the use of modern electronic technology for the intelligent control of indoor farms.

The indoor farm collects outdoor sunlight through the use of the Finel lens, plane mirror, and other combinations, collecting as much outdoor sunlight as possible, distributing it evenly throughout the indoor farm [2]. For instance, it prioritizes the use of the Finel lens mounted on the wall or the roof to gather as distribute as much sunlight as possible. The Finel lens is divided into light and scattering two. The Finel lens converges the outdoor sunlight and focuses it onto the astigmatic lens. Then, homogeneous astigmatic scattering of this sunlight over the crops throughout the indoor farm occurs.

The indoor farm simulates the outdoor sunlight through the indoor artificial natural sunlight device as shown in Fig. 2. Sunlight is collected by 11 solar panels through diode D1 to E1 battery charging. When the battery voltage E1 is large enough, a reverse voltage input voltage is applied through input end of a voltage comparator IC1. This voltage is higher than the end of a voltage comparator IC1. When the output voltage of the comparator IC1 is high, through a resistor R3 and a resistor divider R4, R5 through the resistor connected to a base of a triode Q1, a triode Q1 forces saturation conduction. When the collector of the triode Q1 is low, leading to the first input terminal A3B3 third or IC2-3 circuit for low level Y3 third, the output end of the OR gate circuit IC2-3 goes to a low level. Similarly, when the B4 first input A4, fourth or IC2-4 circuit for low output fourth Y4 gate circuit IC2-4 is low, the resistance forces the R10 unidirectional thyristor SCR2 cutoff, extinguishing the artificial sun light LP2 lamp; in addition, the output voltage of the comparator IC1 is at a high level, through a resistor R3 and a resistor R4 partial pressure, leading to the first input terminal through the resistance R6 A1 B1, the first gate circuit IC2-1 is at a high level, and the output end of the first gate circuit Y1 IC2-1 is at a high level. When the B2 first input A2, second or IC2-2 circuit is at a high level, the output second Y2 gate circuit IC2-2 is at a high level, forcing the conduction of unidirectional thyristor SCR1 through a resistor R9, lighting the artificial sun light infrared LP1 lamp and sending infrared light to the indoor space radiation plant required for growth.

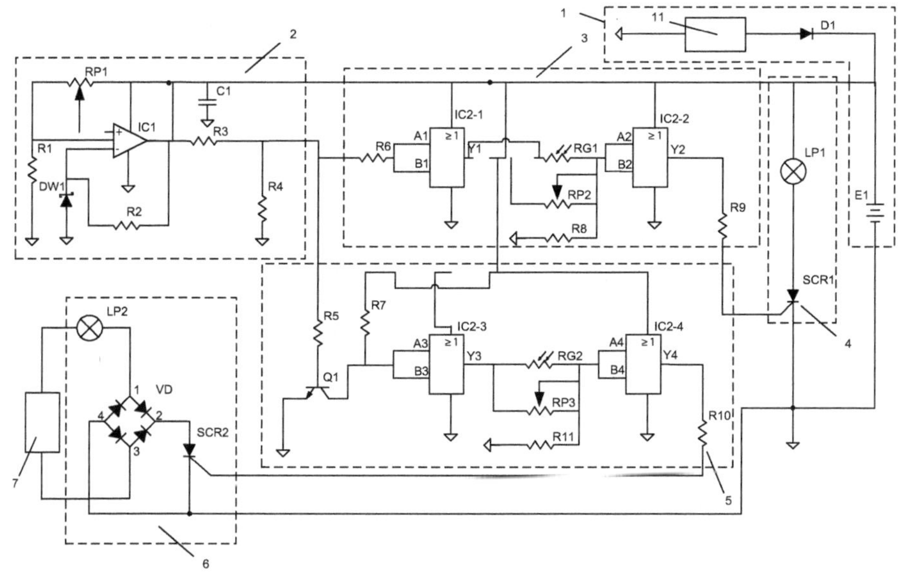

Fig. 2. Indoor man-made natural sunlight device.

Similarly, when the battery voltage of E1 is reduced, the artificial sun light and LP1 lamp is extinguished. The artificial solar light LP2 lamp is lit, providing infrared light radiation to the indoor space required for plant growth.

The output end of a voltage comparator IC1 is high when the B1 first input A1, the first gate circuit IC2-1, is also high, resulting in the output Y1 of the first gate circuit to the high level of IC2-1. When the sun sets, photosensitive resistor RG1 resistance increases with the potentiometer RP2 increasing the parallel resistance, and then in series with the resistance of R8, by adjusting the potentiometer RP2 until the first cause input A2, the B2 s IC2-2 gate circuit is low, while the output second Y2 gate circuit IC2-2 is at a low level, resulting in the unidirectional thyristor SCR1 cutoff and the extinguishing of the LP1 lamp. Similarly, when the sun sets, the one-way thyristor SCR2 is cut off. Artificial sunlight lamp extinguishes LP2. The device preferentially uses solar energy to supply power to the red light LP1. When the voltage of the battery is not enough, the device is switched to 220 V AC power supply, which supplies power for the red light LP2, and realizes the synchronous sunrise and sunset indoors and outdoors.

Water Supply and Drainage System. Inside the storage tank used to collect outdoor rainwater, when the water level reaches a certain threshold, water is automatically discharged to sewers, through the control of microcontroller MCU instruction, which can also carry out drip irrigation, watering, and planting operations.

Ventilation System. Through the illuminance sensor information on the wind, temperature, and humidity, the microcontroller MCU can comprehensively deal with the data, within the range of presupposition, to achieve the desired parameters. The system will automatically open a fan, such as miniature blower ventilation device, when necessary, operate it for a period of time and then automatically stop it. After a certain period of time, the system can carry out automatic detection, all according to how it is programmed.

Remote Control Management System. The design of the remote control system is mainly based on the 32-bit microcontroller serving as the data processing center and on data transmission through the Internet to a server 6, remote monitoring, data storage, and analysis. The structure diagram is shown in Fig. 3, displaying the main part of the software, including the computer browser or mobile phone browser and server end equipment, main control module, 32-bit microcontroller, output drive module, a variety of sensors, infrared indoor lighting, and the components of ventilation, drip irrigation, watering, and farming.

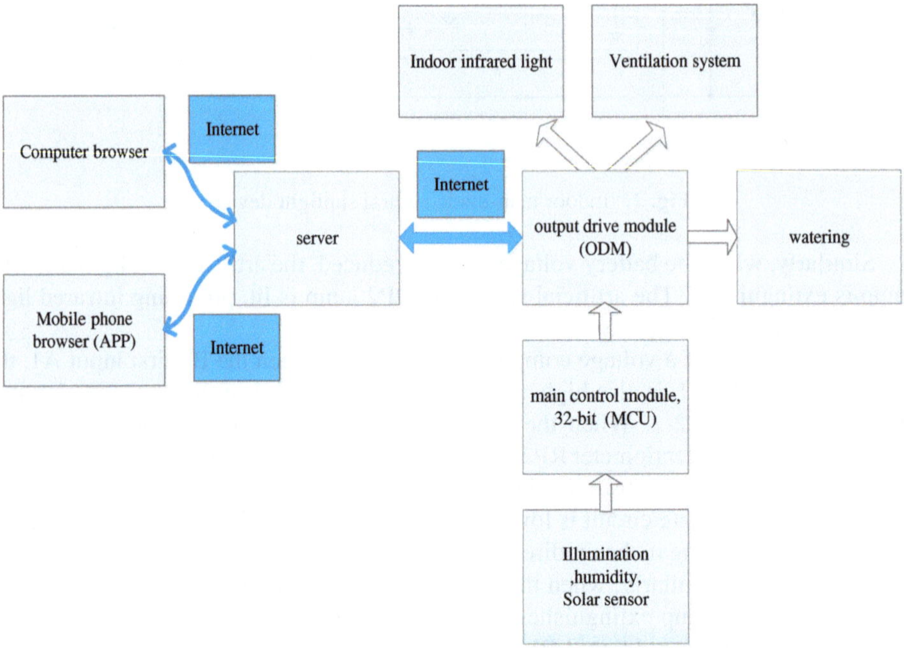

Fig. 3. Block diagram of remote control system.

The remote control computer or mobile phone users manipulate the browser through the function of the variable name access server to achieve an indoor environment data query and farming process. It also can be accessed through the web input control signal, as the microcontroller module server sends the control signal to the terminal equipment for analysis. The MCU control module control signal output module is the driving force to carry out the corresponding work, achieving the purpose of the remote control indoor farm.

The server returns the corresponding web page (HTML page) to the user's browser by parsing the commands requested by the user, and the browser displays the corresponding page to achieve the simple interaction. If the request received by the server is to control the farming command, the server sends the control command to the master module at the farm device end via the network.

The main control module receives the control commands to the server. After being resolved, the control command is transmitted to the output control module, and the output control module executes the corresponding control commands. The output driver circuit then drives the corresponding equipment. At the same time, farm environmental data can also be detected and transmitted to the terminal equipment of the main control module. The main control module sends environmental data information through the network TCP/IP protocol to send data to the server for tasks such as saving data or remote calling a client.

3 Exploration of Application Extension

3.1 The Significance of Interior Farms

The ratio of the ocean to the land area is about 7:3. However, with limited land resources, arable land area used is decreasing, the capacity per unit area on the earth to support the population is decreasing, and the resulting social contradictions have intensified, plaguing government leaders. If the excess solar energy collected by the five transmission modes can be allocated to the desired indoor farm, creating a man-made indoor solar system and restoring cultivated land resources, it is possible to resolve the land occupied social contradictions, making the world a more beautiful place.

3.2 Construction and Promotion of Indoor Farm

Now, the construction of indoor farms may not be accepted, because the land occupied may not be a danger to human beings at this time. A period of time may pass before this is a threat, so this paper suggests that government departments focus on the future, and that developers should collaborate and pioneer indoor farms, actively devoting more manpower and material resources, with the development and improvement of further financial resources in the future.

Building Community Air Corridor Protection System. Test area planning can involve the community in the first test, as part of the new community planning, that involves such activities as the construction of indoor air corridors for farming and the scaling of the construction of indoor farms, as shown in the schematic in Fig. 4, in which the community combines roofs through the air corridor protection system into a single one. For farm building in the floor and indoor roof top, the ultimate goal is to restore the building-occupied land resources, developing modern farm planting base in the building.

Interior Farm Architecture. The indoor farm architecture can be varied, one of the schematic as shown in Fig. 5, it can be used as the roof and the top farm, occupied by this building land area will recover more land resources.

Fig. 4. Aerial corridor of indoor farm.

Fig. 5. Sketch map of indoor farm building.

4 Conclusion

We have put forward a research plan of indoor planting, using modern electronic technology to subvert the traditional farming mode and achieve indoor plant growth and growth outside the natural growth environment, indoor artificial sunshine, ventilation, water supply and drainage. Through this effort, a number of technologies are integrated to achieve an intelligent indoor farming system that includes the use of personal mobile phone or computer browser through the Internet or wireless data transmission technology to achieve indoor farming and remote monitoring and management of a new combination of farming.

References

1. Yao, L.D.: To create artificial indoor land required for imitation of natural sunlight device. Invention Patent Application Number: 201710225452.3
2. Yao, L.D.: A technology recycling air intelligence land system. Invention Patent Number: 201510707967.8
3. Yao, L.D.: High frequency power amplifier, a C has authorized invention patent number: ZL 2012141362.6
4. Zhang, Y.: Principle and interface technology of single chip microcomputer. People's Post and Telecommunications Publishing House, Beijing (2011)
5. Cheng, J.: Equality: Solar Photovoltaic Power Generation System. Publishing House of Electronics Industry, Beijing (2016)
6. Infineon technologies Inc.: Wireless Components ASK/FSK Transmitter 868/433MHz TDA5100 v1.0, 25 July 2017. www.infineno.com

Design for Intelligent Control System of Curtain Based on Arduino

Sheng-Hui Meng[1,2](✉), An-Chi Huang[3], Chia-Jung Lee[2], Tian-Jiun Huang[4], and Jian-Nan Dal[2]

[1] Fujian Provincial Key Lab of Big Data Mining and Applications, Fujian University of Technology, Fuzhou 350118, China
menghui@fjut.edu.cn

[2] School of Information Science and Engineering, Fujian University of Technology, Fuzhou 350118, Fujian, China

[3] Department of Electrical Engineering, National Sun Yat-Sen University, Kaohsiung, Taiwan
abc94299@gmail.com

[4] Kaohsiung Municipal ChungShan Senior High School, Kaohsiung, Taiwan
abc45562@gmail.com

Abstract. In this research, a design for smart home curtains was proposed. The design is based on an Arduino UNO R3 development board equipped with an ESP8266-01 Wi-Fi module that communicates with an Android client application (app) using HTTP protocols to control the sensor and motor modules.

Keywords: Application (app) · Automatic curtains · Arduino · Hypertext Transfer Protocol (HTTP) · Smart home · Wi-Fi

1 Introduction

The demand for "smart devices" is rapidly rising, therefore, smart homes will provide a market for cross applications for artificial intelligence and the Internet of things (IoT) [1]. The construction of a reasonable, economic, and easily controllable smart home architecture has become an increasingly important topic for research and development [2]. Technologies may be assimilated into buildings by leveraging improvements in smartization and automation technologies and the integration of security, IT, the Internet, and service related industries [3]. Therefore, we have proposed a method for remotely controlling automated curtains in smart homes through local-area networks (LAN) and light sensors controls [4].

J.-S. Pan et al. (eds.), *Advances in Smart Vehicular Technology, Transportation, Communication and Applications*, Smart Innovation, Systems and Technologies 86,
https://doi.org/10.1007/978-3-319-70730-3_22

2 Materials and Methods

The system consists of the client app, an Arduino board, an ESP8266-01 Wi-Fi module, a light-sensing module, direct current (DC) motor module, and peripheral circuit, as shown in Fig. 1. The modules are described below:

The client app: The app was designed in Android Studio and written in Java. It uses the HTTP protocol to transmit data to the Wi-Fi module (which acts as a server) via URL queries, through a LAN. The Arduino Uno development board The Arduino board serves as the core of the system. It receives, processes, and transmits the data. The ESP8266-01 Wi-Fi module: The Wi-Fi module acts as a server in the LAN. It facilitates real-time communication with the client-end app, and serial communication with the Arduino board. Light-sensing module: This module uses photoresistors to sense the ambient light intensity, and it transmits data to the Arduino board via an analog port.

Light-sensing module: This module uses photoresistors to sense the ambient light intensity, and it transmits data to the Arduino board via an analog port. Motor module: This module consists of an L293D driver chip and a small DC motor. The logic level at the input port controls the forward and reverse DC motor rotations. Peripheral circuit: This module is composed of light-emitting diodes (LEDs), and it functions as a status indicator.

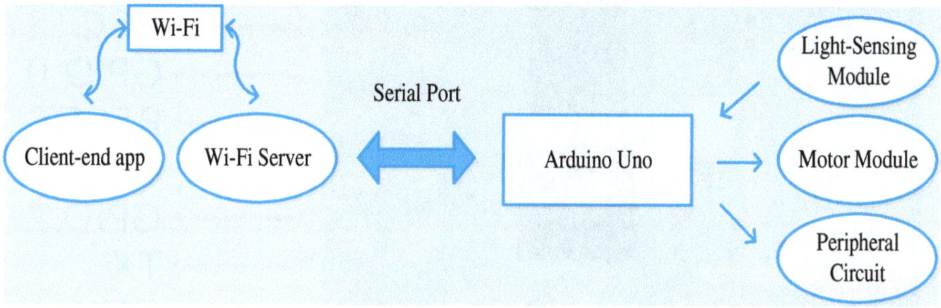

Fig. 1. Modular system.

3 Description of Hardware Modules and the Overall Electronic Circuitry

3.1 The L293D Motor Driver Chip

The L293D chip can simultaneously control two DC motors, and the direction of the motor (reverse or forward rotation) can be controlled simply by varying the logic level at its input port [5].

3.2 Circuit Connection of ESP8266-01Wi-Fi Module

TX pin:	The arduino digital I/O port of the second pin for the output pin.
RX pin:	The arduino digital I/O port of the third pin for the input pin.
GND pin:	The GND is connect the Arduino to the GND pin.
VCC pin:	Access to 3.3 V power for the power pin.
CH_PD pin:	Connect the 3.3 V from the power source.

An example of ESP8266-01 is shown in Fig. 2.

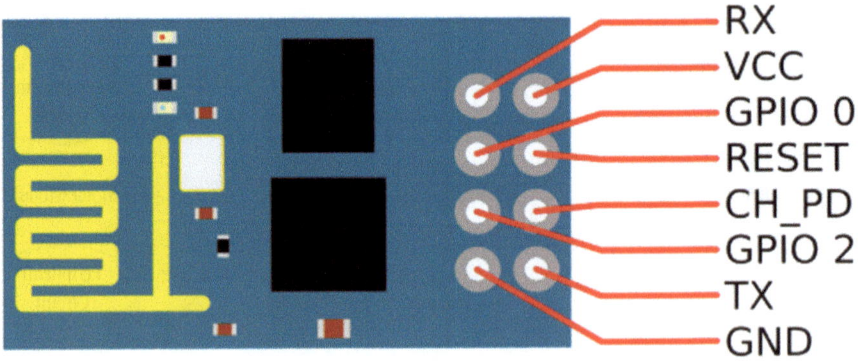

Fig. 2. ESP8266-01.

The circuit interconnection of the Arduino development board [6] and a schematic diagram of the overall circuit interconnection are illustrated in Figs. 3 and 4.

Fig. 3. a schematic diagram of the overall circuit interconnection.

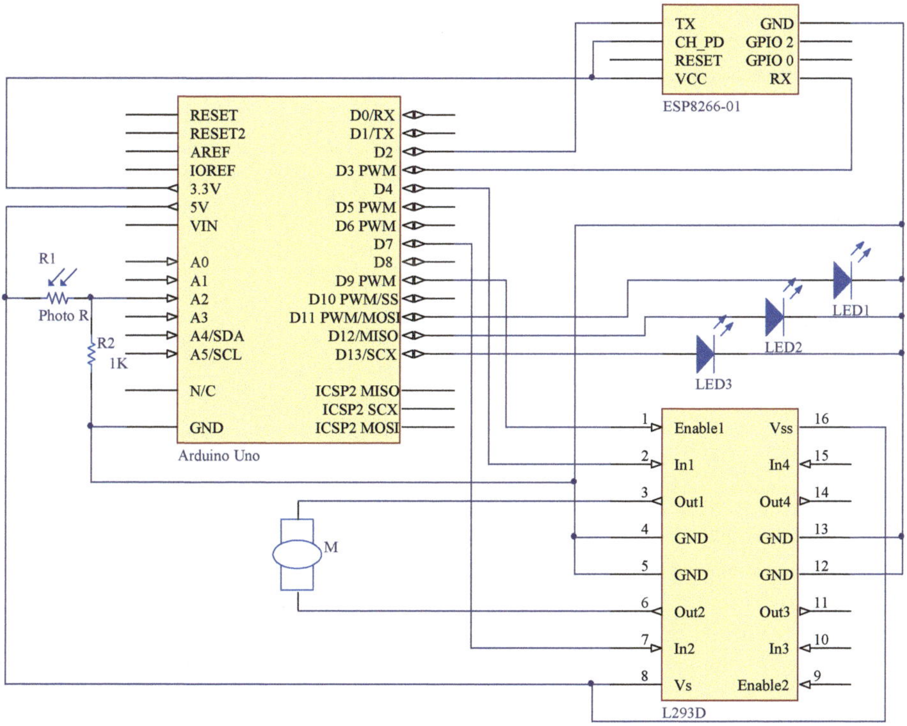

Fig. 4. Schematic diagram of the system's overall circuit interconnection.

4 Arduino Control Program-App Transfer Data Function Setting

Sample codes of the app's button function program:

```
if(digitalRead(pinNumber))
  {
   content += "ON";
   switch(pinNumber){
    case 11:
    z=1;
    break;
    case 12:
     if(curtainState == 1){
    analogWrite(enablepin,150);
    digitalWrite(in1a,HIGH);
    digitalWrite(in2a,LOW);
    delay(2000);
    curtainState = 0;
    }
    break;
    case 13:
     if(curtainState == 0){
    analogWrite(enablepin,150);
    digitalWrite(in1a,LOW);
    digitalWrite(in2a,HIGH);
    delay(2000);
    curtainState = 1;


        break;
        default:
        break;
       }
      }
      else
      {
       content += "OFF";
        switch(pinNumber){
        case 11:
        z=0;
        break;
          break;
        case 1
```

This code defines the functions of the data transmitted by the app. case11 is the button that activates the automated light-sensing control, and it corresponds to the INTELLIGENCE button of the app. The case12 button determines the state of the curtains and activates the motor driver chip if the curtains are closed, so that the motor rotates forwards and opens the curtains; thus, case12 corresponds to the OPEN button of the App. The case13 determines the state of the curtains and activates the motor driver chip if the curtains are open, so that the motor rotates in reverse and closes the curtains; thus, case13 corresponds to the CLOSE button of the app.

5 Results and Discussion

The required basic functionalities have been achieved in this design. However, note that the app's control buttons should not be pressed in quick succession since the "communication gap" between the app and Wi-Fi module is approximately 4.75 s, as shown in Fig. 5. Pressing the buttons too quickly results in data transmission failures.

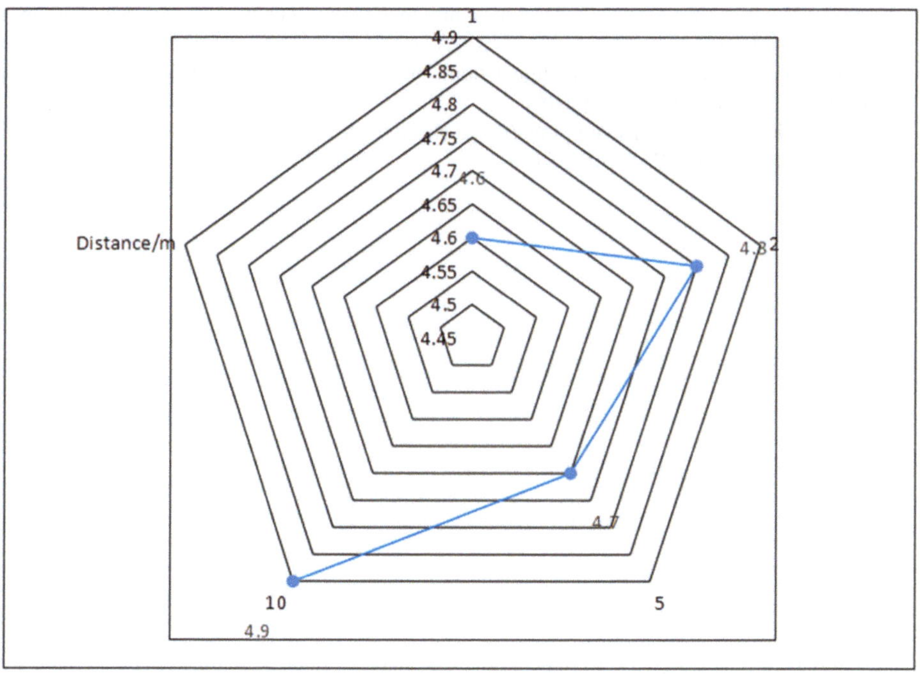

Fig. 5. Distance versus communication time line chart.

6 Conclusion

A light-sensing module was used in this system to detect the intensity of ambient light, which transmits the data back to the Arduino board. Programs were then set up to automatically close the curtains at night and open them in the morning using the transmitted data. An Android app was used to facilitate the remote and intelligent control of these automated curtains that possess ambient intelligence and autonomous control. In summary, this system improves the comfort and overall quality of a home.

References

1. Minling, Z., Li, N.: Intelligent home development status quo and future analysis. Telev. Technol. **39**(4) (2015)
2. Alam, M., Reaz, M., Ali, M.: A review of smart honest-past, present, and future. IEEE Trans. Syst. Man Cybern. **42**(6), 12–16 (2012)
3. Xiaoyu, T., Bingyi, F., Yunyong, Z.: Analysis on the development of Internet of Things smart homes. Mob. Commun. **34**(9), 18–19 (2010)
4. Yong, S., Wenyue, Y., Yuxin, Z.: Design of automatic curtain control system. Microcomput. Appl. **13**, 8 (2010)
5. Xinglong, F.: ESP8266 in the intelligent home monitoring system application. Single-Chip Embed. Syst. Appl. **09**, 52–56 (2016)
6. Yu, Z.: Android Network Development from Basic to Proficient. Tsinghua University Press, Beijing (2014). ISBN 978-7-302-34192-5

Graphene Saturable Absorption and Applications in Fiber Laser

Hancong Wang[1,2,3(✉)], Jinyang Lin[3], and Shihao Huang[3]

[1] Key Lab of Automotive Electronics and Electric Drive Technology of Fujian Province, Fujian University of Technology, Fuzhou 350108, China
whcuser@163.com

[2] Fujian Provincial Key Laboratory of Digital Equipment, Fujian University of Technology, Fuzhou 350108, China

[3] School of Information Science and Engineering, Fujian University of Technology, Fuzhou 350108, China

Abstract. Graphene as an ideal ultra-thin two-dimensional (2D) form of carbon has not only unique electronic but also wonderful broadband nonlinear optics. Particularly, under strong laser radiation, graphene can modulate laser field in amplitude due to saturable absorption (SA), and in phase due to large Kerr nonlinearity. Herein, we review the nonlinear optics saturable absorption in graphene and its application for fiber laser photonics. The SA in graphene has already led to intensive research advancements on graphene mode-locking/Q-switching ultrafast lasers while the large Kerr nonlinearity in graphene may also result in potential applications for graphene based Kerr photonics. Moreover, to analyze its evolving mechanism, two carrier relaxation times with distinct scales have been measured in graphene using ultrafast optical pump-probe spectroscopy. In addition, graphene-metal hybrid nanomaterials attract tremendous interest and show exceptional tunable and enhanced nonlinearity due to the surface plasmons in metal nanostructures.

Keywords: Graphene · Nonlinear optics · Saturable absorption · Carrier dynamics · Plasmonics

1 Introduction

The emergence of new materials could promote the development of other fields. One representative material is graphene, a two-dimensional sheet of sp^2-hybridized carbon atoms. Since its first isolation by A.K. Geim etc. through the "scotch tape technique" in 2004, graphene, as the thinnest known semi-metal show wonderful mechanical, electronic and optical properties [1]. Electrons in graphene have a linear *E-k* dispersion relation and behave like massless relativistic particles, which induce the near-ballistic electronic transport and a number of peculiar electronic properties. Due to its unprecedented high carrier mobilities, graphene has a vast range of practical nanoelectronic applications, from all carbon-based nanoelectronic devices, which could rival and even potentially substitute Silica, to ultra-capacitors and bio-devices.

J.-S. Pan et al. (eds.), *Advances in Smart Vehicular Technology, Transportation, Communication and Applications*, Smart Innovation, Systems and Technologies 86,
https://doi.org/10.1007/978-3-319-70730-3_23

Equally fascinating, graphene also exhibits unique optical property, which was relatively less investigated [2, 3]. The Dirac point type of energy band structure of graphene endows that it readily absorbs photons ranging from the visible to the infrared with the record strong inter-band transition efficiency. This broadband optics property provokes the realization of an ultra-fast graphene photodetector with very broad wavelength detection range. Under intensive illumination, SA of graphene will decrease, as a natural consequence of the Pauli blocking [4]. Taking full advantage of this SA property, the photonics application of graphene saturable absorber in passively mode locking and Q-switching laser had already been widely investigated [5–9].

The combination of its unique optical and electronic properties, and the linear band structure of the Dirac electrons renders graphene with ultra-broadband SA, delivering the flexibility of graphene based saturable absorbers to mode lock lasers without cut-off wavelength limitation, which constitutes the first example of "universal" saturable absorber. This unique nonlinear optical property in graphene could not only "overturn" the way how we look at the physics of SA but also offer new insight to explore the nonlinear optical property of other 2-dimensional materials (e.g. MoS_2 and phosphorene). In addition to its fundamental importance, owing to the special morphology and broadband SA property, graphene also shows remarkable advantages including: (i) weaker non-saturable scattering loss owing to its 2D morphology; (ii) easier to modify its SA through chemical functionalization or hybrid with polymer composite; (iii) more difficult to form bundled or entangled sate than SWCNTs; (iv) easy to be fabricated with various approaches; (v) ultrafast saturation recovery time in the range of 100 fs; (vi) isotropic optical absorption for each polarization of normally incident light. Graphene SA had already attracted considerable interests within laser community, particularly, because graphene laser photonics application belongs to the first few practice applications of graphene. The primary goal of this review is to summarize the experimental progress and also offers a guide to the potential development of graphene laser photonics.

2 Physics of Saturable Absorption (SA)

Figure 1a shows excitation processes responsible for absorption of light in monolayer graphene, in which electrons from the valence band (VB) are excited into the conduction band (CB). Shortly after photo-excitation (within 10–150 fs), these hot electrons thermalize and cool down to form a hot Fermi-Dirac distribution (Fig. 1b) with electronic temperature T_e. These created electron-hole pairs could block some of the originally possible interband optical transitions in a range of $k_B T_e$ (k_B is the Boltzmann constant) around the Fermi energy E_F and decrease the absorption of photons $\hbar\omega \sim k_B T_e$. In the following $\sim$1 ps, intraband phonon scattering further cools the thermalized carriers. After that, electron-hole recombination will dominate the process until the equilibrium electron and hole distribution is restored (Fig. 1a). However, these only describe the linear optical transition under low excitation intensity. As the excitation is increased, the photogenerated carriers increase in concentration (much larger than the intrinsic electron and hole carrier densities of $\sim 8 \times 10^{10}\ \mathrm{cm}^{-2}$ at room temperature) and cause the states near the edge of the CB and VB to fill, blocking further absorption (Fig. 1a), and thus imparting transparency to light at photon energies just above the band-edge. Band-filling occurs because no two electrons can fill the same state. Thus, SA or absorption bleaching is achieved due to this Pauli blocking process.

The nonlinear absorption properties could also be measured by balanced twin-detectors apparatus is shown in Fig. 1b. Stable and standard soliton mode locking fiber laser with output pulse width 1 ps and repetition rate 100 MHz is used as input seed pulse. Then, the seed pulse can be amplified through the EDFA (erbium doped fiber amplifier). To reduce the errors caused by the fluctuation of input power, the output powers (with or without passing through graphene) were determined simultaneously through two set of power meters with equal optoelectronic responses and parameter setups including sweeping time, as shown in Fig. 1b. Similar to the measurement of polarization independent SA of graphene, however, the incident laser source was passed through a polarizer firstly in order that a linearly polarized light with orthogonal polarization states were used as the input light, respectively. The SA feature of the graphene sample under the illumination of each of the orthogonally polarized light was measured, respectively. To reduce any artificial discrepancy caused by the fluctuation of the input power, output powers (with or without passing through the graphene sample) were detected simultaneously through two set of power meters. It thus demonstrated that graphene could be reasonably regarded as a polarization insensitive saturable absorber.

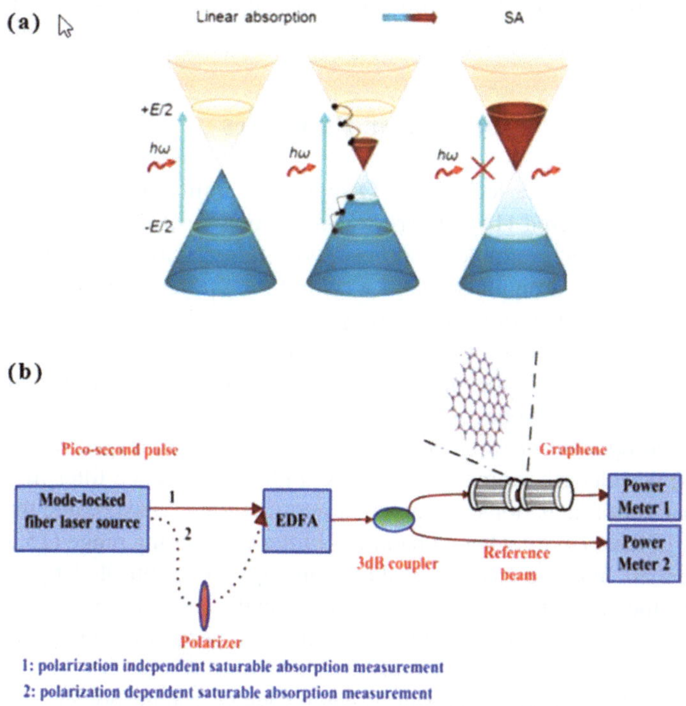

Fig. 1. (a–c) Excitation processes responsible for absorption of light in monolayer graphene. (d) Experimental setup for measuring nonlinear power-dependent absorption of graphene samples.

We also experimentally investigated the nonlinear optical response in few-layered graphene/silver nanowires (G/NWs) hybrid film that consists of two-dimensional graphene and one-dimensional silver nanowires [10]. The NWs can operate as a means of modulating the Fermi level of graphene and can be coupled to graphene for activating electron transfer. The enhanced SA based on the open-aperture femtosecond Z-scan measurement in G/NWs is obviously not the linear superposition of SA in pristine graphene [11] and reverse SA [12] in NWs. The enhancement of SA is attributed to interfacial nanowires network on graphene for large local field enhancement [13] and exciton-plasmon-photon conversion channel in which the propagating SPs couple out as photons at the ends of NWs [14]. In addition, accelerated carrier relaxation times in G/NWs are also observed by pump-probe spectroscopy, in which nanowires might supply an additional relaxation channel for graphene. The nanowire-induced defects and the increase of either electrons or holes induced by charge transfer, may also contribute to the ultrafast dynamics in G/NWs. These obtained results are important not only for understanding the coupling mechanism between graphene and plasmonic metal nanostructures, but also for developing graphene-based plasmonic devices [15]. More importantly, this work offers a new strategic approach to modulate the nonlinear optical properties of two-dimensional materials (such as topological insulator and molybdenum disulfide) or heterostructures beyond graphene, through mixing with one-dimensional metallic nanowires, which may boost the development and applications of 2D materials based nonlinear optics, especially the optical Kerr effect-based photonic devices, such as laser mode locker, Q-switcher, etc.

3 Graphene-Based Fiber Laser Photonics

To make full use of mode locking or Q-switching ability of graphene, laser cavity design must be considered [16]. After the demonstration of different methods in fabricating graphene laser saturable absorber devices, laser cavity design must be considered because cavity design also plays a central role in the laser output performance. In an Erbium-doped fiber ring laser cavity as illustrated in Fig. 2a, a piece of high doping EDF with normal group velocity dispersion (GVD) functions as the gain medium while a piece of Dispersion compensation fiber (DCF) with tiny normal GVD or a segment of single mode fiber (SMF) with abnormal GVD was spliced as the passive fiber component or to balance the cavity dispersion, respectively.

Whereas, wavelength tuning achieved through birefringence filter might be sensitive to temperature fluctuations and external perturbation. To overcome this drawback, another type of laser cavity where a band-pass filter with tuning range (1530–1555 nm) was incorporated into the cavity, as shown in Fig. 2b. Controlled by a micrometer screw, mode locking pulses with central wavelength from 1555 to 1559 nm, pulse width ~ 1.15 ps, average output power of 1 mw and per-pulse energy of 125 pJ, had been observed.

Another graphene mode locked laser cavity, where a high power EDFA replaced the combination of WDM, EDF and laser pump, is shown in Fig. 2c. Through depositing the reduced graphene oxide onto a side-polished fiber as the passive mode

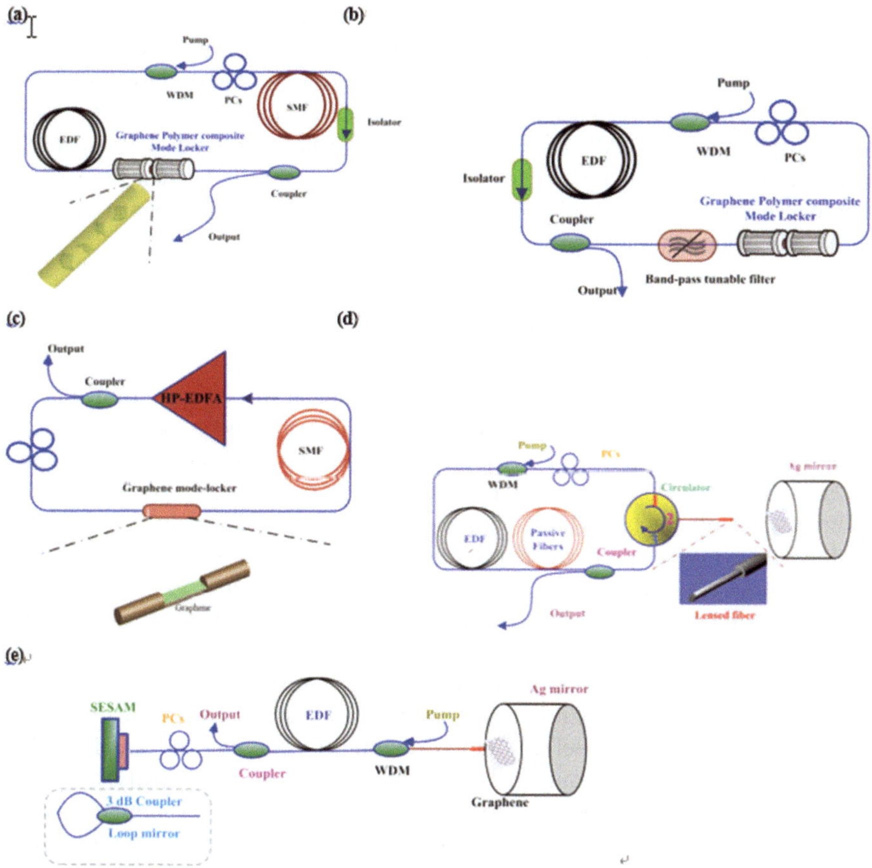

Fig. 2. Different types of graphene mode-locked erbium doped fiber lasers. (a) Graphene polymer composite mode locked fiber laser; (b) Wavelength tunable graphene polymer composite mode locked fiber; (c) Graphene on side polished fiber mode locked fiber laser; (d) GSAM mode locked ring cavity fiber laser, (e) GSAM mode locked linear cavity fiber laser.

locker, stable pulse with a repetition rate of 6.99 MHz and a central wavelength of $\sim$1561.6 nm, was formed.

To incorporate reflection type graphene into fiber laser ring cavity, an optical circulator was used to coupe light between the circulator and the graphene saturable absorber mirror (GSAM). Similar to the SESAM mode locking fiber laser, Fig. 2d is the GSAM mode locked ring cavity. Lensed fiber, fixed on the XYZ translation stage, was used to focus the output light and collect the reflected light with a minimum loss down to 3 dB. Through positioning the XYZ translation stage, the output light from the port 2 of the circulator was normally incident onto the surface of few layer graphene. In the following, mode locking operation was initiated through further optimizing the laser cavity parameters, such as: increasing the pump power and finely tuning the polarization controllers.

Besides the ring cavity configuration, GSAM could also be operated in a linear cavity fiber laser, as illustrated in Fig. 2e. At one side of the cavity, laser emission was reflected back by GSAM while at the other side, either a SESAM or a loop mirror made of a perfect 3 dB coupler, could function as the reflection mirror of cavity. If a SESAM is used instead, as SESAM has a relative slower relaxation time $\sim$10 ps than graphene, this fiber laser cavity has a lower self-starting threshold than pure graphene mode-locked fiber laser. Hybrid SA from both quantum well and graphene can result in mode locking operation and new soliton dynamics as well.

4 Conclusions and Prospective

So far, graphene laser photonics have been heralded in this review: from the SA of graphene to the characterization of graphene saturable absorbers, to their applications in different kinds of fiber lasers at different wavelengths. Beyond saturable absorber devices, other graphene photonics devices also show attractive laser applications. Another aspect of graphene laser photonics arises from its widely controllable and tunable nonlinear optics property, making graphene competent with silicon and gallium arsenide. There are several ways to modulate SA in graphene: varying SA by controlling the number of layers and polymerization and functionalization of graphene. Additionally, we can also engineer the band structure by electron doping, in which an external electric gate field is applied upon graphene surface to shift its Fermi Level (also called as chemical potential) and change the carrier density in graphene; by chemical doping: inflicting defects and *N*-doping or *P*-doping; by graphene hybrid polymer and graphene heterojunction structure and so on. Principally, as-proposed approaches will allow researchers to design laser saturable absorber devices with customized optics property. To date, however, direct experiments were not yet provided. It is anticipated that the wide control and tuning of the nonlinear optics in graphene may lead to the birth of new graphene laser photonics devices.

Acknowledgements. National Natural Science Foundation of China (61604041); Fujian Provincial Natural Science Foundation of China (2016J05147, 2017J05097); Mid-youth Project of Education Bureau of Fujian Province (JAT160331, JAT160334); Startup Foundation of Fujian University of Technology (GY-Z160049); Fujian Provincial Major Research and Development Platform for the Technology of Numerical Control Equipment (2014H2002).

References

1. Novoselov, K.S., Geim, A.K., Morozov, S.V., Jiang, D., Zhang, Y., Dubonos, S.V., Grigorieva, I.V., Firsov, A.A.: Electric field effect in atomically thin carbon films. Science **306**(5696), 666–669 (2004)
2. Bonaccorso, F., Sun, Z., Hasan, T., Ferrari, A.C.: Graphene photonics and optoelectronics. Nat. Photonics **4**(9), 611–622 (2010)
3. Avouris, P., Freitag, M.: Graphene photonics, plasmonics, and optoelectronics. IEEE J. Sel. Top. Quantum Electron. **20**(1), 72–83 (2014). Article ID 6000112

4. Bao, Q., Zhang, H., Wang, Y., Ni, Z., Yan, Y., Shen, Z.X., Loh, K.P., Tang, D.Y.: Atomic-layer graphene as a saturable absorber for ultrafast pulsed lasers. Adv. Funct. Mater. **19**(19), 3077–3083 (2009)
5. Bao, Q., Loh, K.P.: Graphene photonics, plasmonics, and broadband optoelectronic devices. ACS Nano **6**(5), 3677–3694 (2012)
6. Zheng, Z., Zhao, C., Lu, S., Chen, Y., Li, Y., Zhang, H., Wen, S.: Microwave and optical saturable absorption in graphene. Opt. Express **20**(21), 23201–23214 (2012)
7. Miao, L., Jiang, Y., Lu, S., Shi, B., Zhao, C., Zhang, H., Wen, S.: Broadband ultrafast nonlinear optical response of few-layers graphene: toward the mid-infrared regime. Photonics Res. **3**(5), 214 (2015)
8. Yamashita, S.: A tutorial on nonlinear photonic applications of carbon nanotube and graphene. IEEE J. Lightwave Technol. **30**(4), 427–447 (2012)
9. Sun, Z., Hasan, T., Torrisi, F., Popa, D., Privitera, G., Wang, F., Bonaccorso, F., Basko, D.M., Ferrari, A.C.: Graphene mode-locked ultrafast laser. ACS Nano **4**(2), 803–810 (2010)
10. Wang, H., Miao, L., Jiang, Y., Lu, S., Li, Z., Li, P., Zhao, C., Zhang, H., Wen, S.: Enhancing the saturable absorption and carrier dynamics of graphene with plasmonic nanowires. Phys. Status Solidi B **252**(10), 2159–2166 (2015)
11. Zhang, H., Virally, S., Bao, Q., Kian Ping, L., Massar, S., Godbout, N., Kockaert, P.: Z-scan measurement of the nonlinear refractive index of graphene. Opt. Lett. **37**(11), 1856–1858 (2012)
12. Kalanoor, B.S., Bisht, P.B., Akbar Ali, S., Baby, T.T., Ramaprabhu, S.: Optical nonlinearity of silver-decorated graphene. J. Opt. Soc. Am. B **29**(4), 669–675 (2012)
13. Xu, H., Bjerneld, E.J., Käll, M., Börjesson, L.: Spectroscopy of single hemoglobin molecules by surface enhanced raman scattering. Phys. Rev. Lett. **83**(21), 4357–4360 (1999)
14. Wei, H., Li, Z., Tian, X., Wang, Z., Cong, F., Liu, N., Zhang, S., Nordlander, P., Halas, N.J., Xu, H.: Quantum dot-based local field imaging reveals plasmon-based interferometric logic in silver nanowire networks. Nano Lett. **11**(2), 471–475 (2011)
15. Qian, H., Ma, Y., Yang, Q., Chen, B., Liu, Y., Guo, X., Lin, S., Ruan, J., Liu, X., Tong, L., Wang, Z.L.: Electrical tuning of surface plasmon polariton propagation in graphene-nanowire hybrid structure. ACS Nano **8**(3), 2584–2589 (2014)
16. Martinez, A., Sun, Z.: Nanotube and graphene saturable absorbers for fibre lasers. Nat. Photonics **7**(11), 842–845 (2013)

Power Supply Loss Riding Through Method for High-Voltage Great-Power Cascaded H-Bridge Multilevel Inverters

Xinjian Cai(✉), Tianjian Li, and Jiaxin Chen

School of Information Science and Engineering,
Fujian University of Technology, Fuzhou, China
cxj2001@fjut.edu.cn

Abstract. In present, adjustable speed drive systems are widely used in industrial production and ship propelling. One of the type of inverters that are commonly applied to high-voltage great-power adjustable speed drive system is Cascaded H-bridge multilevel inverters. In practical applications, many factors may result in power supply losses. Therefore, the capability of riding through power supply losses is required to improve the reliability. Unfortunately, few researcher pay attention to the riding through method for cascaded H-bridge multilevel inverters. This paper proposes a riding through method and designs the controller. The proposed method is verified by simulation.

Keywords: Adjustable speed drive system · Power supply loss riding through · Cascaded H-bridge multilevel inverter

1 Introduction

In present, adjustable speed drive system is applied to ship propelling, vehicle and industrial production to improve the performances [1–5]. Unfortunately, many factors result in power supply losses that usually lead to inverters shutting down. The shutting down of inverters usually brings about economic losses and accidents [6–9]. Therefore, the capability of power supply losses riding through (PSLRT) is required to improve the performance and reliability.

Many researchers are interested in PSLRT method for inverters. In [3], a riding through method using the ultra-capacitor is proposed. The energy is stored in the ultra-capacitor when the inverter operates under the normal mode or braking mode. The stored energy is applied to riding through the grid power supply loss. The other method of stabilizing inverter DC voltages is feeding electrical energy, which is generated by utilizing the inertia energy of the rotor and the connected mechanical loads, to inverters. Scholars who take interest in this area usually focus on designing and improving the control method of the motor that is working as a generator.

Most of the riding through method are designed for two-level or three-level inverters, in which only one DC voltage needs to be stabilized during the riding through process. At present, three-phase cascaded H-bridge multilevel inverters that consists of numerous H bridge cells, are widely used in high-voltage great-power ASDs.

J.-S. Pan et al. (eds.), *Advances in Smart Vehicular Technology, Transportation, Communication and Applications*, Smart Innovation, Systems and Technologies 86,
https://doi.org/10.1007/978-3-319-70730-3_24

Unfortunately, few researcher pays attention to the PSLRT method of cascaded H-bridge multilevel inverters.

This paper focuses on the PSLRT method for cascaded H-bridge multilevel inverters. Since this type inverters usually includes three legs and each leg consists of numerous H-bridge cells. The PSLRT method includes three parts: (1) stabilizing the sum of all cell DC voltages ($U_{DC\Sigma}$); (2) balancing three leg DC voltages (U_{DCA}, U_{DCB}, and U_{DCC}), achieving $U_{DCA} = U_{DCB} = U_{DCC} = U_{DC\Sigma}/3$; (3) balancing the cascaded cell DC voltages in one leg, achieving $U_{DC\_1} = U_{DC\_2} = \ldots = U_{DC\_N} = \mathrm{U}_{DCp}/\mathrm{N}$ (p = A, B and C). In this paper, three parts of the PSLRT method are designed. This paper is organized as follow. Section 2 introduces the simplified topology of cascaded H-bridge multilevel inverters during riding through peroids. Section 3 proposes the PSLRT method. Section 4 offers simulation results.

2 Simplified Topology of Cascaded H-Bridge Multilevel Inverters Under Power Supply Loss Riding Through Condition

Cascaded H-bridge multilevel inverters that are applied to adjustable speed drive systems, usually consists of three legs. Each leg is composed of N cascaded H bridge cells. Each cell includes an isolated DC source. The DC source usually consists of DC capacitor and three-phase rectifier that is connected to the output ports of a multi-winding transformer. Electrical energy flows into H-bridge cells by way of the multi-winding transformer and the rectifier when the grid voltage is normal. Under the condition of power supply losses, the rectifier usually shuts down.

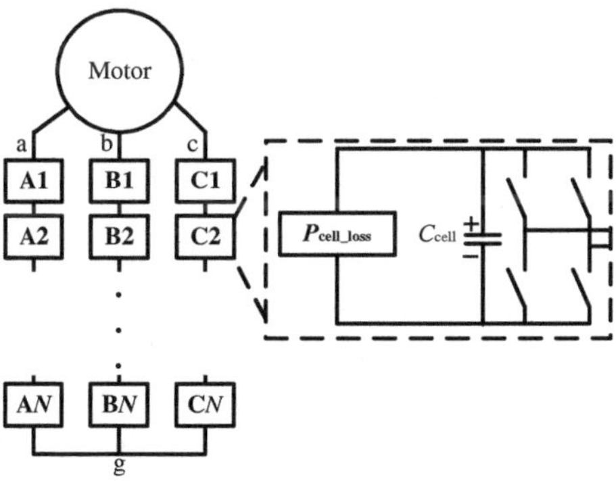

Fig. 1. The simplified topology of cascaded H bridge multilevel inverters under the condition of riding through power supply loss

On the other hand, the electrical power consumption of power electronic devices is mainly related to the magnitude of the output current, the output power factor, the carrier waveform frequency, the cell DC voltage. In practical applications, the carrier waveform frequency is constant. During the riding through period, the output current magnitude is nearly constant and the power factor is nearly equal to zero. Therefore, under the condition that the inverter DC voltages a little deviate from the required value, the electrical power consumption is almost constant.

The simplified topology of cascaded H-bridge multilevel inverters is shown in Fig. 1, during riding through periods. In Fig. 1, $P_{\text{cell-loss}}$ is the power consumption of the H-bridge cell.

3 The Proposed Method of Riding Through Power Supply Loss for Cascaded H-Bridge Multilevel Inverters

Since cascaded H-bridge multilevel inverters consists of three legs that includes several cascade-connected H-bridge cells, not only should the sum of all H-bridge cell DC voltages be stabilized, but also three leg DC voltages and the H-bridge cell DC voltages during the riding through period.

3.1 The Method of Stabilizing the Sum of All H-Bridge Cell DC Voltages

Since cascaded H-bridge multilevel inverters consists of numerous H-bridge cell that needs an isolated DC voltage, riding through power supply by using storage device or specially designed circuits is very complex and costly. Therefore, utilizing the inertia energy of the rotor and the mechanical loads and feeding the energy back to the inverter is a feasible method to stabilizing the sum of all cell DC voltages. This method is inexpensive because it needs no additional devices and circuits. During the riding through period, the induction motor operate as a generator and the electrical energy is feeding back to the inverter. The relationship among the sum of all cell DC voltages and the feed-back power should be analyzed to design the controller at first. When the field orientated control (FOC) method is utilized, the feed-back power can be expressed as:

$$P_{\Sigma} = -1.5 n_{\text{p}} i_{\text{sq}} \varphi_{\text{sd}} \omega_{\text{r}} - P_{\text{e\_loss}} \tag{1}$$

In Eq. (1), P_{Σ} is the feed-back power, n_{p} is the number of pole pair, i_{sq} is the current component of q axis, φ_{sd} is the stator magnetic field component of d axis, ω_{r} is the rotor speed, $P_{\text{e\_loss}}$ is the electrical power consumption of the motor. Since the magnitude of the motor current is nearly constant, $P_{\text{e\_loss}}$ can be treated as a constant value, during the riding through period. Under the condition that the sum of all cell DC voltages ($U_{\text{DC}\Sigma}$) is a little deviating from the static value ($U_{\text{DC}\Sigma 0}$). The relationship among P_{Σ}, $U_{\text{DC}\Sigma}$ and the inverter power consumption ($\Sigma P_{\text{cell\_loss}}$) can be expressed as:

$$P_{\Sigma} - \sum P_{\text{cell\_loss}} = C_{\Sigma} U_{\text{DC}\Sigma 0} \frac{\text{d}U_{\text{DC}\Sigma}}{\text{d}t} \tag{2}$$

In Eq. (2), C_Σ is the equivalent capacitor of the inverter. The value of C_Σ is equal to all H-bridge cell capacitors are connected in series. Based on Eqs. (1) and (2), the relationship between i_{sq} and $U_{DC\Sigma}$ can be expressed as:

$$G_{isq\_UDC\Sigma}(s) = -\frac{1.5n_p\varphi_{sd}\omega_r}{C_\Sigma U_{DC\Sigma0}s} \tag{3}$$

It can be observed from Eq. (3) that a proportional controller can be used to stabilize $U_{DC\Sigma}$. But the static error is inevitable when the proportional controller is utilized without compensation. Considering that the electrical power consumption of the motor and the inverter can be measured or calculated easily, a compensation (i_{sq_com}) can be added to the controller to reduce the static error. The picture that describes the control system are shown in Fig. 2.

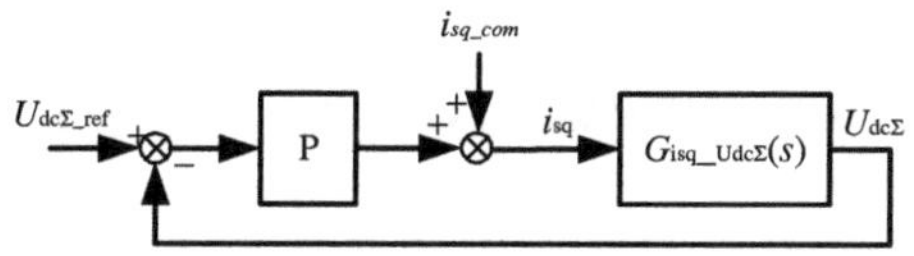

Fig. 2. The control system for stabilizing $U_{DC\Sigma}$

3.2 The Method of Balancing Three Leg DC Voltages

Since there are usually no zero sequence current way between the inverter and the motor, injecting a zero sequence voltage in the inverter output AC voltage little affects the motor. Regulating the leg feed-back electrical energy and balancing three leg DC voltages by injecting a zero sequence voltage in the inverter output AC voltage is a valuable method. The positive sequence voltage (fundamental waveform) and the zero sequence voltage of the inverter output voltages can be expressed as:

$$u_A = U_{pm}\sin\omega_m t + u_0 \tag{4}$$

$$u_B = U_{pm}\sin\left(\omega_m t - \frac{2\pi}{3}\right) + u_0 \tag{5}$$

$$u_C = U_{pm}\sin\left(\omega_m t + \frac{2\pi}{3}\right) + u_0 \tag{6}$$

$$u_0 = U_{0m}\sin(\omega_m t + \theta_0) \tag{7}$$

In Eqs. (4)–(7), *U*pm is the magnitude of the positive sequence phase-voltage waveform, ω_m is the frequency of the output voltage fundamental waveform, θ is the difference between the voltage phase and the current phase, U_{0m} is the magnitude of the zero sequence voltage waveform, θ_0 is the difference between the zero sequence voltage phase and u_A phase. In common application, there is no way for zero sequence current between inverters and motors. The inverter output current can be expressed as:

$$i_A = I_{pm} \sin(\omega_m t - \theta) \tag{8}$$

$$i_B = I_{pm} \sin\left(\omega_m t - \theta - \frac{2\pi}{3}\right) \tag{9}$$

$$i_C = I_{pm} \sin\left(\omega_m t - \theta + \frac{2\pi}{3}\right) \tag{10}$$

In Eqs. (8)–(10), I_{pm} is the magnitude of the phase-current waveform. The active-power adjustments by injecting the zero sequence voltage can be expressed as:

$$\Delta P_A^0 = 0.5 U_{0m} I_{pm} \cos(\theta_0 + \theta) \tag{11}$$

$$\Delta P_B^0 = 0.5 U_{0m} I_{pm} \cos\left(\theta_0 + \theta + \frac{2\pi}{3}\right) \tag{12}$$

$$\Delta P_C^0 = 0.5 U_{0m} I_{pm} \cos\left(\theta_0 + \theta - \frac{2\pi}{3}\right) \tag{13}$$

Based on the principle of the capacitor storing electrical energy, the relationship of the active-power adjustment and the leg DC voltage can be expressed as:

$$G_{\Delta P_p\_U_{DCp}}(s) = \frac{1}{C_p U_{DCp0} s} \quad (p = A, B, C) \tag{14}$$

In Eq. (14), C_p is the equivalent capacitor of the leg. The value of C_p is equal to the leg H-bridge cell capacitors are connected in series. U_{DCp} is the leg DC voltage, and U_{DCp0} is the static value. On the other hand, under the condition that $U_{DC\Sigma}$ is stabilized, if U_{DCA} and U_{DCB} are adjusted to $U_{DC\Sigma}/3$, U_{DCC} is $U_{DC\Sigma}/3$. Therefore only two controller are required to balancing three leg DC voltages. It can be observed from Eq. (14) that proportional controllers can be used to control U_{DCA} and U_{DCB}. But the static error is inevitable when the proportional controller is utilized. In practical applications, small static error is permitted. The picture that describes the control system are shown in Fig. 3. The zero sequence voltage is calculated by (11), (12) and (13).

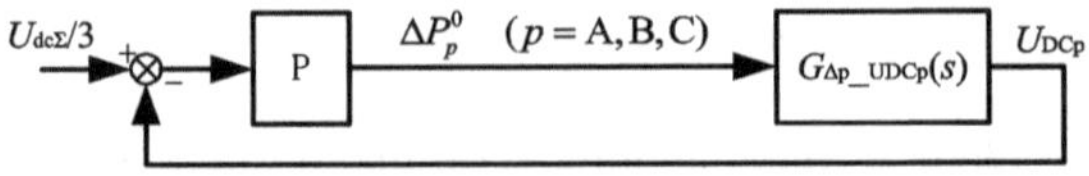

Fig. 3. The control system for controlling U_{DC_p}

3.3 The Method of Balancing the H-Bridge Cell DC Voltage in One Leg

It is well known that the cell DC voltage can be controlled by regulating the active power that is fed in the cell. The cell active power without regulation is expressed as:

$$P_{\mathrm{cell}} = U_{\mathrm{cell}} I_{\mathrm{cell}} \cos\theta \tag{15}$$

In Eq. (15), cosθ is the power factor, U_{cell} and I_{cell} are the RMS of the cell output AC voltage and current without regulations, respectively. If an AC voltage is injected, the active power can be expressed as:

$$P_{\mathrm{cell}} + \Delta P_{\mathrm{cell}} = U_{\mathrm{cell}} I_{\mathrm{cell}} \cos\theta + \Delta U_{\mathrm{cell}} I_{\mathrm{cell}} \cos\theta_1 \tag{16}$$

Since the riding through is required to operate well under the condition that the rotor speed is high. In some special conditions, the regulation of magnetic field is forbidden. The magnitude of output voltage is proportional to rotor speed and the magnitude of the magnetic field. Therefore, the method of controlling and balancing the H-bridge cell DC voltages is required to operate well when the cell output voltage is high. Unfortunately, the higher the output voltage magnitude, the smaller the AC voltage magnitude that is injected to regulate the cell feed-back active power without over-modulation. The method of this paper needs smaller injecting AC voltage and achieves better performance. It can be observed that ΔU_{cell} is smallest when cosθ_1 is zero, under the condition that ΔU_{cell} and I_{cell} is constant. Therefore, injecting an AC voltage of which the phase is equal to the current, can regulate the active power and the magnitude is smallest. This method fits to the condition that small magnitude AC voltage is permitted to inject without over-modulation.

Based on the principle of the capacitor storing electrical energy, the relationship of the active-power adjustment and the cell DC voltage can be expressed as:

$$G_{\Delta U_{cell}\_U_{DCcell}}(s) = \frac{1}{C_{cell} U_{\mathrm{DC}cell0} I_{\mathrm{cell}} s} \tag{17}$$

In Eq. (17), C_{cell} is cell capacitor. $U_{\mathrm{DC}cell}$ is the cell DC voltage, and $U_{\mathrm{DC}cell0}$ is the static value. It can be observed from Eq. (17) that proportional controllers can be used to control U_{DCcell} because small static error is permitted in practical applications. The picture that describes the control system are shown in Fig. 4.

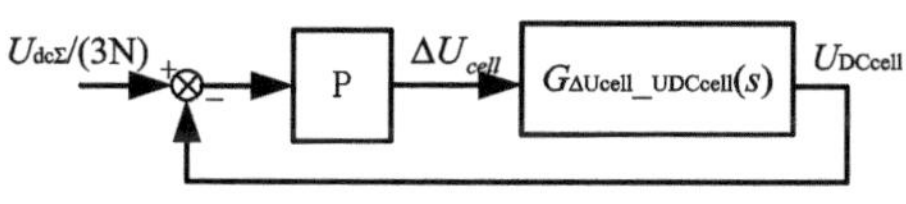

Fig. 4. The control system for controlling $U_{\mathrm{DC}cell}$

4 Simulation

To verify the proposed PSLRT method, a mathematic model of the cascaded H-bridge multilevel inverter and the induction motor is established by means of MATLAB/Simulink. Each H-bridge cell is mainly composed of an H-bridge inverter, a capacitor (9.8 mF) and a three-phase rectifier, of which the input ports are connected to a multi-winding transformer. Each leg includes five H-bridge cells. The inverter consists of three legs, of which the topologies are identical. The nominal RMS value of the inverter output voltage is 6000 V. The nominal output power of the inverter is 1000 kW. Each cell DC voltage is approximately 1000 V, and each leg DC voltage is nearly 5000 V when the grid voltage is normal. Since the cell DC voltage is one of the criteria of detecting the power supply loss and the recovery of the power supply, the stabilized cell DC voltage should be lower than 850 V (0.85 times the nominal value) during the riding through period. On the other hand, because the cell is required to output nominal AC voltage during the riding through period, the cell DC voltage should be greater than 800 V (0.80 times the nominal value). Therefore, the stabilized value of the cell DC voltage is nearly 820 V (0.82 times the nominal value). The stabilized value of the leg DC voltage that is the sum of five cascaded cell DC voltages, is approximately 4100 V. The stabilized value of $U_{DC\Sigma}$ is 12300 V. To simulate the cell power consumption the resisters are placed in the cells and connected with the capacitor (10 mF) in parallel. The resisters of three legs are not equal to simulate the differences of three leg power consumptions. The resisters placed in the cells of A phase leg are 1000 Ω. The resisters placed in the cells of B and C phase leg are 1075 Ω and 925 Ω respectively.

The nominal voltage and power of the motor are 6000 V and 300 kW respectively. The nominal rotor speed is 750 r/min. The load of the motor is a synchronous generator of which the output ports are connected to the great-power resisters. When the rotor speed is 500 r/min and the proposed method of stabilizing $U_{DC\Sigma}$ is utilized, the waveform of $U_{DC\Sigma}$ is shown in Fig. 5. Because the inverter cannot ride through power supply losses without stabilizing $U_{DC\Sigma}$, the simulation result of no stabilizing $U_{DC\Sigma}$ is invaluable and is not provided in this paper.

The waveform of U_{DCA}, U_{DCB} and U_{DCC} are shown in Fig. 6 when the method of balancing three leg DC voltage is not used. The simulation result of the proposed balancing three leg DC voltages is shown in Fig. 7. The waveform of five cell DC voltages in A phase is shown in Fig. 8 without the proposed method of balancing cascaded H-bridge cell DC voltages. The simulation results of the proposed method of controlling and balancing cell DC voltages is shown in Fig. 8.

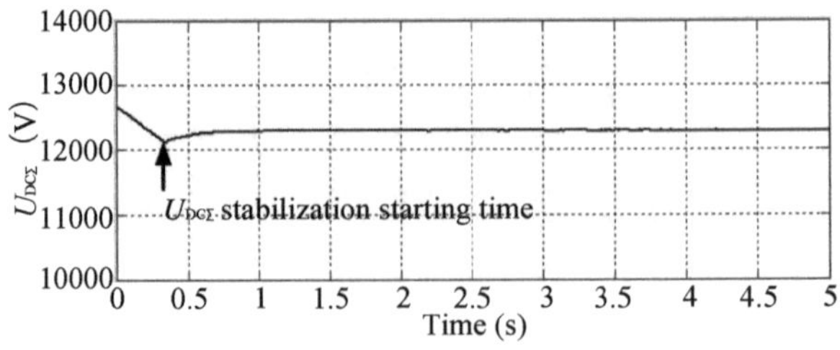

Fig. 5. The waveform of $U_{DC\Sigma}$ when the proposed method of stabilizing $U_{DC\Sigma}$ is utilized

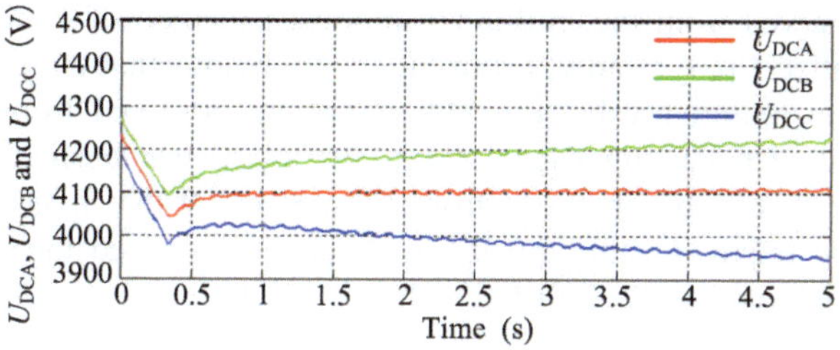

Fig. 6. The waveform U_{DCA}, U_{DCB} and U_{DCC} without the proposed method of balancing three leg DC voltages

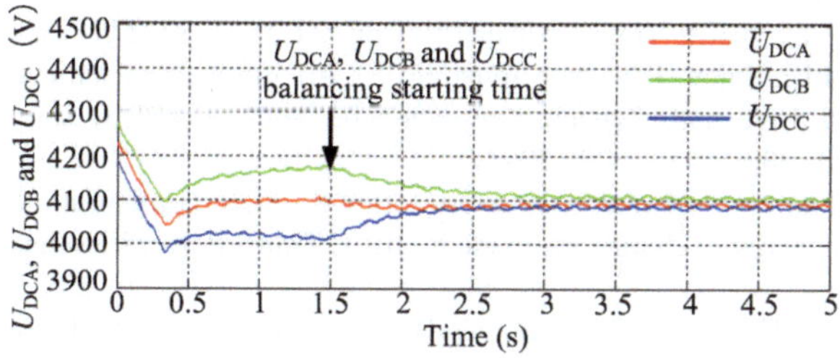

Fig. 7. The waveform U_{DCA}, U_{DCB} and U_{DCC} with the proposed method of balancing three leg DC voltages

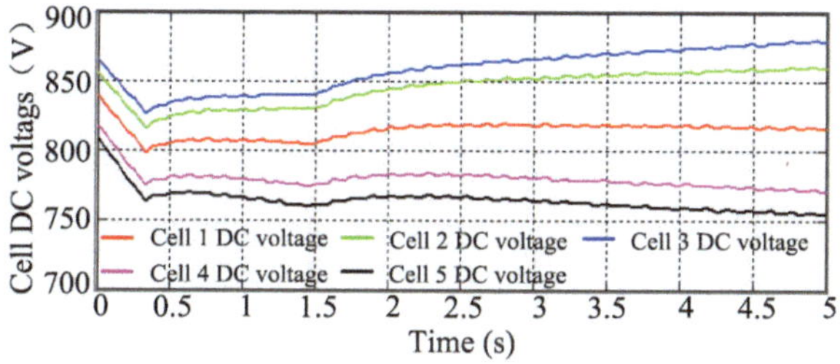

Fig. 8. The waveform of five cell DC voltages without the proposed method of balancing cell DC voltages in one leg

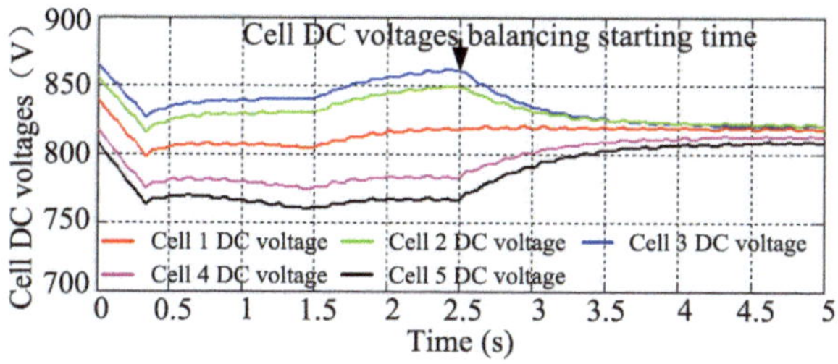

Fig. 9. The waveform of five cell DC voltages with the proposed method of balancing cell DC voltages in one leg

It can be observed from Fig. 5 that the proposed method of stabilizing $U_{DC\Sigma}$ can control and stabilize $U_{DC\Sigma}$ rapidly. From Fig. 6, it obvious that three leg DC voltages deviates from the required value quickly. It can be observed from Fig. 7 that the proposed method of balancing U_{DCA}, U_{DCB} and U_{DCC} can balancing three leg DC voltages quickly. It can be observed from Figs. 8 and 9 that five cell DC voltages deviates from the required value quickly without the proposed method and the proposed method can balance five cell DC voltages quickly.

References

1. Khamphakdi, P., Nitta, M., Hagiwara, M., Akagi, H.: Zero-voltage ride-through capability of a transformerless back-to-back system using modular multilevel cascade converters for power distribution systems. IEEE Trans. Power Electron. **31**(4), 2730–2741 (2016)
2. Duran-Gomez, J.L., Enjeti, P.N., von Jouanne, A.: An approach to achieve ride-through of an adjustable-speed drive with flyback converter modules powered by super capacitors. IEEE Trans. Ind. Appl. **38**(2), 514–522 (2002)
3. Orser, D., Mohan, N.: A matrix converter ride-through configuration using input filter capacitors as an energy exchange mechanism. IEEE Trans. Power Electron. **30**(8), 4377–4385 (2015)
4. von Jouanne, A., Enjeti, P.N., Banerjee, B.: Assessment of ride-through alternatives for adjustable-speed drives. IEEE Trans. Ind. Appl. **35**(4), 908–916 (1999)
5. Klumpner, C., Boldea, I., Blaabjerg, F.: Limited ride-through capabilities for direct frequency converters. IEEE Trans. Power Electron. **16**(6), 837–845 (2001)

Prediction of Hourly Power Consumption for a Central Air-Conditioning System Based on Different Machine Learning Methods

Si-qi Gao[1,2,3], Fu-min Zou[2(✉)], Xin-hua Jiang[3], Lyuchao Liao[3], and Yun Chen[3]

[1] Beidou Navigation and Smart Traffic Innovation Center of Fujian Province Fuzhou, Fuzhou 350118, Fujian, China
784912077@qq.com
[2] Fujian University of Technology, Fuzhou 350118, Fujian, China
fmzou@fjut.edu.cn
[3] Fujian Key Laboratory for Automotive Electronics and Electric Drive, Fujian University of Technology, Fuzhou 350118, Fujian, China
achao@fjut.edu.cn

Abstract. This paper uses a variety of machine learning methods to predict the hourly power consumption of a central air-conditioning system in a public building. It is found that the parameters of the central air-conditioning system are different at different times, so is the corresponding power consumption. The paper applies the time series to predict the power consumption on account of the time, which predicts the hourly power consumption based on historical time series data. Comparing the prediction accuracy of multiple machine learning methods, we find that the Gradient Boosting Regression Tree (GBRT), one of the ensemble learning methods, has the highest prediction accuracy.

Keywords: Central air-conditioning system · Machine learning methods · Ensemble learning · Time series prediction

1 Introduction

With the information technology shooting up, intelligent buildings and energy issues have become the focus of increasing attention. Central air conditioning, as an indispensable part of intelligent building, often accounts for a large proportion of the energy consumption in large public buildings. Energy is largely lost in that the system works at low temperature and large flow rate for a long time in the case of low load and low temperature. Therefore, if the system power consumption of each moment can be accurately predicted, the optimization plan could be designed in advance to reduce the energy consumption.

Conventional prediction models are based on physical principles to calculate thermal dynamics and energy behavior at the building level. Some of them include models of space systems, natural ventilation, air conditioning system,

J.-S. Pan et al. (eds.), *Advances in Smart Vehicular Technology, Transportation, Communication and Applications*, Smart Innovation, Systems and Technologies 86,
https://doi.org/10.1007/978-3-319-70730-3_25

passive solar, photovoltanic systems, financial issue, occupants behavior, climate environment, and so on [1]. There are also statistical methods [2–4]. These methods are used to predict energy consumption and that affecting variables such as weather and energy cost. In addition, in order to influence the development of future building systems, there are also some hybrid methods [5,6], which combine the models above to optimize performance. Conventional methods are advantages in that they are easily implemented and interpreted, however, their disability to handle non-linearity in short-term load series instead encourages the use of machine learning methods [7]. In recent years, a large number of machine learning algorithms have been used to predict energy consumption [8–10].

The paper aims to predict the system hourly power consumption based on historical data of the running parameters of the various system components, which utilizes the four machine learning algorithms: Random Forests, K-Nearest Neighbour, Gradient Boosting Tree Regression and Support Vector Regression. Moreover, seeking a model that is easy to be implemented with minimal input requirements and maximum accuracy.

2 Data Exploratory Analysis

The data source used in the paper has 51 variables (including timestamp). There are totally 1750 instances (from October 4, 2016 at 10:00 to December 29, at 13:00 once every hour to collect) that is an actual data set collected from sensors of central air conditioning system parameters. The following Table 1 shows the partial variables description and the remaining variables are specified in the paper.

Table 1. Partial variables description

Variables	Description	Unit
chwsfhdr	The velocity difference of the cooler	gal/min
$ch_ikw(i = 1, 2, 3)$	The power consumption of the cooling device	kw
chwrhdr	The water temperature that flows into the cooler	°C
rh	Relative humidity	%
drybulb	Drybulb temperature	°C
effsys	Effect of the system	kw/RT
systotpower	Total power consumption of the system	kw
loadsys	Load of the system	RT
cwgpmrt	Speed of load flow	gal/min·RT
ct_pc	Speed of cooling tower fan (25 Hz–50 Hz)	Hz

2.1 Data Preprocessing

Before digging into the characteristics and regularities of the data changes, we first utilize the data preprocessing, which consists of data cleaning, data integration and data reduction.

Data Cleaning. According to the principle of heat balance of the system, we filter out the data points when the system is stable (when the heat balance $> 5\%$ means the system is unstable). Then the system power is drawn by time series and it is found that there are several missing values. Here we directly eliminate in that the small amount and the final effect is shown in Fig. 1.

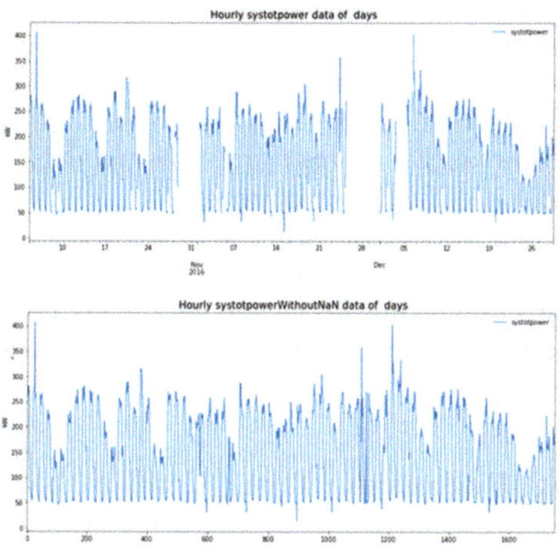

Fig. 1. Time series diagram of hourly system power consumption

Data Integration and Data Reduction. Since the data set of the central air conditioning system has many attributes, in view of the integration of each other, the correlation coefficients between attributes are obtained, and the irrelevant attributes are removed by dimension reduction, thus reducing the amount of data mining.

The Pearson correlation coefficient reflects the correlation between two variables. The closer the absolute value is closer to 1, the closer the correlation between the operating parameters. Similarly, the closer the absolute value is closer to 0, the lower the correlation is. After analysis of the data integration and the Pearson correlation coefficient between the attributes and systotpower, operating parameters that are significantly related (the absolute value of Pearson correlation coefficient greater than 0.5) to systotpower are selected. A total of 24 variables were eliminated and 26 variables remained.

3 Mathematical Background

3.1 Ensemble Learning Methods (Random Forests and GBRT)

The core of ensemble learning is to build multiple different models and aggregate these models to improve the performance of the final model. According to the generation strategy of base learner, ensemble learning can be divided into two categories: (1) parallel methods, take the bagging as the main representative; (2) sequential methods, take the boosting as the main representative.

Random Forests. Random Forests is a typical application algorithm of bagging. Bagging is abbreviated as Bootstrap AGGregatING, that is, a number of different training sets are constructed by means of bootstrap sampling (repeatable sampling). Then the corresponding base learner is trained on each training set. Finally, these base learners are aggregated to obtain the final model. Compared with a single model, the disadvantages of ensemble learning are: (1) computational complexity, in that multiple models need to be trained. (2) the resulting model is difficult to interpret.

However, the random forest can estimate importance of the variable. Therefore, a random forest model directly tells us the importance of each variable, which is beneficial for eliminating irrelevant variables and redundant data. Not only could it improve the performance of the model but also reduce the computational complexity.

Gradient Boosting Regression Tree. Gradient boosting algorithm was fathered by FreidMan in 2000, which is another method of ensemble learning methods. The core is that each tree learns from the residuals of all previous trees. The negative gradient of the loss function $L(y_i, f(x_i)$ is used in the current model as the approximate value of the residuals in the boosting algorithm to fit a regression (classification) tree. The negative gradient of the loss function has the form:

$$r_{mi} = -[\frac{\partial L(y_i, f(x_i))}{\partial f(x_i)}]_{f(x)=f_{m-1}(x)}.$$

For the Gradient Boosting algorithm, multiple learners are sequentially build. Using the correlation of the base learner, the existing shortcomings of weak learners is overcome, which the characterization is the gradient. Gradient Boosting selects the descent direction of gradient during iteration to improve the performance after polymerization. The loss function is used to describe the reliable degree of the model. If the model is not over fitting, the greater the loss function, the higher the error rate of the model. If our model allows the loss function to continue to decline, it indicates that our model is constantly improving, and the best way is to let the loss function fall in its gradient direction.

3.2 K-Nearest Neighbor Regression

K-Nearest Neighbor is a simple algorithm that stores all available cases and predicts digital targets based on similarity (e.g. distance function). KNN, as a non-parametric technique, has been used in statistical estimation and pattern recognition in early 1970. KNN regression is a non-parametric learner, namely, you do not need to give a function such as linear regression and logistic regression, and then to fit parameters. Non-parametric learner is a function without guesswork, and it eventually fail to get an equation, but it could acquire fine prediction. The advantage of non-parametric learner is that it's easy to change the model and the training speed is fast. The disadvantage is that all the points need to be stored so that the space is consumed and the query is slow.

3.3 Support Vector Regression

Support vector regression (SVR) is one of the most popular algorithms for machine learning and data mining. SVR is derived from support vector machines (SVM) and mainly used to fit the sample data and predict the unknown data.

The input data is first mapped to a high-dimensional feature space by kernel functions, and then the linear regression functions in high-dimensional feature space are computed. In contrast to the expression of both Linear Regression and SVR, the latter is a constrained optimization problem that allows the model to find a tube instead of a line. For the $\varepsilon - SVR$ (Epsilon insensitive support vector regression), our goal is to find a function, such as linear function $f(x) = W^{\mathrm{T}}x+b$, to minimize generalization errors:

$$\min \frac{1}{2}\|w\|^2 + c\sum_{i=1}^{l}(\xi_i + \xi_i^*)$$

$$s.t\{\begin{matrix} |y_i - W^{\mathrm{T}} - b| < \varepsilon \\ \xi, \xi_i^* \geq 0 \end{matrix}$$

A loss function is introduced here, which is used to ignore the fluctuation error in the range of the predicted value and the true value, namely, if the deviation of the data from the regression function subjects to the constraint: $|y_i - W^{\mathrm{T}} - b| < \varepsilon$, the data can be ignored. This constraint guarantees the fitting of the best linear regression function so that more points fall within the range of accuracy accepted. But it can be found that there are still some points whose deviations are quite large, so relaxation factors ξ, ξ_i^* are introduced. SVR is suitable for solving small sample, nonlinear and high-dimensional problems, and also has great potential to overcome overfitting and curse of dimensionality.

4 Experiments and Results

In order to achieve the goal of hourly power consumption prediction, four machine learning models were evaluated and compared with a set of measured

Table 2. Feature importance

Variables	Importance	Variables	Importance	Variables	Importance	Variables	Importance	Variables	Importance
chwsfhdr	0.980011	ch1kw	0.000740	ct1kw	0.000264	loadsys	0.000187	ch1stat	0.000097
chwrhdr	0.006654	ch2kw	0.000528	ct2kw	0.000262	cwp_pc	0.000163	chwp1stat	0.000094
cwshdr	0.004497	cwp1kw	0.000372	chwp4kw	0.000257	chwp_pc	0.000126	cwp1stat	0.000085
cwrhdr	0.003368	cwp2kw	0.000324	drybulb	0.000199	ct_pc	0.000122	cwp3stat	0.000025
cwsfhdr	0.001028	cwp3kw	0.000288	chwp1kw	0.000190	dch	0.000104	ct2stat	0.000017

data. In the data set, operating parameters collected show the evolution of the overall system power over time in a central air-conditioning system. The data set contains 1750 instances, nearly 3 months' data, which is collected hourly. Instances from October 4, 2016 at 10:00 to December 1, 2016 at 13:00, 1127 time series data in total, are as the train set, 623 instances remained as the test set (Table 2).

4.1 Experiments Details

We have implemented data fitting and prediction models in Python2.7 using four machine learning algorithms described in Sect. 3. We first trained a random forest containing 1000 decision trees to assess the importance of 25 dimensional features preprocessed in Sect. 2, results are shown in Table 3.

From the result analysis, we can see that the state of air conditioning water system is the independent variable, so we can eliminate these 5 variables: ch1stat, chwp1stat, cwp1stat, cwp1stat, cwp3stat, ct2stat. The remaining 20 variables in Table 4 are used for learning of the latter three models.

Table 3. Variables used in the training model

chwrhdr	cwshdr	loadsys	chwp_pc	ct1kw	ch1stat	chwsfhdr
ch1kw	cwsfhdr	chwp4kw	chwp1stat	ch2kw	cwp3kw	cwp_pc
cwp1stat	cwp2kw	cwp3stat	cwrhdr	cwp1kw	drybulb	dch
ct2kw	ct_pc	chwp1kw	systotpower	ct2stat		

Table 4. MSE and R^2 score of the four models

Model	MSE	R^2	
		Train set	Test set
RF	32.33	0.99934	0.99452
GBRT	4.49	0.99999	0.99925
KNN	253.70	0.98265	0.95697
SVR	7.14	0.99513	0.99879

4.2 Prediction of Hourly Power Consumption

Predicting the power consumption of office buildings will be directly affected by human behavior and other factors. All these factors lead to a nonlinear time series [1]. The estimation and prediction accuracy of the four models used in this paper show good results, Table 4 and Figs. 2, 3, 4, 5, 6, and 7 give detailed results for all models.

In this paper, we use combination of grid search and cross validation to seek the k with the highest R^2 score from $\{1, 2, 3, 4, 5, 6, 7, 8, 9, 10\}$ as the optimal parameter in KNN regression model, and then the k is used to prediction. Experiment result shows that cross validation with K-fold ($K = 10$) returns the best estimator and result is shown in Fig. 4. The best $k = 3$ found by 10-fold cv, and the predictive values are plotted with a scatter chart for the next Fig. 5.

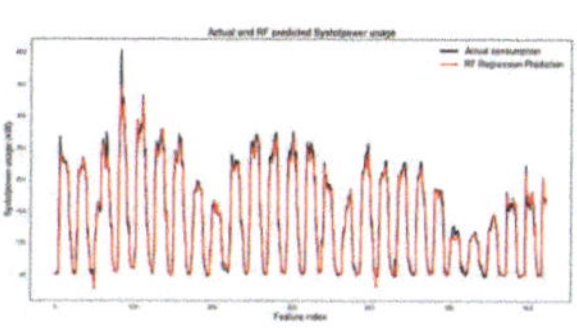

Fig. 2. RF

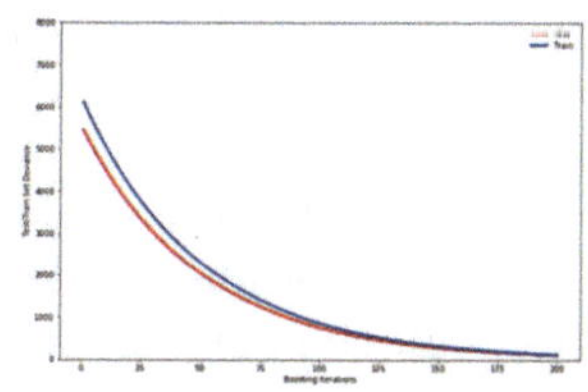

Fig. 3. GBRT

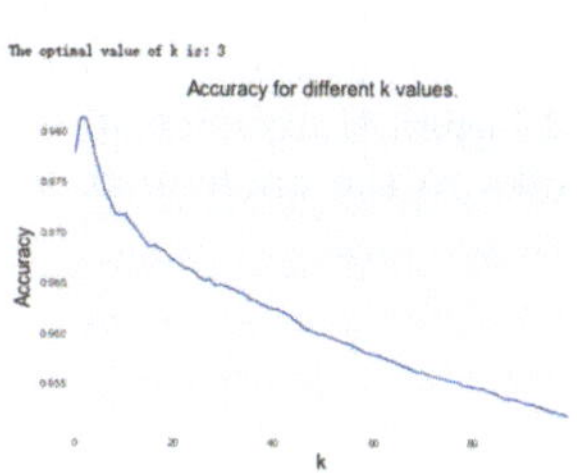

Fig. 4. Seeking the best k

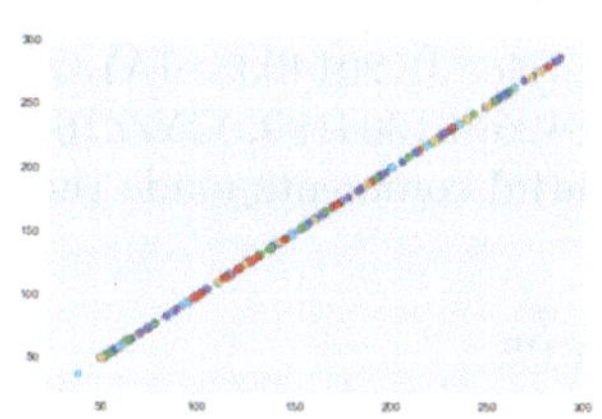

Fig. 5. Scatter plot of predicted systotpower (k = 3)

Finally, we will train a SVR model, where the ε-SVR is used for regression prediction. During the design, we also used grid search and cross validation to fine tune the SVR parameters. And find the optimal parameters: Optimum parameters $C = 1$ and $\gamma = 0.01$ for SVR. Optimum ε and kernel for SVR: {'ε':0.1, 'kernel':'linear'}. Figure 7 shows the prediction of hourly systotpower using SVR.

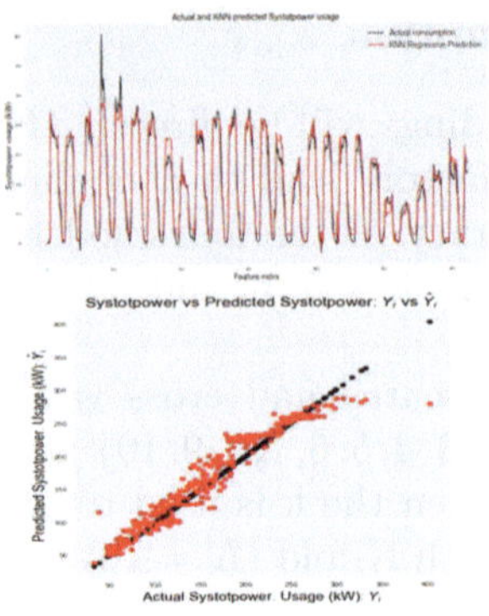

Fig. 6. KNN

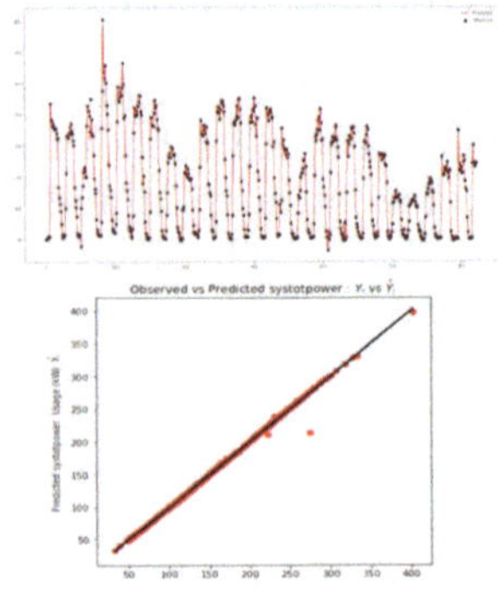

Fig. 7. SVR

5 Conclusion

Our goal is to predict the hourly system power consumption based on historical data and find the optimal model. Experiment result shows that the winner is GBRT. Although the machine learning algorithms are very successful in fitting, when the data is complex and mess, the choose of features and parameters are very important. We have to repeatedly adjust the features and parameters to improve the model.

Acknowledgments. This work is supported by National Natural Science Foundation of China (No. 61304199), Fujian Science and Technology Department (No. 2014H0008), Fujian Transportation Department (No. 2015Y0008), Fujian Education Department (No. JK2014033, JA14209, JA1532), and Fujian University of Technology (No. GYZ13125, 61304199, GY-Z160064). Many thanks to the anonymous reviewers, whose insightful comments made this a better paper.

References

1. Mocanu, E., Nguyen, P.H., Gibescu, M., et al.: Comparison of machine learning methods for estimating energy consumption in buildings. In: International Conference on Probabilistic Methods Applied To Power Systems, pp. 1–6. IEEE (2014)
2. Williams, K.T., Gomez, J.D.: Predicting future monthly residential energy consumption using building characteristics and climate data: a statistical learning approach. Energy Buildings. **128**, 1–11 (2016)
3. Braun, M.R., Altan, H., Beck, S.B.M.: Using regression analysis to predict the future energy consumption of a supermarket in the UK. Appl. Energy **130**(5), 305–313 (2014)
4. Takeda, H., Tamura, Y., Sato, S.: Using the ensemble Kalman filter for electricity load forecasting and analysis. Energy **104**, 184–198 (2016)
5. Li, X., Ding, L., Lv, J., et al.: A novel hybrid approach of KPCA and SVM for building cooling load prediction. In: International Conference on Knowledge Discovery and Data Mining, pp. 522–526. IEEE (2010)

6. Vonk, B.M.J., Nguyen, P.H., Grand, M.O.W., et al.: Improving short-term load forecasting for a local energy storage system. In: Universities Power Engineering Conference, pp. 1–6. IEEE (2012)
7. Hedn, W.: Predicting hourly residential energy consumption using random forest and support vector regression? An analysis of the impact of household clustering on the performance accuracy, Dissertation (2016)
8. Abdelkader, S.S., Grolinger, K., Capretz, M.A.M.: Predicting energy demand peak using M5 model trees. In: IEEE, International Conference on Machine Learning and Applications, pp. 509–514. IEEE (2016)
9. Liu, D., Chen, Q., Mori, K.: Time series forecasting method of building energy consumption using support vector regression. In: IEEE International Conference on Information and Automation, pp. 1628–1632. IEEE (2015)
10. Chen, Y., Tan, H.: Short-term prediction of electric demand in building sector via hybrid support vector regression. Appl. Energy (2017)
11. Shuang, H., Linlin, Z., Xiaohua, W.: The study of mechanical state prediction algorithm of high voltage circuit breaker based on support vector machine. High Voltage Apparatus **7**, 155–159 (2015)

Innovative Intelligent Computation Technology

Compact Evolutionary Algorithm Based Ontology Meta-matching

Xingsi Xue[1,2(✉)] and Shijian Liu[1,2]

[1] College of Information Science and Engineering, Fujian University of Technology, Fuzhou 350118, Fujian, China
[2] Fujian Provincial Key Laboratory of Big Data Mining and Applications, Fujian University of Technology, Fuzhou 350118, Fujian, China
jack8375@gmail.com

Abstract. Since different ontology matchers do not necessarily find the same correct correspondences, usually several competing matchers are applied to the same pair of ontology entities to increase evidence towards a potential match or mismatch. How to select, combine and tune various ontology matchers, so-called ontology meta-matching, is one of the main challenges in ontology matching domain. In recent years, Evolutionary Algorithm (EA) based ontology meta-matching technique has become the state-of-the-art methodology to solve the ontology meta-matching problem, but it suffers from some defects like the slow convergence, premature convergence and the huge memory consumption. To overcome these drawbacks, in this paper, a Compact EA (CEA) based ontology meta-matching technique is proposed, which makes use of a probabilistic representation of the population to perform the optimization process. In particular, we construct an optimal model for the ontology meta-matching problem, propose a problem-specific CEA to optimize the aggregating weights of various matchers, and utilize a Cross Sum Quality Measure (CSQM) to adaptively extract the final alignment. The experimental results show that our approach outperforms other EA based ontology matching techniques and Ontology Alignment Evaluation Initiative (OAEI 2016)'s participants.

Keywords: Compact Evolutionary Algorithm · Ontology meta-matching · Cross sum quality measure

1 Introduction

Ontology describes the domain concepts and their relationships, which is recognized as an effective solution to data heterogeneity on the semantic web. However, different ontology developers might use different ways to describe a concept, which causes the ontology heterogeneity problem and raises the heterogeneous problem to a higher level. There is an urgent need to develop the automatic ontology matching techniques, so called ontology matcher, to identify the correspondences between the same ontology entities. Since different ontology matchers do

J.-S. Pan et al. (eds.), *Advances in Smart Vehicular Technology, Transportation, Communication and Applications*, Smart Innovation, Systems and Technologies 86,
https://doi.org/10.1007/978-3-319-70730-3_26

not necessarily find the same correct correspondences, usually several competing matchers are applied to the same pair of entities in order to increase evidence towards a potential match or mismatch [9]. How to select, combine and tune various ontology matchers, so called ontology meta-matching problem, is one of the main challenges in ontology matching domain [11]. In recent years, Evolutionary Algorithm (EA) based ontology meta-matching technique has become the state-of-the-art methodology to solve the ontology meta-matching problem. GOAL [6] is the first ontology matching system that utilizes EA to determine the weight configuration for a weighted average aggregation of several matchers by considering a reference alignment. More recently, similar idea of combining multiple ontology matchers is also developed in [13,14].

However, slow convergence, premature convergence and the huge memory consumption are three main drawbacks that limit the application of EA based ontology meta-matching techniques. To overcome these problems, in this paper, a Compact EA (CEA) based ontology meta-matching technique is proposed to automatically select, combine and tune basic ontology matchers. In particular, we construct an optimal model for the ontology meta-matching problem, propose a problem-specific CEA to optimize the aggregating weights of various matchers, and utilize a Cross Sum Quality Measure (CSQM) to adaptively extract the final alignment.

The rest of the paper is organized as follows: Sect. 2 describes the ontology meta-matching problem; Sect. 3 presents CEA based ontology meta-matching technique; Sect. 4 gives the detailed of CSQM; Sect. 6 presents the experimental studies and analysis; finally, Sect. 7 draws the conclusions.

2 Ontology Meta-matching

In this work, we define an ontology as a 9-tuple $O = (C, P, I, A, \leq_C, \leq_P, \phi_{CP}, \phi_{CI}, \phi_{PI})$ [16], where:

- C is a nonempty set of classes,
- P is a nonempty set of properties,
- I is a set of instances (it can be empty),
- A is a set of axioms which should not be empty,
- $\leq_C$ is a partial order on C, called class hierarchy or taxonomy,
- $\leq_P$ is a partial order on P, called property hierarchy,
- $\phi_{CP} : P \rightarrow C \times C$ is a function which associates a property $p \in P$ with two linked classes through the property p,
- $\phi_{CI} : C \rightarrow \mathcal{P}(I)$ is a function which associates a concept $c \in C$ with a subset of I representing the instances of the concept c,
- $\phi_{PI} : P \rightarrow \mathcal{P}(I^2)$ is a function which associates a property $p \in P$ with a subset of cartesian product $I \times I$ representing the pair of instances related through the property p.

An ontology alignment A between two ontologies is a correspondence set and each correspondence inside is a 3-tuples $(e, e', =, confidence)$, where e and e' are respectively the entities of two ontology, the relation of the correspondence is the equivalence (=), and *confidence* is the trust extent of the equivalence holds between e and e'.

Assume one entity in source ontology is matched with only one entity in target ontology and vice versa, we define the ontology meta-matching problem as follows:

$$\begin{cases} max \quad f(X) \\ s.t. \quad X = (x_1, x_2, \cdots, x_n)^T \\ \qquad \sum_{i=1}^{n} x_i = 1 \\ \qquad x_i \in [0,1], i = 1, 2, \cdots, n \end{cases} \tag{1}$$

where n is the number of ontology matchers, $x_i, i = 1, 2, \cdots, n$ is the i-th ontology matcher's aggregating weight, and function $f()$ calculates the MatchFmeasure value [17] of the alignment which is determined by X.

3 Compact Evolutionary Algorithm

CEA simulates the behaviour of population-based EA by employing the probabilistic representation of the population, and a run of CEA is able to highly improve the performance of traditional EA in terms of both memory consumption and runtime. In this work, given n ontology matchers, we first generate the similarity matrix M_i for i-th ontology matcher, whose row and column are respectively entities of the source ontology and the target ontology and the element m_{pq} inside are the similarity value of p-th source ontology entity and q-th target ontology entity. Given the aggregating weight set W, an aggregated similarity matrix can be obtained by $M^{agg} = \sum_{i=1}^{n} w_i \times M_i$, after that, we further utilize the CSQM to extract the final alignment.

In the next, we respectively present the probability vector, adaptive alignment extraction and the pseudo-code of CEA.

3.1 Probability Vector

We use the binary encoding where the parameter set is encoded in a solution. Since the weighted average approach is used to combine various similarity measures where the sum of the weights is equal to 1, our encoding mechanism indirectly represents them through the numbers in the interval [0,1]. Specifically, assume that n is the number of weights required, the number set can be represented as $\{c_1, c_2, \ldots, c_n\}$, and the chromosome decoding is carried out by dividing the numbers with their sum, then we get the parameter set $\{\frac{c_1}{\sum_{i=1}^{n} c_i}, \frac{c_2}{\sum_{i=1}^{n} c_i}, \ldots, \frac{c_n}{\sum_{i=1}^{n} c_i}\}$.

On this basis, we use one Probability Vector (PV) [10] to characterize the entire population in population-based EA. The number of elements in PV is equal to the number of individual's gene bits and each element's value is in [0,1]. Since each element's value in PV represents the probability of being one, we can use PV to generate various solutions. In addition, PV can be updated based on the better solution in terms of its fitness value, with the aim to move the PV toward the better solution. Here is an example of generating a new solution through PV $(0.3, 0.7, 0.6, 0.9)^T$. First, generate four random numbers, such as 0.4, 0.5, 0.8 and 0.1. Then compare the numbers with the elements in PV accordingly to determine the new generated individual's gene values, e.g., since $0.4 > 0.3$, the first gene bit's value of the new solution is 0, and in this way, we can generate a new solution 0101. If 0101 is the elite solution, PV will be updated according to its information, e.g. if the gene value of the elite is 0, the corresponding element of PV will decrease by 0.1, otherwise increase by 0.1. So, the updated PV is $(0.2, 0.8, 0.5, 1.0)^T$.

4 Adaptive Alignment Extraction

After determining the aggregating parameters, we utilize CSQM to automatically extract the final ontology alignment. CSQM assigns a higher quality score to a mapping that does not conflict with other mappings, a conflict occurring when there exists another mapping for the same source or target concept. This measure takes into account the similarity score of the mappings, assigning a lower quality to mappings that conflict with mappings of higher similarity [3]. In particular, for each element M_{ij}^{agg}, CSQM first sums all the similarity values in the same i-th row and j-th column of M^{agg}, and then the sum is normalized by the maximum sum of the similarity values per row and column:

$$CSQM(M_{ij}^{agg}) = 1 - \frac{\sum_{v=1}^{|O_2|} M_{iv}^{agg} + \sum_{k=1}^{|O_1|} M_{kj}^{agg}}{MaxRowSum(M^{agg}) + MaxColumnSum(M^{agg})} \quad (2)$$

where $|O_1|$ and $|O_2|$ are respectively the number of entities in source ontology O_1 and O_2. After calculation, for the i-th row of M^{agg}, we pick up the element with highest cross sum quality value, say the one in the j-th column, and the correspondence $(e_i, e_j, =, CSQM(M_{ij}^{agg}))$ is extracted into the final alignment.

5 Pseudo-code of Compact Evolutionary Algorithm

In the following, we give the pseudo-code of CEA:

Step (1) Initialization:

1. generation=0;
2. for($i = 0$; $i < PV.length$; i++)
3. $PV[i] = 0.5$;
4. end for
5. generate an individual ind_{elite} through PV;

Step (2) Update PV:

```
generate an individual ind_temp through PV;
//Execute the local search on ind_temp to improve its quality
ind_new = elite.copy();
generate i = round(rand(0, solution.length));
ind_new[i] = ind_temp[i];
 while (rand(0, 1) < 0.8)
   ind_new[i] = ind_temp[i];
   i = i + 1;
     if (i == len)
       i = 0;
     end if
 end while
//Update PV based on the elite
 [winner, loser] = compete(ind_elite, ind_new);
 if (winner == ind_new)
   ind_elite = ind_new;
end if
for(i = 0; i < solution.length; i++)
  if (winner[i]==1)
    PV[i] = PV[i] + 0.1;
    if (PV[i] > 1)
      PV[i] = 1;
    end if
  else
    PV[i] = PV[i] - 0.1;
    if (PV[i] < 0)
      PV[i] = 0;
    end if
  end if
end for
//Initial the PV to prevent the premature convergence
if (each bit of PV is either 1 or 0)
  for(i = 0; i < PV.length; i++)
    PV[i] = 0.5;
  end for
end if
```

Step (3) Stopping Criteria:

```
if (maxGeneration = 3000 is reached)
  stop and output ind_elite;
else
  generation=generation+1;
  go to Step 2;
end if
```

6 Experimental Studies and Analysis

In the experiment, four kinds of basic similarity matchers are utilized to evaluate the confidence of the entity pair: (1) syntactic-based matcher [12], which computes a string distance or edit distance between the ontology entities; (2) linguistic-based matcher [8], which calculates the similarity between ontology entities by considering linguistic relations such as synonymy, hypernym, and so on; (3) structure-based matcher [7], which utilizes the specialization relation, e.g. the subsumption relation, in ontologies to calculate the similarity between ontology entities; (4) instance-based matcher [15], which exploits the similarity between classified instances to discover the same entities in ontologies.

The testing cases we utilize is the well-known bibliographic track provided by Ontology Alignment Evaluation Initiative (OAEI 2016)[1], which consists of a set of ontologies that are built around a seed ontology and many variations of it. Table 1 shows a brief description on the bibliographic track, Tables 2 and 3 show the statistical comparison among three EA based ontology matching approaches and our approach, and Table 4 compares our approach with OAEI 2016's participants on the bibliographic track. Our results are the average values over 30 independent runs.

6.1 Experimental Results

We first carry out the statistical comparison on the alignment's quality in terms of f-measure among Genetic Algorithm (GA) based [6], Memetic Algorithm (MA) based [1], Particle Swarm Optimization (PSO) based [2] ontology matching approaches and our approach. The statistical comparison is formally carried out by means of a multiple comparison procedure which consists of two steps: in the first one, a statistical technique, i.e. the Friedmans test [4], is used to determine whether the results provided by various approaches present any difference; in the second one, which method is outperformed is determined by carrying out a post-hoc test, i.e. Holms test [5], when in the first step an difference is found.

Table 1. Brief description on bibliographic track. 1XX, 2XX and 3XX stands for the test case whose ID beginning with the prefix digit 1, 2 and 3, respectively.

ID	Description
1XX	The ontologies under alignment are the same
2XX	The ontologies under alignment have different lexical, linguistic or structure features
3XX	The ontologies under alignment are real world cases

[1] http://oaei.ontologymatching.org/2016/benchmarks/index.html.

Table 2. Friedman's test on bibliographic track. Each value represents the f-measure, and the number in round parentheses is the corresponding computed rank.

ID	GA	MA	PSO	CEA
101	1.00 (2.5)	1.00 (2.5)	1.00 (2.5)	1.00 (2.5)
103	0.99 (4)	1.00 (2)	1.00 (2)	1.00 (2)
104	0.99 (4)	1.00 (2)	1.00 (2)	1.00 (2)
201	0.50 (3)	0.62 (2)	0.42 (4)	0.66 (1)
203	0.97 (3)	0.96 (4)	1.00 (1.5)	1.00 (1.5)
204	0.94 (4)	0.97 (3)	0.98 (1.5)	0.98 (1.5)
205	0.83 (1.5)	0.79 (3)	0.73 (4)	0.83 (1.5)
206	0.84 (4)	0.88 (2)	0.85 (3)	0.92 (1)
221	0.99 (3.5)	0.99 (3.5)	1.00 (1.5)	1.00 (1.5)
222	0.99 (2.5)	0.99 (2.5)	0.99 (2.5)	0.99 (2.5)
223	0.99 (2.5)	0.99 (2.5)	0.99 (2.5)	0.99 (2.5)
224	1.00 (2.5)	1.00 (2.5)	1.00 (2.5)	1.00 (2.5)
225	0.99 (4)	1.00 (2)	1.00 (2)	1.00 (2)
228	0.99 (2.5)	0.99 (2.5)	0.99 (2.5)	0.99 (2.5)
230	0.93 (3.5)	0.93 (3.5)	0.98 (2)	1.00 (1)
231	0.99 (3)	0.99 (3)	0.99 (3)	1.00 (1)
301	0.70 (2.5)	0.70 (2.5)	0.64 (4)	0.73 (1)
302	0.61 (3)	0.63 (2)	0.04 (4)	0.66 (1)
304	0.83 (3)	0.87 (2)	0.72 (4)	0.89 (1)
Average	0.89 (3.07)	0.91 (2.57)	0.85 (2.68)	0.92 (1.65)

Table 3. Holm's test on bibliographic track.

i	Approach	z value	Unadjusted p-value	$\frac{\alpha}{k-i}, \alpha = 0.05$
3	MA	2.57	0.0285	0.05
2	PSO	2.68	0.0142	0.02
1	GA	3.07	0.0006	0.01

As can be seen from Table 2, in the Friedmans test, the computed $\mathcal{X}_r^2$ value is 10.65, which is greater than $\mathcal{X}_{0.05}^2 = 7.81$. In the Holm's test, as shown in Table 3, our approach statistically outperforms other EA based ontology matching approaches on the alignment's quality at 0.05 significance level.

Finally, as can be seen from Table 4, our approach's result also outperform all the participants of OAEI 2016 in terms of f-measure and recall, which shows the effectiveness of our approach.

Table 4. Comparison of our approach with OAEI 2016's participants on bibliographic track.

Systems	*recall*	*precision*	*f − measure*
edna	0.51	0.35	0.41
AML	0.24	1.00	0.38
CroMatcher	0.83	0.96	0.89
Lily	0.83	0.97	0.89
LogMap	0.39	0.93	0.55
LgoMapLt	0.50	0.43	0.46
PhenoMF	0.01	0.03	0.01
PhenoMM	0.01	0.03	0.01
PhenoMP	0.01	0.02	0.01
XMap	0.40	0.95	0.56
LogMapBio	0.24	0.48	0.32
CEA	0.85	0.97	0.90

7 Conclusion

Research on ontology meta-matching is mainly concerned about how to select, combine and tune different ontology matchers to obtain the high quality ontology alignment, which is one of the main challenges in ontology matching domain is. To face this challenge, in this paper, we present a CEA based ontology meta-matching technique, which can automatically determine the optimal weights of aggregating the ontology matchers and determine the high quality ontology alignment. The experimental results show that our approach outperforms other EA based ontology matching techniques and OAEI 2016's participants.

Acknowledgment. This work is supported by the National Natural Science Foundation of China (Nos. 61503082 and 61403121), Natural Science Foundation of Fujian Province (No. 2016J05145), Scientific Research Startup Foundation of Fujian University of Technology (No. GY-Z15007), Fujian Province outstanding Young Scientific Researcher Training Project (No. GY-Z160149) and Fundamental Research Funds for the Central Universities (No. 2015B20214).

References

1. Acampora, G., Loia, V., Vitiello, A.: Enhancing ontology alignment through a memetic aggregation of similarity measures. Inf. Sci. **250**, 1–20 (2013)
2. Bock, J., Hettenhausen, J.: Discrete particle swarm optimisation for ontology alignment. Inf. Sci. **192**, 152–173 (2012)
3. Cruz, I.F., Loprete, F., Palmonari, M., Stroe, C., Taheri, A.: Pay-as-you-go multi-user feedback model for ontology matching. In: International Conference on Knowledge Engineering and Knowledge Management, pp. 80–96. Springer (2014)

4. Friedman, M.: The use of ranks to avoid the assumption of normality implicit in the analysis of variance. J. Am. Stat. Assoc. **32**(200), 675–701 (1937)
5. Holm, S.: A simple sequentially rejective multiple test procedure. Scand. J. Stat. 65–70 (1979)
6. Martinez-Gil, J., Alba, E., Montes, J.F.A.: Optimizing ontology alignments by using genetic algorithms. In: Proceedings of the First International Conference on Nature Inspired Reasoning for the Semantic Web, vol. 419. pp. 1–15. CEUR-WS.org (2008)
7. Melnik, S., Garcia-Molina, H., Rahm, E.: Similarity flooding: a versatile graph matching algorithm and its application to schema matching. In: 2002 Proceedings of the 18th International Conference on Data Engineering, pp. 117–128. IEEE (2002)
8. Miller, G.A.: WordNet: a lexical database for english. Commun. ACM **38**(11), 39–41 (1995)
9. Nguyen, T.T.A., Conrad, S.: Ontology matching using multiple similarity measures. In: 2015 7th International Joint Conference on Knowledge Discovery, Knowledge Engineering and Knowledge Management (IC3K), vol. 1, pp. 603–611. IEEE (2015)
10. Parsopoulos, K.E.: Cooperative micro-differential evolution for high-dimensional problems. In: Proceedings of the Conference on Genetic and Evolutionary Computation, Montreal, Canada, pp. 531–538, July 2009
11. Shvaiko, P., Euzenat, J.: Ontology matching: state of the art and future challenges. IEEE Trans. Knowl. Data Eng. **25**(1), 158–176 (2013)
12. Stoilos, G., Stamou, G., Kollias, S.: A string metric for ontology alignment. In: International Semantic Web Conference, pp. 624–637. Springer (2005)
13. Xue, X., Ren, A.: An evolutionary algorithm based ontology alignment extracting technology. J. Netw. Intell. **2**(2), 205–212 (2017)
14. Xue, X., Wang, Y.: Optimizing ontology alignments through a memetic algorithm using both matchfmeasure and unanimous improvement ratio. Artif. Intell. **223**, 65–81 (2015)
15. Xue, X., Wang, Y.: Ontology alignment based on instance using NSGA-II. J. Inf. Sci. **41**(1), 58–70 (2015)
16. Xue, X., Wang, Y.: Using memetic algorithm for instance coreference resolution. IEEE Trans. Knowl. Data Eng. **28**(2), 580–591 (2016)
17. Xue, X., Wang, Y., Hao, W., Hou, J.: Optimizing ontology alignments through NSGA-II without using reference alignment. Comput. Inform. **33**(4), 857–876 (2015)

Transfer Knowledge Based Evolution of an External Population for Differential Evolution

Zhenyu Meng[1], Jeng-Shyang Pan[1,2](✉), and Xiaoqing Li[3]

[1] Innovative Information Industry Research Center, Department of Computer Science and Technology, Harbin Institute of Technology Shenzhen Graduate School, Shenzhen, China
mzy1314@gmail.com, jengshyangpan@gmail.com
[2] College of Information Science and Engineering, Fujian University of Technology, Fuzhou, China
[3] Shenzhen Institute of Advanced Technology, Chinese Academy of Sciences, Shenzhen, China

Abstract. Population size plays an important role in the optimization performance of Differential Evolution. Researches in earlier literature usually employed constant population size, and these recommended settings of different population sizes usually varied from one DE variant to another. As we know, smaller population size settings perform better on some objective functions while bigger settings perform better on the other within the same number of function evaluations. Therefore, adaptive schemes for population size became much more popular recently and performed very well on a large number of benchmark functions. These schemes dynamically changed the population size either in increasing or decreasing approaches during the evolution. Moreover, most of these adaptive schemes mainly focused on decreasing population size. Nevertheless, this paper reveals an approach to diversify the individuals (increase the population size) by employing an external population without increasing number of function calls. This approach employs transfer knowledge learned from the target population in the evolution of an external population for Differential Evolution. CEC2013 test suite for real-parameter single objective optimization is employed in the verification of our approach and experiment results show that the proposed approach is very useful in maintaining a better diversity of individuals without increasing function calls.

Keywords: Benchmark functions · Differential evolution · External population · Global optimization · Transfer knowledge

1 Introduction

Differential evolution is a simple but powerful stochastic algorithm for optimization problems, and it is originated from Genetic Annealing algorithm [1] which

J.-S. Pan et al. (eds.), *Advances in Smart Vehicular Technology, Transportation, Communication and Applications*, Smart Innovation, Systems and Technologies 86,
https://doi.org/10.1007/978-3-319-70730-3_27

can be considered as a hybrid algorithm of Genetic Algorithm and Simulated Annealing [2]. Operations, such as mutation, crossover and selection employed in Genetic Algorithm, is inherited into DE as well. As its simplicity and excellent performance, many researchers have learnt about it, advanced many variants to enhance the performance of it and employed its variants in many practical applications since the inception in 1995 [3–16].

There are three control parameters, including population size ps, scale factor F and crossover rate Cr in the canonical DE algorithm. The inventors of DE algorithm also proposed 5 mutation strategies and two crossover schemes, including binomial crossover and exponential crossover, of it [1], and this constituted the 10 basic trial vector generating strategies of DE algorithm. There were other mutation strategies proposed by researchers in literature, e.g. DE/rand/dir [17], DE/target-to-pbest/1/bin [18]. Though there were benchmark functions on which a certain mutation strategy performed better than other mutation strategies, the DE/target-to-pbest/1/bin had an overall better performance on a large number of benchmark functions. Moreover, there were DE variants employing hybrid mutation strategy or a strategy pool containing several selected mutation strategies for optimization problems.

When given a fixed mutation strategy, control parameters played very important roles in the optimization performance. Early DE researches often employed fixed control parameters, and a common setting was $ps = 100$, $F = 0.5$, and $Cr = 0.1$ for uni-modal objective functions, $Cr = 0.9$ for multi-modal objective functions. Later, researchers found that adaptive control parameters F and Cr performed even better for a large number of objective functions [18–21]. Most adaptive schemes for parameters control in literature mainly focused on scale factor F and crossover rate Cr, and the adaptive scheme for population size ps in literature only took up a very small proportion of all these adaptive schemes. Moreover, almost all adaptive schemes for population size ps decreased its value with increased generation number [22,23]. As we know, smaller population size setting may performs better on some objective functions while bigger population size setting performs better on the other objective functions. Therefore, for a fixed initial population size setting, maybe we need to increase the population size during the evolution.

In this paper, we reveal an approach to diversify the individuals (increase the population size) by employing an external population without increasing number of function calls. The proposed approach employs transfer knowledge learnt from the target population in the evolution of external population. The external population maintains an individual diversity pool from which individuals can be selected and added into the target population. Then we can increase the target population size when needed. The rest of the paper is organized as follows. Section 2 gives a brief review of Differential Evolution. Section 3 makes a detailed description on the transfer knowledge and evolution guidance on the external population. Section 4 shows the experiment analysis under CEC2013 test suit for real-parameter single objective optimization. Finally, the conclusion is given in Sect. 5.

2 Differential Evolution

Differential Evolution is a stochastic optimization algorithm, and it employs ps individuals to tackle complex optimization problems. These individuals, i.e., the vector of ith individual in the Gth generation, $X_{i,G} = (x_{i1,G}, x_{i2,G}, ..., x_{iD,G}), i = 1, ..., ps$, are also called candidate solutions, which have the same dimension number as objective function. A general definition of minimizing an objective function can be represented like this:

$$\Omega^* \equiv \arg\min_{X \in \Omega} f(X) = \{X^* \in \Omega : f(X^*) \le f(X), \forall X \in \Omega\} \tag{1}$$

where X is the D-dimensional vector of parameters, also the coordinate of an individual in the solution space $\Omega \subseteq \mathbb{R}^D$. All these individuals are initialized by uniformly randomizing ps vectors in the solution space. For a hyper-cube solution space, the jth parameter of the ith vector in 0th generation can be initialized according to Eq. 2. $x_{max,j}$ and $x_{min,j}$ are the upper and lower bound on the jth parameter respectively.

$$x_{ij,0} = x_{min,j} + rand_{ij}(0,1) \cdot (x_{max,j} - x_{min,j}) \tag{2}$$

After initialization, the donor vector can be generated and crossover operation is consequently done to generate trial vector for the target vector. In the stage of trial vector generation, mutation strategy takes great effect on it. A general convention DE/x/y/op is used to differ different mutation strategies. x denotes the base vector, y denotes the difference pairs and op denote crossover scheme, including binomial crossover and exponential crossover. Here we use DE/target-to-pbest/1/bin [18] mutation strategy in our paper because this mutation strategy performs very well on a large number of objective function. Equation 3 gives the detailed equation of DE/target-to-pbest/1/bin. $X^p_{gbest,G}$ denotes one individual in a handful of best individuals of the population in the Gth generation, and p is the proportion of the handful top superior individuals. $X_{i,G}$, $X_{r_1,G}$, $\widetilde{X}_{r_2,G}$ have the same meanings as the ones in algorithm JADE [18].

$$\begin{aligned} &DE/target-to-pbest/1: \\ &V_{i,G} = X_{i,G} + F_i * (X^p_{best,G} - X_{i,G}) + F_i \cdot (X_{r_1,G} - \widetilde{X}_{r_2,G}) \end{aligned} \tag{3}$$

Selection is finally made between the target vector and trial vector. Equation 4 illustrates the selection operation of a minimizing optimization.

$$X_{i,G+1} = \begin{cases} X_{i,G}, & if(f(X_{i,G}) \le f(U_{i,G}) \\ U_{i,G}, & otherwise \end{cases} \tag{4}$$

3 Transfer Knowledge Based Evolution for External Population

In this part, we mainly discuss the new proposed transfer knowledge based evolution for external population. The external population can be considered as a satellite population in which the individuals evolve according to transfer knowledge learnt from the target population. We use η to define the ratio of the external population size ps_e to the target population size ps, and Eq. 5 shows the definition.

$$\eta = \frac{ps_e}{ps} \tag{5}$$

For $X^e_{i,G}$, the ith individual of the Gth generation in the external archive, it is evolved according to Eq. 6. $X^p_{best,G}$, $X_{r_1,G}$ and $X_{r_2,G}$ are the knowledge learned from the target population of the Gth generation. We can see from Eq. 6 that individuals in the external population evolve with disabled crossover and selection operation in comparison with the individuals in the target population.

$$X^e_{i,G} = X^e_{i,G} + F * (X^p_{best,G} - X^e_{i,G}) + F \cdot (X_{r_1,G} - \widetilde{X}_{r_2,G}) \tag{6}$$

For the target population, we employ a new recently proposed PALM-DE [5] variant for evolution. PALM-DE employed an enhance DE/target-to-pbest/1/bin mutation strategy and a totally new adaptive scheme for control parameters. PALM-DE tackles the mis-interaction of control parameter F and Cr by separating them into different groups and it achieves a great success. F in PALM-DE algorithm obeys Cauchy distribution, $F \sim C(\mu_F, \sigma_F)$ and Cr obeys normal distribution, $Cr \sim randn(\mu_{Cr}, \sigma_{Cr})$. F and Cr is renewed according to Eqs. 7–11. $P(Cr_j)$ denotes the selection probability of jth group, $P(Cr_j) = P(j)$. ns_j and nf_j denote the number of success individuals and failure individuals in j^{th} group. $nSucc$ and $nFail$ denote the number of success individuals and failure individuals of the whole population respectively. r_j is the proportion of individuals of the jth group. ϵ is a small positive value to avoid possible null value of probability. k is the number of groups. F_{ji} denotes the scale factor value of ith individual in the jth group. $w_{F_{ji}}$ denotes the fitness weight of F_{ji}. S_{F_j} denotes the set of success scale factor values in the jth group.

$$\begin{cases} ps \cdot P(Cr_j) = ns_j + nf_j, \\ nSucc = \sum_{j=1}^{k}(ns_j), \\ nFail = \sum_{j=1}^{k}(nf_j), \\ ps = nSucc + nFail. \end{cases} \tag{7}$$

$$\begin{cases} r_j = \begin{cases} \dfrac{ns_j^2}{nSucc \cdot (ns_j + nf_j)}, & if\ ns_j > 0, \\ \epsilon, & otherwise. \end{cases} \\ P(j) = \dfrac{r_j}{\sum_{j=1}^{k}(r_j)}. \end{cases} \tag{8}$$

$$\mu_{Cr} = \frac{\sum_{j=1}^{k}(P(j) \cdot Cr_j^2)}{\sum_{j=1}^{k}(P(j) \cdot Cr_j)} \tag{9}$$

$$\begin{cases} w_{F_{ji}} = \dfrac{\Delta f_i}{\sum_{F_{ji} \in S_{F_j}} \Delta f_i} \\ F_j = \dfrac{\sum_{F_{ji} \in S_{F_j}} w_{F_{ji}} \cdot F_{ji}^2}{\sum_{F_{ji} \in S_{F_j}} w_{F_{ji}} \cdot F_{ji}} \end{cases} \quad (10)$$

$$\Delta f_i = f(U_{i,G}) - f(X_{i,G}) \quad (11)$$

4 Experiment Analysis

In this part, we mainly analyze the performance of transfer knowledge based evolution of external population for Differential Evolution. Benchmark functions in Congress of Evolutionary Computation (CEC) Competition 2013 for real-parameter single objective optimization are employed as the test-beds for algorithm verification. These 28 benchmarks can be separated into three groups, the unimodal function group, the basic multimodal function group, and even some composition functions, and they can reflect some real world complex optimization problems.

In order to validate and evaluate the proposed algorithm, comparisons are made between the target population and external population. We conduct 51 runs because of the stochastic nature of DE variants. The best, mean and standard deviation are collected and calculated from the total 51 runs. For the parameters in the paper, most parameter settings are the same as PALM-DE algorithm, $F_j = 0.5, F_{ji} \sim C(F_j, 0.2), \mu_{Cr} = 0.5, Cr \sim N(\mu_{Cr}, 0.1), ps = 100, k = 8, p = 0.1, a = 1.6, T_0 = 70$, the extra parameter $\eta = 10\%$. The comparison results are shown in Table 1. Symbols $>$, $=$, $<$ in the right column of the table mean "Better Performance", "Similar Performance" and "Worse Performance" respectively. Wilcoxon's signed rank test with significant level $\alpha = 0.05$ is employed for experiment result evaluation. All these experiments are conducted on a PC with Intel(R) Core(TM) $i5-3470$ CPU @ 3.2 Hz on RedHat Linux Enterprise Edition 5.5 Operating System and Matlab software with 2011b Unix version. The values of fitness error that smaller than "*eps*" ($eps = 2.2204e-016$) are considered as zeros herein.

In order to maintain a better diversity, individuals in the external population should obey two basic rules. First, vectors of individuals in the external population should not be the equivalents calculated by arithmetical operations of the target individuals. Second, the performance of the external individuals should be similar to ones of the target individuals. As the external individuals are generated with disabled crossover and mutation operation, the target individuals are generated with enabled crossover and selection operation, so they are arithmetically different. Moreover, from Table 1, we can see that the optimization performance of external population is similar to the performance of the target population on the 28 benchmark functions test suite except benchmark function f_8, f_{14}, f_{17} and f_{22} with the significant $\alpha = 0.05$. When we enlarge the value of η from 10% to 100% (the external population size $ps_e = \eta \times ps$),

Table 1. Best (minimum), mean and standard deviation of 51 runs fitness error comparisons between target population and external population. The population size $ps = 100$ and the total number of function evaluations is $nfe = 10000 \times D$, generation number is $gen = \frac{nfe}{ps}$.

10D	Target population			External population			Performance
No.	Best	Mean	Std	Best	Mean	Std	>≈<
f_1	0	0	0	0	1.3370E-014	5.4032E-014	≈
f_2	0	0	0	0	8.9166E-015	4.4574E-014	≈
f_3	0	1.2802E-001	8.8384E-001	0	1.2802E-001	8.8384E-001	≈
f_4	0	0	0	0	1.3375E-014	5.4032E-014	≈
f_5	0	0	0	0	1.5604E-014	3.9511E-014	≈
f_6	0	5.3872E+000	4.9312E+000	0	5.3872E+000	4.9312E+000	≈
f_7	2.2737E-013	1.2624E-004	4.7767E-004	5.6843E-013	1.2784E-004	4.7742E-004	≈
f_8	2.0000E+001	2.0089E+001	1.6331E-001	2.0000E+001	2.0102E+001	2.1410E-001	>
f_9	0	6.1008E-001	6.5943E-001	0	6.1008E-001	6.5943E-001	≈
f_{10}	0	7.4830E-003	1.0598E-002	0	7.4830E-003	1.0598E-002	≈
f_{11}	0	0	0	0	1.1146E-015	7.9597E-015	≈
f_{12}	9.9496E-001	4.0189E+000	1.7113E+000	9.9496E-001	4.0189E+000	1.7113E+000	≈
f_{13}	9.9496E-001	4.3515E+000	2.6362E+000	9.9496E-001	4.3515E+000	2.6362E+000	≈
f_{14}	6.2455E-002	6.1690E-001	9.3505E-001	6.2455E-002	7.6272E+001	1.9219E+002	>
f_{15}	1.4703E+002	5.9191E+002	2.2392E+002	1.4703E+002	5.9191E+002	2.2392E+002	≈
f_{16}	2.8422E-014	1.2172E-001	1.3590E-001	5.6843E-014	1.2172E-001	1.3590E-001	≈
f_{17}	1.0122E+001	1.0126E+001	8.0088E-003	1.0122E+001	1.2202E+001	4.4275E+000	>
f_{18}	1.1348E+001	1.6355E+001	2.9638E+000	1.1348E+001	1.6355E+001	2.9638E+000	≈
f_{19}	2.3375E-001	5.1972E-001	1.5432E-001	2.3375E-001	5.1972E-001	1.5432E-001	≈
f_{20}	1.0957E+000	1.9611E+000	5.2254E-001	1.0957E+000	1.9611E+000	5.2254E-001	≈
f_{21}	4.0019E+002	4.0019E+002	0.0000E+000	4.0019E+002	4.0019E+002	4.5475E-014	≈
f_{22}	3.4106E-012	2.4150E+001	2.7503E+001	5.9117E-012	3.0814E+001	4.9210E+001	>
f_{23}	2.7766E+001	3.9394E+002	2.2584E+002	2.7766E+001	3.9394E+002	2.2584E+002	≈
f_{24}	2.0000E+002	2.0000E+002	1.0853E-005	2.0000E+002	2.0000E+002	1.0853E-005	≈
f_{25}	1.0245E+002	1.9627E+002	1.8629E+001	1.0245E+002	1.9627E+002	1.8629E+001	≈
f_{26}	1.0100E+002	1.2132E+002	3.6870E+001	1.0099E+002	1.2132E+002	3.6870E+001	≈
f_{27}	3.0000E+002	3.0000E+002	0.0000E+000	3.0000E+002	3.0000E+002	1.1139E-013	≈
f_{28}	3.0000E+002	3.0000E+002	0.0000E+000	3.0000E+002	3.0000E+002	3.2155E-014	≈

the optimization performance of external population is similar to the target population on the 28 benchmark functions except f_{14} and f_{22}. When we enlarge η to 1000%, performance exception occurs only on function f_{14} (Performances of $\eta = 100\%$ and $\eta = 1000\%$ are not given here as page limit). In a word, we can say that the individuals in the external population are effective, and our approach is very useful in maintaining a better diversity of individuals without increasing function calls.

5 Conclusion

In this paper, we proposed transfer knowledge based evolution of an external population for Differential Evolution. Our approach can maintain a better diversity of individuals without increasing number of function calls. This can tackle the obstacle on the road when we need to add more individuals into the target population but don't know how to generate these extra individuals of an adaptive scheme for population size. Experiment results revealed that the external population are arithmetically different from the target population, and the optimization performances are roughly equal. When the individuals in the target population converged into the global optima, so did the individuals in the external population. This means that our approach is useful in maintaining a better diversity of individuals during the evolution process.

Acknowledgement. This work was supported by Shenzhen Innovation and Entrepreneurship Project (GRCK20160826105935160) and National Natural Science Foundation of China (61371178).

References

1. Price, K., Storn, R.M., Lampinen, J.A.: Differential Evolution: A Practical Approach to Global Optimization. Springer, Heidelberg (2006)
2. Kirkpatrick, S., Vecchi, M.P.: Optimization by simmulated annealing. Science **220**(4598), 671–680 (1983)
3. Storn, R., Price, K.: Differential evolutional simple and efficient adaptive scheme for global optimization over continuous spaces. International Computer Science Institute, Berkeley, CA (1995)
4. Storn, R., Price, K.: Differential evolution - a simple and efficient heuristic for global optimization over continuous spaces. J. Global Optim. **11**(4), 341–359 (1997)
5. Meng, Z., Pan, J.-S.: Parameters with adaptive learning mechanism (PALM) for the enhancement of differential evolution, submitted to Knowledge-Based Systems
6. Meng, Z., Pan, J.-S.: A simple and accurate global optimizer for continuous spaces optimization. In: Genetic and Evolutionary Computing, pp. 121–129. Springer (2015)
7. Meng, Z., Pan, J.-S., Alelaiwi, A.: A new meta-heuristic ebb-tide-fish-inspired algorithm for traffic navigation. Telecommun. Syst. **62**, 1–13 (2015)
8. Meng, Z., Pan, J.-S.: Monkey king evolution: a new memetic evolutionary algorithm and its application in vehicle fuel consumption optimization. Knowl. Based Syst. **97**, 144–157 (2016)
9. Meng, Z., Pan, J.-S., Xu, H.: QUasi-Affine TRansformation Evolutionary (QUATRE) algorithm: a cooperative swarm based algorithm for global optimization. Knowl. Based Syst. **109**, 104–121 (2016)
10. Pan, J.S., Meng, Z., Xu, H., et al.: QUasi-Affine TRansformation Evolution (QUATRE) algorithm: a new simple and accurate structure for global optimization. In: International Conference on Industrial, Engineering and Other Applications of Applied Intelligent Systems, pp. 657–667. Springer (2016)
11. Meng, Z., Pan, J.S., Quasi-affine transformation evolutionary (QUATRE) algorithm: a parameter-reduced differential evolution algorithm for optimization problems. In: 2016 IEEE Congress on Evolutionary Computation (CEC), pp. 4082–4089 (2016)
12. Meng, Z., Pan, J.S.: A competitive QUasi-Affine TRansformation Evolutionary (C-QUATRE) algorithm for global optimization. In: 2016 IEEE International Conference on Systems, Man, and Cybernetics (SMC), pp. 1644–1649. IEEE (2016)

13. Meng, Z., Pan, J.S.: QUasi-Affine TRansformation Evolutionary (QUATRE) algorithm: the framework analysis for global optimization and application in hand gesture segmentation. In: 2016 IEEE 13th International Conference on Signal Processing (ICSP), pp. 1832–1837 (2016)
14. Pan, J.S., Meng, Z., Chu, S.C., et al.: Monkey king evolution: an enhanced ebb-tide-fish algorithm for global optimization and its application in vehicle navigation under wireless sensor network environment. Telecommun. Syst. **65**(3), 351–364 (2017)
15. Pan, J.S., Meng, Z., Xu, H., et al.: A matrix-based implementation of DE algorithm: the compensation and deficiency. In: International Conference on Industrial, Engineering and Other Applications of Applied Intelligent Systems, pp. 72–81. Springer, Cham (2017)
16. Cai, D.: A new evolutionary algorithm based on uniform and contraction for many-objective optimization. J. Netw. Intell. **2**(1), 171–185 (2017)
17. Feoktistov, V., Janaqi, S.: Generalization of the strategies in differential evolution. In: 18th International Parallel and Distributed Processing Symposium, Proceedings, pp. 165–170. IEEE (2014)
18. Zhang, J., Sanderson, A.C.: JADE: adaptive differential evolution with optional external archive. IEEE Trans. Evol. Comput. **13**(5), 945–958 (2009)
19. Brest, J., Greiner, S., Bošković, B., Mernik, M., Zumer, V.: Self-adapting control parameters in differential evolution: a comparative study on numerical benchmark problems. IEEE Trans. Evol. Comput. **10**(6), 646–657 (2006)
20. Qin, A.K., Huang, V.L., Suganthan, P.N.: Differential evolution algorithm with strategy adaptation for global numerical optimization. IEEE Trans. Evol. Comput. **13**(2), 398–417 (2009)
21. Tanabe, R., Fukunaga, A., Success-history based parameter adaptation for differential evolution. In: 2013 IEEE Congress on Evolutionary Computation, pp. 71–78, June 2013
22. Brest, J., Maučec, M.S.: Population size reduction for the differential evolution algorithm. Appl. Intell. **29**(3), 228–247 (2008)
23. Tanabe, R., Fukunaga, A.S.: Improving the search performance of shade using linear population size reduction. In: 2014 IEEE Congress on Evolutionary Computation (CEC), pp. 1658–1665, July 2014

Research on Ships Collision Avoidance Based on Chaotic Particle Swarm Optimization

Lisang Liu[1,2(✉)], Dongwei He[1,2], Ying Ma[1], Tianjian Li[1], and Jianxing Li[1]

[1] Fujian University of Technology, Fuzhou, China
liulisang@fjut.edu.cn
[2] Key Laboratory of Digital Equipment, Fuzhou, Fujian, China

Abstract. Reasonable ways and suitable decision of ships collision avoidance are the guarantee of the safety navigation. Based on analysis of the existing collision avoidance strategy, this paper studies on the effectiveness of chaotic particle swarm optimization (CPSO) algorithm for ships collision avoidance, by introducing the improved Tent mapping to solve the premature convergence problem of PSO. The mathematical model and objective function of collision avoidance are put forward and the simulation is carried out to compare the CPSO with PSO in various encounter situations of ships. The results shows that the CPSO is feasible and practical in conducting an optimal steering collision-avoidance scheme.

Keywords: Collision avoidance · Chaotic · PSO algorithm · Tent mapping

1 Introduction

The ships collision avoidance problem is one of the key research field of marine science and technology. In recent years, the collision accidents occur continuously and bring great losses to people, property and environment, as the increasing number and increasing speed of ships, poor environmental conditions and maneuvering experience. In 1995, the international maritime organization (IMO) made it clear that ship collision accidents accounted for the largest proportion of all kinds of maritime accidents, and 80% of the incidents on the water were related to human factors. The United States studied 13191 ship crashes in American waters during 1970 to 1974, and concluded that the main cause of the crashes was human error, accounting for 66.5%. Among the water traffic accidents in China's Yangtze River Basin between 2005 to 2009, collision accidents accounted for 63.8% of the total number of accidents. All the evidence show that human error, wrongly judgments or violation of International Regulations for Preventing Collision at Sea (COLREGS), is a great factor to the ships collision accidents.

The COLREGS has standardized the collision avoidance behavior of the ship drivers though, it cannot offer specific collision avoidance scheme when ships encountering. Therefore, how to develop a reasonable, efficient and perfect collision avoidance scheme to realize intelligent ship is still an essential research field.

J.-S. Pan et al. (eds.), *Advances in Smart Vehicular Technology, Transportation, Communication and Applications*, Smart Innovation, Systems and Technologies 86,
https://doi.org/10.1007/978-3-319-70730-3_28

2 Research Status

2.1 A Subsection Sample

In 1960s, some experts and scholars made quantitative research on some important qualitative concepts by using collision avoidance geometry method [1], which is not only the foundation of mathematical model establishment, but also the foundation of collision avoidance automatic research and system establishment. There were lots of achievements [2] and some important concepts were proposed, such as distance to closest point of approach (DCPA), time to closest point of approach (TCPA), moving field and ship domain. In 1980s, the artificial intelligence was used to study automatic ships collision avoidance, for example, the expert system was developed by Tokyo University of Merchant Marine and Liverpool Polytechnic University [3]. Since 21 Century, University of Southampton and other units have been engaged in the field of ship intelligent collision avoidance decision-making system [4–7], many Chinese scholars also have done a lot of research in this area [8–10]. The simulated annealing algorithm and neural network are applied to the determination of ship collision risk. However, these research has not been applied in practical yet.

The mathematical model of collision avoidance decision and the automation of collision avoidance have been studied extensively and deeply, however, some problems still exist. For example, the evaluation of the ship collision risk is limited to DCPA and TCPA, which is not comprehensive enough, no consensus has been reached on the methods of automatic collision avoidance and the establishment of mathematical models, the collision avoidance decision system is either not perfect to applied to real ships or not strictly tested in real ships, in addition, how to effectively fuse ARPA (Automatic Radar Plotting Aid), AIS (Automatic Identification System) and other navigation information to help ship collision avoidance is seldom studied [11].

According to the COLREGS, sailing habits and automatic collision avoidance method, a multi-objective optimization function with constraints is established in this paper, taking influential factors such as DCPA, TCPA, speed ratio and relative distance of the ships, relative bearing into account to judge ship collision risk. Particle swarm optimization (PSO) algorithm is used for solving the optimal problem, and the chaotic Tent mapping method is introduced to solve problems of premature convergence of PSO so as to get the best steering angle in anti-collision scheme. This algorithm is called chaotic particle swarm optimization (CPSO). Simulation results show that the CPSO algorithm is effective and feasible for dealing with the optimal steering angle of the ship in case of multi ships encountering.

3 Ship Collision Avoidance

3.1 Collision Cases

The collision zone is divided into 6 areas according to the position of the own ship and the approaching ship, shown as Fig. 1.

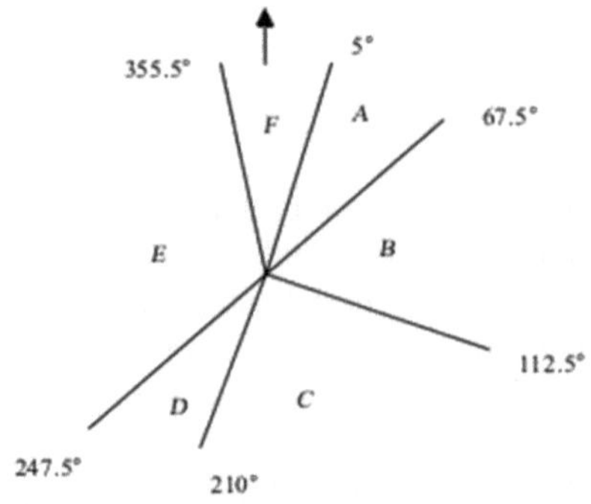

Fig. 1. The division of collision zone.

The own ship takes different measures due to the different collision areas. In particular,

(1) In area A (5°–67.5°), own ship should be rightwards to avoid collide the approaching ship.
(2) In area B (67.5°–112.5°), assume the own ship speed is V_o and the approaching ship speed is *Va*, speed ratio *K* is the criterion of the measure, the own ship should be leftwards to collide the approaching ship if $K = V_a/V_o \leq 0.95$, otherwise, take rightwards.
(3) In area C (112°–210°), the own ship keeps sailing with current direction and speed while the approaching ship should take measure to avoid collision. If no measure is taken, then DCPA need to be calculated, own ship turns leftwards if DCPA larger than 0, otherwise rightwards.
(4) In area D (210°–247.5°), the own ship keeps sailing with current direction and speed while the approaching ship should take measure to avoid collision. If no measure is taken, then DCPA need to be calculated, own ship turns rightwards if DCPA larger than 0, otherwise leftwards.
(5) In area E (247.5°–355.5°), no collision risk.
(6) In area F (355.5°–0°), both of own ship and approaching ship turn rightwards.

The process of ship collision avoidance is shown in Fig. 2.

The collected necessary information includes the AIS information, the approaching ships motion state. Then, determine whether there is any collision risk or not, whether to take action or not and how to act.

There are several measures to avoid collision, such as right turn, left turn, half deceleration, left-handed circle, reversing, parking, left-handed deceleration, etc. Generally speaking, three kinds of steering, deceleration, steering & transmission are three main actions. According to the statistical results of the collision avoidance at sea, over 90% of the collision avoidance operations are steering. Moreover, in accordance with the rule eighth of COLREGs, steering is the most common operations. Thus, this paper mainly considers steering collision avoidance.

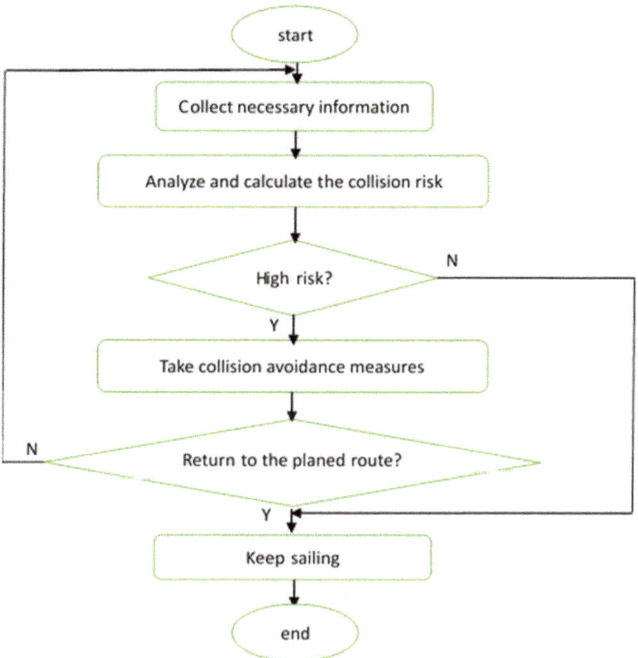

Fig. 2. Collision avoidance process.

3.2 Mathematical Model of Collision Avoidance

Lots factors should be taken into account in making collision avoidance decision. Assuming the own ship speed is V_O and heading is C_O. The approaching ship speed is V_a and heading is C_a. Radar observations are shown in Fig. 3.

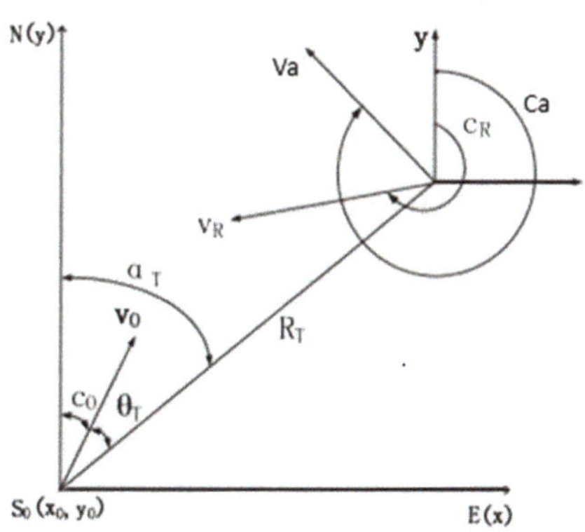

Fig. 3. Relative moving parameters of two ships.

Where, R_T is relative distance, α_T is relative bearing, V_R is relative velocity. θ_T is relative heading, DCPA is short for distance to closest point of approach while TCPA is short for time to closest point of approach. $\left(x_O, y_O\right)$ and $\left(v_{xo}, v_{yo}\right)$ are the position and velocity of the own ship respectively, $\left(x_a, y_a\right)$ and $\left(v_{xa}, v_{ya}\right)$ are the position and velocity of the approaching ship respectively.

Assuming the own ship should take measures, the approaching ship keeps the original heading and speed. We can obtain:

(1) Calculating the relative bearing α_T and relative distance RT.

Assuming the position of the own ship $\left(x_O, y_O\right)$ is (0, 0), the initial relative distance is R_0, the initial relative bearing is α_0 then the position of the approaching ship $\left(x_a, y_a\right)$ will be:

$$\begin{cases} x_a = R_0 cos\left(c_o + \alpha_0\right) \\ y_a = R_0 sin\left(c_o + \alpha_0\right) \end{cases} \tag{1}$$

After time t, the own ship speed is $V_o(t)$ and heading is $C_o(t)$, the position is

$$\begin{cases} x_o(t) = \int_0^t v_o cosc_o(t)dt \\ y_o(t) = \int_0^t v_o sinc_o(t)dt \end{cases} \tag{2}$$

The relative displacement between the two ships is

$$\begin{cases} \Delta x = x_a + \int_0^t \left(v_a cosc_a(t) - v_o cosc_o(t)\right)dt \\ \Delta y = y_a + \int_0^t \left(v_a sinc_a(t) - v_o sinc_o(t)\right)dt \end{cases} \tag{3}$$

As a result, the relative bearing α_T and relative distance R_T will be

$$\alpha_T(t) = arctan\frac{\Delta y}{\Delta x} \tag{4}$$

$$R_T(t) = \sqrt{\Delta x^2 + \Delta y^2} \tag{5}$$

(2) Calculating the relative velocity V_R and relative heading θ_T.

$$\theta_T(t) = arctan\frac{v_a cosc_a - v_o cosc_o}{v_a sinc_a - v_o sinc_o} \tag{6}$$

$$V_R(t) = \sqrt{v_a^2 + v_o^2 - 2v_a v_o cos\left(c_o - c_a\right)} \tag{7}$$

(3) Calculating the DCPA and TCPA

$$DCPA = R_T(t) \cdot \left|sin\left(\theta_T - \alpha - 180\right)\right| \tag{8}$$

$$TCPA = R_T(t)cos\left(\theta_T - \alpha - 180\right)/V_R \tag{9}$$

3.3 Objective Function of Collision Avoidance

The ships encountering situations are complicated. As a result, an appropriate steering is very important. This paper considers the steering range of collision avoidance as a multi-objective optimization problem, and finds the optimal range of collision avoidance that satisfied the target function and the constraint condition in the feasible solution space by particle swarm optimization (PSO) algorithm. After steering, the own ship can: ① reduce the risk of collision with other ships; ② minimize the steering angle; ③ return to the original heading and speed in the shortest time [12]. The objective function can be written as

$$\begin{cases} minf_i^1(x),\ i = 1, 2, \cdots, n \\ minf^2(x) \\ minf^3(x) \end{cases} \tag{10}$$

With the following constraint condition:

$$\begin{cases} 10 \le x \le 180 \\ f^3(x) - 40 \le 0 \\ f_i\left(\mu_A\right) - f_i^1(x) > 0, i = 1, 2, \cdots, n \end{cases} \tag{11}$$

Where, x is the steering angle and decision variable, $f_i^1(x)$ is the objective function of collision risk of the own ship and the i_{th} approaching ship. $f^2(x) = x$ is also an objective function. $f^3(x)$ is the time-object function with maximum limit 40 min that return to the original route. $f_i\left(\mu_A\right)$ is the collision risk of the own ship and the i_{th} approaching ship, which follows that

$$f_i\left(\mu_A\right) = f_i(DCPA, TCPA) = DCPA \oplus TCPA \tag{12}$$

Where, a $\oplus$ b = $|min(min(a/10, b), 1)|$.

4 Chaotic Particle Swarm Optimization

4.1 A Subsection Sample

Particle Swarm Optimization (PSO) is one of the intelligent optimization method proposed by Professor Kennedy and Eberhart originally in 1995. PSO is a metaheuristic as it makes few or no assumptions about the problem being optimized and can search very large spaces of candidate solutions. However, metaheuristics such as PSO do not guarantee an optimal solution is ever found. More specifically, PSO depends on the selection of parameters, is easy to fall into the local optimum, and the convergence speed is slow when approaching the best solution area.

The system is initialized with a population of random solutions and searches for optima by updating generations. However, unlike GA, PSO has no evolution operators such as crossover and mutation. In PSO, the potential solutions, called particles, fly through the problem space by following the current optimum particles. Each particle keeps track of its coordinates in the problem space which are associated with the best solution (that is, fitness) it has achieved so far. (The fitness value is also stored.) This value is called *pbest*. Another best value that is tracked by the particle swarm optimizer is the best value, obtained so far by any particle in the neighbors of the particle. This location is called *lbest*. When a particle takes all the population as its topological neighbors, the best value is a global best and is called *gbest*. Usually *gbest* is more popular than *lbest*.

Assume a group consists of m particles is flying at a certain speed in D-dimension search space. The position in the space is the solution to the problem. The i_{th} particle $(i = 1, 2, \cdots, m)$ has three vectors: current position $x_i = (x_{i1}, x_{i2}, \cdots, x_{iD})$, historical best position $p_i = (p_{i1}, p_{i2}, \cdots, p_{iD})$ and velocity $v_i = (v_{i1}, v_{i2}, \cdots, v_{iD})$. $p_g = (p_{g1}, p_{g2}, \cdots, p_{gD})$ is the historical best position of the group. Each particle is evaluated by the defined fitness function, which is related to the problem to be solved, and iterates through the following formula:

$$v_{id}^{k+1} = v_{id}^{k} + c_1 rand(0, 1)(p_{id}^{k} - x_{id}^{k}) + c_2 rand(0, 1)\left(p_{gd}^{k} - x_{id}^{k}\right) \tag{13}$$

$$x_{id}^{k+1} = x_{id}^{k} + v_{id}^{k+1}, d = 1, 2, \cdots, D \tag{14}$$

Where, c_1 and c_2 are learning factor, usually $c_1 = c_2 = 2$ or $c_1 = 2.8, c_2 = 1.3$. v_{id}^{k} and x_{id}^{k} are velocity and position for each individual i on dimension d in the k_{th} iteration.p_{id}^{k} and p_{gd}^{k} are personal best position and global best position for each individual i on dimension d in the k_{th} iteration. The only additional rule is the v_{id} is limited by some value v_{max}, that is, $v_{id} \in [-v_{max}, v_{max}]$. In general, if the searching space is limited within $[-x_{max}, x_{max}]$, we can set $v_{max} = l \cdot x_{max}$, $0 \le l \le 1$. The group number m is generally 20–40. For complex problems, 100 or more is recommended.

However, since the PSO is prone to premature and it is hard to tell global convergence or premature convergence when the group fitness is zero, it is expected that the PSO algorithm can jump out of the local optimum when the premature convergence occurs and continue searching until the global optimal solution is found. Therefore, an improved Tent mapping is introduced as chaos search into PSO based on group fitness variance in order to avoid premature convergence. The improved Tent mapping is more ergodic than the basic Tent mapping by introducing the random process [13]. Mathematically,

$$x_{k+1} = \begin{cases} 2(x_k + 0.1 \cdot rand(0, 1)) & 0 \le x_k \le 0.5 \\ 2(1 - (x_k + 0.1 \cdot rand(0, 1))) & 0.5 \le x_k \le 1 \end{cases} \tag{15}$$

The group fitness variance is defined as

$$\sigma^2 = \frac{1}{N}\sum_{i=1}^{N}\left(\frac{f_i - \bar{f}}{f}\right)^2 \tag{16}$$

f is a normalization factor to limit σ^2.

$$f = \begin{cases} max(|f_i - \bar{f}|) & max(|f_i - \bar{f}|) \geq 1 \\ 1 & others \end{cases} \tag{17}$$

The group fitness variance σ^2 reflects the degree of convergence of all particle groups in the particle swarm. The smaller the σ^2 is, the more convergent the particle swarm is.

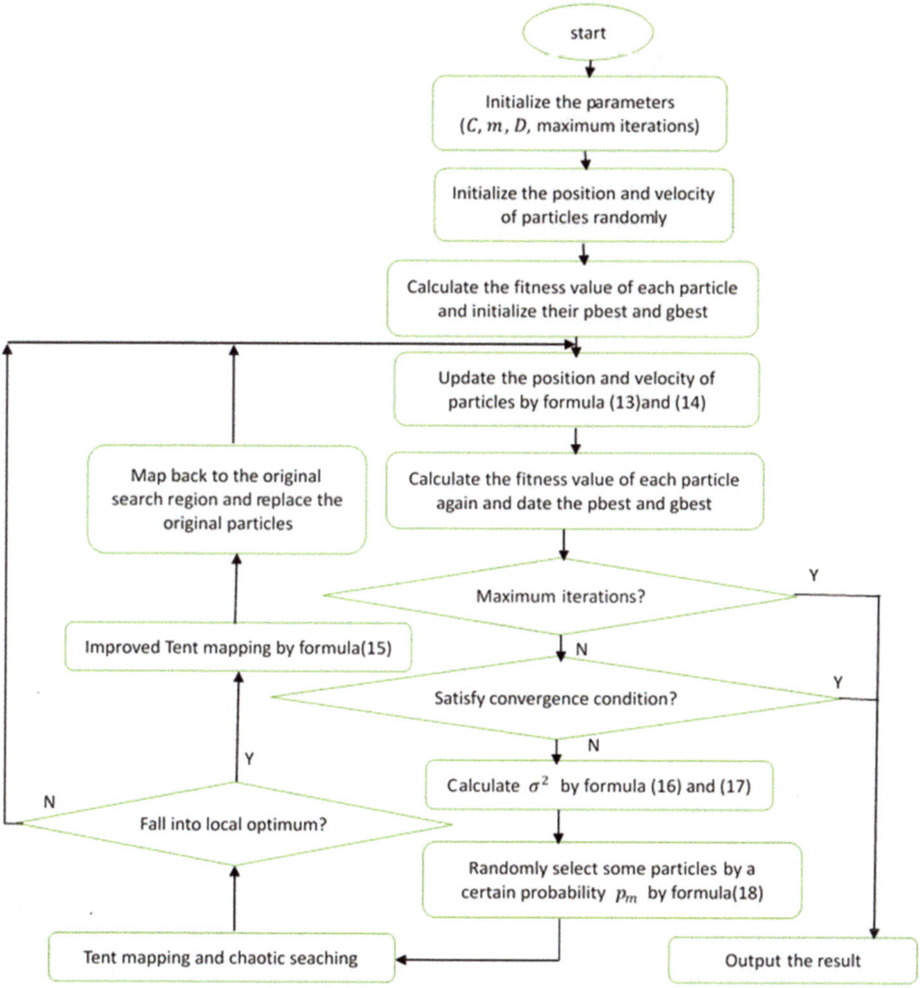

Fig. 4. Flow of the CPSO.

On the contrary, the particle swarm is in a random search phase. If the algorithm falls into premature convergence or reaches global convergence, $\sigma^2 = 0$.

This above-mentioned algorithm, called as chaotic particle swarm optimization (CPSO), chaotic searches for some particles in a particle swarm randomly selected by a certain probability p_m

$$p_m = \begin{cases} k & \sigma^2 \leq C \\ 0 & others \end{cases} \tag{18}$$

Where, k is an arbitrarily value within [0.1, 0.3], C is related to the actual problem and is a small one. The flow of the CPSO algorithm is as follows (Fig. 4):

5 Simulation and Discussion

5.1 A Subsection Sample

Assuming that the crew are experienced, the visibility is good in water area and ship maneuvering is good, too. The own ship's length is 100 m, heading $C_o = 0$, speed $V_o = 20$ Kn (1 Kn = 1.852 Km/h). The particles number $m = 20$. $l = 0.5$, $k = 0.2$. The approaching ships' information and the encountering simulation results are listed in Table 1.

Table 1. Ships encountering simulation.

	One approaching ship-crossing	One approaching ship-overtaking	Two approaching ships-crossing	
α_T	3	10	30	110
$V_a(Kn)$	20	10	15	20
C_a	185	3	235	95
RT (n mile)	7	2	6.5	4
Steering (PSO)	Rightwards 63.54°	Rightwards 72.31°	Leftwards 45.49°	
Steering (CPSO)	Rightwards 62.11°	Rightwards 73.48°	Leftwards 44.37°	
Time to recover (min)	About 10.27 min	About 11.43 min	About 14.08 min	
DCPA before steering (n mile)	0.43	0.54	0.59	0.32
DCPA after steering (n mile)	1.98	2.35	1.33	2.10

In the simulation, different types of approaching ships are designed and different encounter situations are considered, such as crossing and overtaking one approaching ship, crossing two approaching ships and so on. It can be seen from the above table that in the cross case, the steering angle of CPSO is smaller than that of PSO, and the angle is larger than PSO when overtaking. This is reasonable because the resumption of the shortest time and speed decline during the steering process and other factors are considered, thus the results more practicable to reality.

6 Conclusion

A reasonable and accurate collision avoidance decision is an important guarantee for the safety sailing of ships. This paper illustrates the chaotic particle swarm optimization based on the improved Tent mapping in detail. Various influential factors in the actual navigation are taken into account and the mathematical model was established. The simulation of the own ship encountering with one or more approaching ships in various situation by PSO and CPSO are carried out, and the comparison shows that anti-collision schemes using CPSO algorithm is more feasible and more reasonable than PSO.

Acknowledgement. In this paper, the research was supported by Initial Scientific Research Fund of Fujian University of Technology and Pre-research Project of Fujian University of Technology.

References

1. Inaish, M.: Basic research on a collision avoidance system using neural networks. Jpn Inst. Navig. **112**, 22–27 (1992)
2. Coenen, F.P., Smeaton, G.P., Bole, A.G.: Knowledge based collision avoidance. J. Navig. **42**(3), 75–78 (2004)
3. Inaish, M.: Building method and application of neural network system for navigation. In: Lateral Symposium of Sino-Japanese Navigation Institute, vol. 112, pp. 22–27 (1992)
4. Wilson, P.A., Harris, C.J.: A line of sight counteraction navigation algorithm for ship encounter collision avoidance. J. Navig. **56**, 111–121 (2003)
5. Imazuh: Problem in the Future on ship's collision avoidance after the fitting of an advanced navigation system. In: Navigation, IAIN (2000)
6. Hwang, C.: The integrated design of fuzzy collision-avoidance and H-autopilots on ships. J. Navig. **55**, 117–136 (2002)
7. Lamp, W.G.P., Hunt, J.M.: Multiple encounter avoidance manoeuvres. J. Navig. **53**(1), 181–186 (2000)
8. Yian, L., Jing, L., Jie, W., et al.: Application of simulated annealing algorithm in turning angle to avoid collision between ships. Shipbuild. China **48**(4), 53–57 (2007)
9. Lina, L., Zhennan, X., Yansong, G., et al.: Method of building and optimizing intelligent decision of single-ship anti-collision. Navig. China **2**, 49–52 (2002)
10. Baozhang, Y., Aiguo, S., Feng, C., et al.: Automatic collision avoidance system. J. Naval Univ. Eng. **13**(4), 57–66 (2001)
11. Pengpeng, P.: Research on ships collision based on particle swarm optimization. Master Thesis of Jiangsu University of Science and Technology (2010)
12. Deyan, W., Yian, L.: Studies on turning angle to avoid collision between ships with PSO arithmetic. Comput. Eng. Des. **30**(14), 3380–3382 (2009)
13. Hao, Z., Tienan, Z., Jihong, S., et al.: Research on decision-makings of structure optimization based on improved tent PSO. Control Decis. **23**(8), 857–862 (2008)

Design of Gear Reducer Based on FOA Optimization Algorithm

Xiaojia Lin[1], Fuquan Zhang[2,3(✉)], and Lin Xu[4]

[1] Department of Computer Science, Fujian Business University, Fuzhou 350012, China
14759227@qq.com

[2] Fujian Provincial Key Laboratory of Information Processing and Intelligent Control, Minjiang Univeristy, Fuzhou 350121, China
8528750@qq.com

[3] School of Software, Beijing Institute of Technology, Beijing 100081, China

[4] Innovative Information Industry Research Institute, Fujian Normal University, Fuzhou 350300, China
71471418@qq.com

Abstract. In order to optimize the design of gear reducer, gear reducer optimal design to improve reliability and security, slow convergence and local optimum for FOA algorithm is proposed based on the improved type FOA gear reducer optimization design model. To avoid falling into local optimum Drosophila optimization algorithms, improved FOA algorithm by introducing a correction factor. Superior to the gear reducer seven variables optimization design model for the study, to ensure the safety and reliability of the premise, the FOA has the advantage of improved convergence speed and avoid local optimization problems, in order to verify the proposed method and reliability.

Keywords: Flying fruit optimization algorithm · Correction factor · Gear reducer · Optimization design · Fitness function

1 Introduction

Gear reducer as an important independent drive unit, to increase its life and minimize its weight and size hold significant meaning [1]. Standard gear reducer design has a flaw caused by non-optimized parameter input. This research proposed an optimized design scheme, which could realize rapid design procedure and improve gear reducer performance without compromise the security and reliability.

Classic Fruit Fly Optimization Algorithm [2] (FOA) has advantages of fast convergence process. However by introducing a modify factor β to FOA could optimize gear reducer design scheme.

J.-S. Pan et al. (eds.), *Advances in Smart Vehicular Technology, Transportation, Communication and Applications*, Smart Innovation, Systems and Technologies 86,
https://doi.org/10.1007/978-3-319-70730-3_29

2 Fruit Fly Optimization Algorithm (FOA)

Fruit Fly Optimization Algorithm was inspired by fruit fly ranging behavior, it is a group intelligent algorithm. Its advantages include less control parameters and fast convergence, yet it easily become "pre-mature "and fall into local optimum. FOA algorithm is shown as follow:

Step 1: Set fruit fly group size "pop-size" and "iteration", initialize group position "X_begin" and "Y_begin";

Step 2: Formulas (1) and (2) realize single fruit fly optimized direction and distance;

$$x_i = X\_begin + Value \times rand() \tag{1}$$

$$y_i = Y\_begin + Value \times rand() \tag{2}$$

"Value" is search distance, "x_i" and "y_i" both indicate the individual fly positions of next moment.

Step 3: Formulas (3) and (4) are to calculate distance of individual fly position to its original start point (d_i), and concentration of individual fly scent(s_i);

$$d_i = \sqrt{x_i^2 + y_i^2} \tag{3}$$

$$s_i = \frac{1}{d_i} \tag{4}$$

Step 4: Formula (5) is a determining function for scent concentration. It shows individual fly scent concentration value on the current position;

$$Smell_i = Function(s_i) \tag{5}$$

Step 5: Find best scent concentration value and position in the fly group. The "$Smell_b$" is the best scent concentration value, "x_b" and "y_b" are the best positions;

Step 6: Track single fly's best position and scent concentration value while fly group search for best position. "$Smellbest = Smell_b$" indicates best scent concentration value; "$X\_begin = x_b$" and "$Y\_begin = y_b$" are fly original position;

Step 7: Iterative optimization, repeat iterative processes of step 2 to step 5 and if current scent concentration is better than previous value, move to step 6.

3 Improved Fly Optimization Algorithm (IFOA)

Based on the distance of FOA (d_i) and scent concentration value (s_i), "d_i" are randomly distributed in large area. The concentration value "s_i" becomes less after Formula (5). At this point "s_i" is used as determine function and it causes problems of partially optimization and pre-mature [3, 4].

In order to avoid such problems, a modified factor "β" is introduced to improve FOA. The improved formula (Improved Fruit Fly Optimization Algorithm,IFOA) as shown below:

$$d_i = \sqrt{x_i^2 + y_i^2} \tag{6}$$

$$s_{Mi} = \frac{1}{d_i} + \beta \tag{7}$$

In Formula 7, "s_{Mi}" is scent criterion function of IFOA.

$$\beta = \begin{cases} g \times d_i \\ K \times X\_axis \ or \ K \times Y\_axis \end{cases} \tag{8}$$

In Formula 8, "g" is evenly distributed and "K" is a constant value.

4 Optimized Gear Reducer Design Model

4.1 Design Variable

Research aim: designing two-level gear transmission reducer; the internal mechanics are shown below.

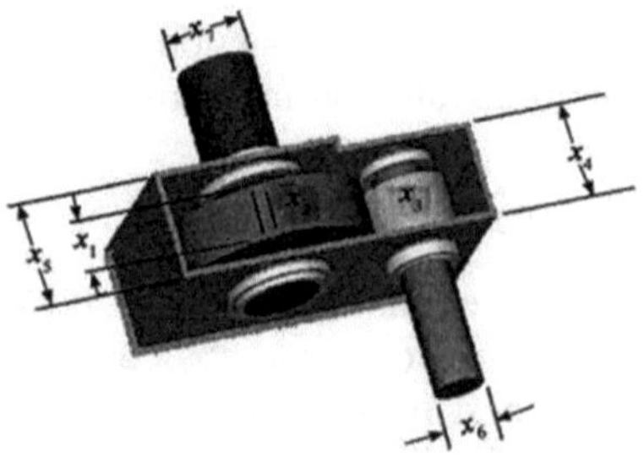

Fig. 1. Reducer internal structure

This optimized design has 7 variables [8, 9]: width x_1, module x_2, pinion teeth number x_3, bearing distance of axle 1 x_4, bearing distance of axle 2 x_5, axle 1 diameter x_6 and axle 2 diameter x_7. The values of variable take from the range below:

$$\begin{aligned} &2.6 \leq x_1 \leq 3.6,\ 0.7 \leq x_2 \leq 0.8,\ 17 \leq x_3 \leq 28,\ 7.3 \leq x_4 \leq 8.3, \\ &7.3 \leq x_5 \leq 8.3,\ 2.9 \leq x_6 \leq 3.9,\ 5.0 \leq x_7 \leq 5.5 \end{aligned} \tag{9}$$

Target function aims to minimize the reducer size [10, 11]:

$$\begin{aligned} \min f1(x) = {} & 0.7854x_1x_2^2(3.3333x_3^2 + 14.933x_3 - 43.0934) \\ & - 1.508x_1(x_6^2 + x_7^2) + 7.477(x_6^3 + x_7^3) \\ & + 0.7854(x_4x_6^2 + x_5x_7^2) \end{aligned} \tag{10}$$

Assume:

$$A_1 = [(745x_2^{-1}x_3^{-1}x_4)^2 + 16.9 \times 10^6]^{0.5} \tag{11}$$

$$B_1 = 0.1x_6^3 \tag{12}$$

$$A_2 = [(745x_2^{-1}x_3^{-1}x_5)^2 + 157.5 \times 10^6]^{0.5} \tag{13}$$

$$B_2 = 0.1x_7^3 \tag{14}$$

In above function, A1 indicates center torsion; B1 indicates coefficients of working condition; A2 indicates coefficients of loads; B2 indicates allowed pressure of contact fatigue.

4.2 Constraints

There are 11 Constraints in this mathematical model [12–14]: size, bending stress, contact stress and axles transverse deviation etc.:

$$g_1(x) = 27x_1^{-1}x_2^{-2}x_3^{-1} - 1 \leq 0 \tag{15}$$

$$g_2(x) = 397.5x_1^{-1}x_2^{-2}x_3^{-2} - 1 \leq 0 \tag{16}$$

$$g_3(x) = 1.93x_2^{-1}x_3^{-1}x_4^3x_6^{-4} - 1 \leq 0 \tag{17}$$

$$g_4(x) = 1.93x_2^{-1}x_3^{-1}x_5^3x_7^{-4} - 1 \leq 0 \tag{18}$$

$$g_5(x) = x_2x_3 - 40 \leq 0 \tag{19}$$

$$g_6(x) = 5 - x_1x_2^{-1} \leq 0 \tag{20}$$

$$g_7(x) = x_1x_2^{-1} - 12 \leq 0 \tag{21}$$

$$g_8(x) = 1.9 - x_4 + 1.5x_6 \leq 0 \tag{22}$$

$$g_9(x) = 1.9 - x_5 + 1.5x_7 \leq 0 \tag{23}$$

$$g_{10}(x) = A_1 B_1^{-1} - 1800 \leq 0 \quad (24)$$

$$g_{11}(x) = A_2 B_2^{-1} - 1800 \leq 0 \quad (25)$$

In above function, g_1 indicates bending stress constraint; g_2 indicates contact stress constraint; $g_3 - g_9$ indicates axles transverse deviation constraint and empirical constraint; $g_{10} - g_{11}$ indicates size constraint.

4.3 Mathematical Model

As mentioned above, the goal of two-level gear transmission reducer design is to minimize its size. The new mathematical model can be illustrated as follow:

$$\min f(x) x = [x_1 x_2 x_3 x_4 x_5 x_6 x_7]^T \in R^7 \quad (26)$$

$$S.t.\ g_j(x) \leq 0\ (j = 1, 2, \cdots, 11) \quad (27)$$

4.4 Optimized Reducer Design with IFOA

The optimized reducer design process is nonlinear with multiple constraints. The new design process base on IFOA is shown as below:

Step 1: A random population with a maximum iteration is initialized based on the upper and lower limit of design variables;
Step2: Calculate single fly's best search direction and distance;
Step3: Base on the Formula (10), calculate the value of fitness function for single fly;
Step4: Base on the Formula (5), calculate fly's current position and scent concentration value;
Step5: Find the best scent concentration value and positions. The best scent concentration value is indicated as $Smell_b$, the best positions are indicated as x_b and y_b;
Step6: Keep records of fly's best position and scent concentration value. The best scent concentration value is indicated as $Smellbest = Smell_b$, fly's initial positions are indicated as $X\_begin = x_b$ and $Y\_begin = y_b$;
Step7: Use iterative optimization method repeat step 2 to step 5, and decide if the concentration value is better than previous value, suppose the concentration value is better, move to step 6.

5 Simulations

In order to prove practical value of the proposed model, the IFOA parameters are set: the iteration is set to 100 and the group size is 30. The simulation results are shown as graphs below:

Figures 2 and 3 indicate optimized reducer design converge process and fruit fly's optimized path respectively. The former one expressed as relationship between numbers of iterations and changes in target function, in simple term it is a relationship between numbers of iteration and changes in reducer volume. The latter expresses fruit fly's search process for optimized path in two-dimensional space.

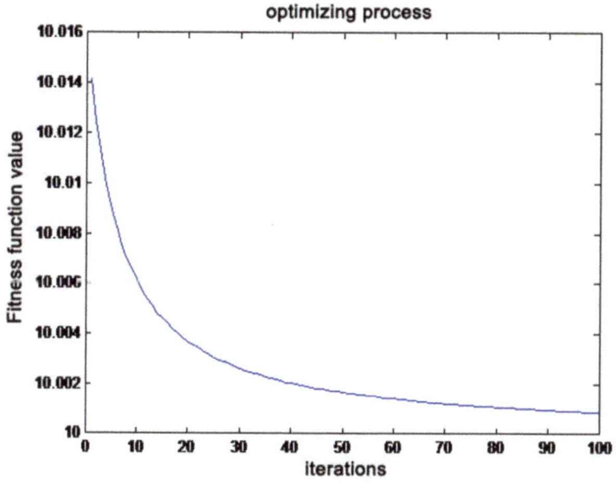

Fig. 2. Fitness function converge process

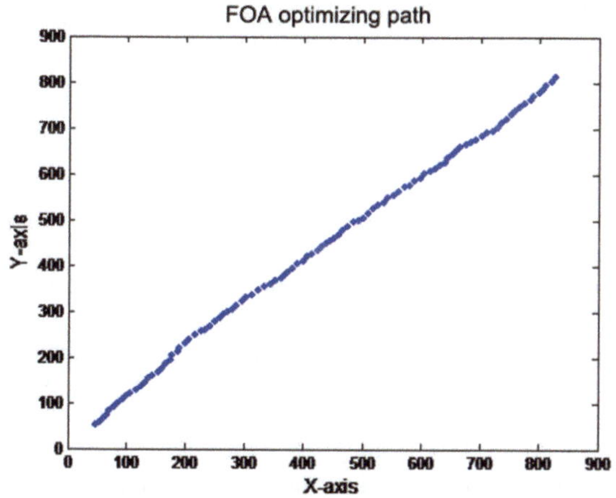

Fig. 3. FOA optimizing path

In order to prove higher reliability and usability of proposed algorithm, the basic FOA algorithm is used for comparison. The result of comparison is shown in Fig. 4. The simulation result in Fig. 4 shows that by introducing modify factor β can avoid local optimum, and improved FOA converges faster compare to traditional FOA.

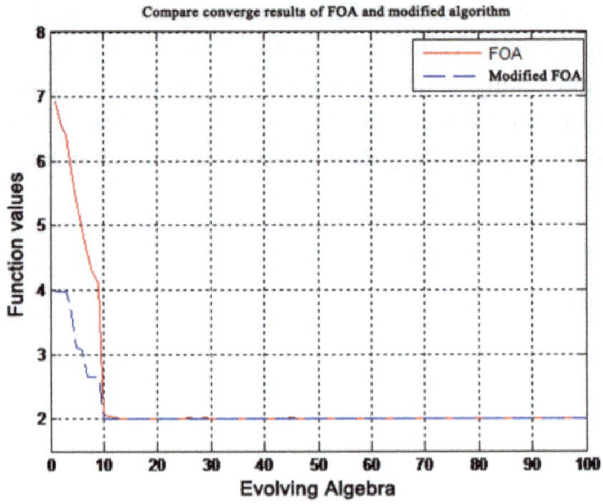

Fig. 4. Compare converge results of FOA and modified algorithm

6 Conclusions

This paper aims to improve gear life and reliability with minimum design cost. Since traditional FOA has slow converge speed and partial optimization problems, this paper proposed a modified gear reducer design algorithm model. This algorithm has 7 variables to ensure its performance and reliability.

Acknowledgement. The work of Fuquan Zhang was supported by the Scientific Research Foundation of Minjiang University (No. MYK17021).

References

1. He, B., Lxian, C., Csheng, L.: Optimized design of gear transmission based on the shuffled frog leaping algorithm. J. Mech. Transm. **7**, 11–15 (2013)
2. Pan, W.T.: A new fruit fly optimization algorithm: taking the financial distress model as an example. Knowl.-Based Syst. **26**, 69–74 (2012)
3. Lxian, C., Zhong, C.: Hybrid discrete differential evolution with a self-adaptive penalty function for constrained engineering optimization. J. Mech. Eng. **3**, 21–25 (2011)
4. Wang, Q., Zhang, W.P., Shi, L.: Improved genetic algorithms' application in reducer optimization design. Mech. Res. Appl. **2**, 18–24 (2009)
5. Hyun, L., Fkai, K.: Planetary gear optimization based on improved genetic algorithm. Appl. Sci. Technol. **12**, 7–12 (2009)
6. Wu, T., Zhang, L.B., Huang, L.: Gear reducer optimal design based on genetic algorithm. Coal Mine Mach. **12**, 17–22 (2009)
7. Hxia, D.: Optimal design of bevel gear reducer based on genetic algorithm. Mod. Manuf. Technol. Equip. **4**, 35–41 (2010)
8. Luo, X.H., Zhang, R.H., Cao, K., Shi, Y.S.: An improved genetic algorithm and its application in optimization design of the gear transmissions. Mach. Des. Res. **2**, 17–20 (2006)

9. Jwei, W., Jming, Z., Xpeng, W.: Multi-objective optimization design of gear reducer based on simulated annealing algorithms. Trans. Chin. Soc. Agric. Mach. **10**, 5–11 (2006)
10. Li, X.Y., Wang, C.X., Guo, Z.Q., Yi, N.: An improved genetic algorithm and its application in engineering optimization. J. Inner Mongolia Univ. Sci. Technol. **3**, 9–12 (2007)
11. Jjun, Y., Hong, Z., Xguo, C.: Optimized design of gear based on hybrid genetic algorithm. Hoisting Conveying Mach. **9**, 12–16 (2008)
12. Gao, Y.G., Wang, G.B., Ding, Y.Z.: Optimized design of helical gear reducer based on genetic algorithm. Hoisting Conveying Mach. **8**, 33–37 (2003)
13. Che, L.X.: Study on differential evolution algorithms orientating analysis and design of mechanisms. J. Chin. Univ. Min. Technol. (2013)
14. Shi, X.: Optimal design of gear based on improved genetic algorithm and its three-dimensional papametric modeing technology. Qingdao University of Technology (2013)

Review of Intelligent Computing Application

Yiou Wang[1], Tianyuan Liu[2], Fuquan Zhang[3,4](✉), Lin Xu[5], Gangyi Ding[3], Rui Xiong[1], and Fei Liu[1]

[1] Beijing Institute of Science and Technology Information, Beijing 100044, People's Republic of China
wangyiou90@163.com, xiongrui_2418@sina.com, liuf@bjstinfo.ac.cn
[2] Carnegie Mellon University, 5000 Forbes Avenue, Pittsburgh, PA 15213, USA
Tianyua2@andrew.cmu.edu
[3] Digital Performance and Simulation Technology Lab., School of Software, Beijing Institute of Technology, Beijing 100081, People's Republic of China
8528750@qq.com, dgy@bit.edu.cn
[4] Fujian Provincial Key Laboratory of Information Processing and Intelligent Control, Minjiang Univeristy, Fuzhou 350121, People's Republic of China
[5] Innovative Information Industry Research Institute, Fujian Normal University, Fuzhou 350300, People's Republic of China
71471418@qq.com

Abstract. Intelligent computing systems can automatically sense environmental changes in the sensor network, make judgments and prediction on the environmental status in time, and provide response strategies in different environments, applying some technologies of pattern recognition, time series prediction and big data analytics. Nowadays, intelligent computing is successfully used in a lot of areas such as transportation, healthcare, home, performance and environmental monitoring. In this paper, the concept of intelligent computing is introduced. Then, the applications in different areas are discussed detailedly. Finally, the future of intelligent computing is analyzed and the conclusion is briefly given.

Keywords: Intelligent computing · Wireless sensor network · Data analysis

1 Introduction

Since 1940s, people are living in an electric information age. The emergence of the Internet, cloud computing, artificial intelligence, and big data are changing people's lives. With the fast development of IoT (Internet of Things), mobile Internet and the related industry, the amount of digital data worldwide is growing at an exponential rate. Statistics show that, in 2015, the overall size of big data market in China is approximately ¥11.06 billion, and will reach ¥501.96 billion in 2020 [1]. Data at such massive scale surpass human's ability to analyze them, however, intelligent computing makes it possible to process the tremendous amount of data. Besides, the fast development of sensing element also generates data with more diversity and heterogeneity. A multimode wireless sensor network (WSN) is constructed by telecommunications equipment, and

J.-S. Pan et al. (eds.), *Advances in Smart Vehicular Technology, Transportation, Communication and Applications*, Smart Innovation, Systems and Technologies 86,
https://doi.org/10.1007/978-3-319-70730-3_30

sensing devices such as video camera, infrared sensor, sound sensor, optical sensor and triaxial accelerometer. In the space of wireless sensor network, the intelligent computing system senses the environmental change of the network automatically, and makes judgments and prediction on the environmental status in time, by applying technology such as pattern recognition, video tracking, time series prediction and big data analytics. The intelligent computing system can also provide response strategy for different environment. Intelligent computing impacts people's daily life in many aspects, including home [2–4], transportation [5, 6], performance [7, 8], healthcare [9–11] and so on, making life more convenient and more comfortable.

2 Concept and Current State of Research

Intelligent Computing refers to applying technologies, such as artificial intelligence, big data analytics and computer networks, to automatically sense, analyze, evaluate, and predict the environmental parameter in the real world, and provide corresponding strategies for different requirements. Intelligent computing system installs the distributed intelligent network devices with the ability of sensing, communicating and information processing on the proper location in the environment. These devices work together and share resources and information with the help of the communication network. Intelligent computing is based on pervasive computing, which is human-orientated. The objective of pervasive computing is to achieve an effective combination of information space and physical space, and make computing devices and computing service everywhere [12]. Intelligent computing constructs an intelligent space, where the system automatically completed the information collection, analysis, processing and evaluation. In 1999, Hashimoto Laboratory of the University of Tokyo proposed the concept of intelligent space [13], which provide a new thinking to solve the complex problem of multivariable system. The objects of intelligent computing is data information. Usually, the computing system achieves a result that is closer to the real value when there are larger amount of data to be computed. Therefore, intelligent computing is very demanding on the processing capacity of computer hardware. In The Road Ahead published in 1995, Bill Gates proposed the concept of Internet of Things (IoT). IoT focuses on information communication and thorough collection on the environment and individuals, while intelligent computing focus on applying advanced technologies, such as neural network and support vector machine, to conduct computing on the massive amount of multimode data intelligently, therefore make some more accurate judgments on the surrounding environment or future status, and then provide appropriate responding strategies according to the judgments.

3 Application of Intelligent Computing

The emergence and development of intelligent computing is having huge impact on many aspects of our everyday life. In this part, the application of intelligent computing will be introduced in detail, including transportation, healthcare, home, performance and environment. Also, the realization methods of intelligent computing in different application scenarios will be analyzed.

3.1 Intelligent Transportation

With the fast growth of the automobile industry, automobile becomes one of the most significant components in the cities' transportation system. However, the serious traffic congestion and environment pollution that come along with it becomes the most urgent problem that needs to be solved, and impeded the economic development of the cities. According to a research in 2015, the loss cost by traffic congestion in 15 cities in China each day is approximately ¥1 billion, and around 500 thousand traffic accident each year results in more than 100 thousand people died in traffic accidents [14]. With the urbanization worldwide and the fast development of technology, the need for a more efficient transportation system is more and more urgent, therefore, the intelligent traffic system are being developed to solve the above problems.

Intelligent transportation system makes use of a group of innovative technology into the transportation management system, including information technology, data communication and data processing, electronic control, computer technology and intelligent vehicle technology [15]. Intelligent transportation system constructed a huge and complicated wireless sensor network that can sense the transportation information in a city in real-time. By applying technologies such as computer vision, pattern recognition, crowd simulation and big data analytics, the intelligent transportation system can provide quick judgments and solutions to the transportation status and traffic accidents. Therefore, the transportation equipment can be fully used, and the use of intelligent transportation system can also improve the efficiency of traffic, and also improve safety, and eventually the transportation service and management will be intelligent. Besides, intelligent transportation is good for reducing traffic congestion, traffic accidents and pollution. Intellectualization, high speed, high precision, and multi-application scope are the development trend of intelligent transportation system. Therefore, intelligent transportation system is a method for sustainable transport.

People's transportation is getting more convenient, comfortable and safe thanks to intelligent transportation management system, transportation information service system, internet of vehicles, self-driving system, urban traffic safety forewarning system, intelligent waiting system. Since 1970s, many major cities in the world are establishing traffic network signal control system, and many large or medium sized cities developed their intelligent transportation management system with signal control system as the key component. For example, Tokyo, Berlin, London, Singapore, Hong Kong and many other large cities in the world have established intelligent transportation management system [16]. Besides, the Interstate 95 highway in the east of USA also established an I-95 corridor coalition [17]. The real-time transportation information service system in Japan, VICS(Vehicle Information and Communication System)overs half of the vehicles in Japan [18]. Europe countries built a radio data system-traffic message channel (RDS-TMC) based on digital radio that covers most area of Europe. Now, dozens of countries and regions in the world are implementing RDS-TMC project [19].

Intelligent transportation system is a high-tech and makes use of a wide range of technologies. The emergence of big data and cloud computing provides significant supports for the fast development of intelligent transportation system.

3.2 Intelligent Healthcare

In the recent years, as the population aging is getting worse, and the people are having more awareness of their health, the needs for healthcare keep increasing. With limited medical resources, how to optimize the allocation of medical resources and improve the usage of medical resources become the key issues that intelligent computing is investigating in the area of medical service [20].

Demirkan Haluk designed a systematic framework for conceptualizing data-driven and mobile-and cloud-enabled intelligent healthcare systems [21]. With adoption of the healthcare systems, healthcare organizations can provide cost-effective quality healthcare services with less IT set-up costs and reduced risk. Beside, telemedicine is a new system of intelligent healthcare. Telemedicine makes use of computer technology, telecommunication technology and modern medical technology, and builds a bridge connecting patients and doctors in the hospital. The patients can receive treatment from the remote hospital, anytime and anywhere, and enjoy efficient treatment and other medical service such as healthcare under the instructions of the doctors [22].

Besides, intelligent wearables which are also known as smart wearables are another research aspects of intelligent healthcare. Intelligent wearables can be put on directly or can integrate with clothing and shoes. The interaction between the built-in softwares and data from the cloud enables the intelligent wearables to analyze the user's health indices [23]. Since Google Glass came out in 2012, Apple, Samsung, Microsoft, Sony and many other technology companies start to explore and research on intelligent wearables in this new domain. According to the report by IHS in October 30, 2013, World Market for Wearable Technology, the total revenue of global intelligent wearable market in 2012 is approximately $8.5 billion, and is expected to reach $30 billion in 2018. The future of intelligent wearables will primarily focus on improving intelligence, convenience and diversity [24]. Intelligent wearables request real-time data transferring to the cloud for big data analytics, and then the digital instructions will be returned to the intelligent wearable devices, and then control the device to conduct human-computer interaction. Therefore, the development of intelligent wearables industry will promote the a lot of industries such as IoT, cloud computing, big data, intelligent home, intelligent grid, healthcare and internet finance.

3.3 Intelligent Home

Intelligent home is also known as smart home. It established an intelligent home space, which connects all the equipments related to information, such as communication devices, home appliances, and home security systems, in order to monitor, control or manage all the devices together in the home or remotely. Also, intelligent home can keep the devices in home and the living environment in harmony [25].

The concept of intelligent home is first proposed in developed countries such as USA, Canada, and Europe countries. The intelligent devices with computing, sensing, and executing abilities are installed in multiple locations in home, which makes a normal physical home becomes an intelligent space that could provide necessary environmental information and technical support for different entities (users or machines) in home, based on knowledge expression and rule-based reasoning. A complete intelligent home system usually includes lighting control system, appliance control system, security system, fire alarm system, remote control system, etc. In the intelligent home industry, the main methods for wireless communication are Zig Bee, Wi-Fi, Zware and Bluetooth. Intelligent home enables all the appliances and devices in home works together by sharing information and applying intelligent computing, and establishes a comfortable and convenient living environment. People can enjoy the convenient intelligent service before realizing it.

In the intelligent home space that intelligent home established information computing are embedded into our daily life. By constructing complex wireless sensing network, intelligent home system can sense the environmental changes and user's needs in time. Intelligent home system achieves the interaction of any information and services, in anytime, anywhere, with anyone, therefore truly realizes the intelligent lifestyle that is people oriented.

3.4 Intelligent Performance

There are many types of performance art. In this part, the application of intelligent performance will be analyzed from two aspects, stage performance and screen performance.

Intelligent stage is based on traditional performance stage, and the mechanical system of the stage is controlled by automatic control system. According to the needs and requirements during the stage performance, the intelligent stage can effectively control the stage automatically, and then improve the quality of performance on stage [26]. As stage lighting has significant effect of light perceptions and easy to arouse strong echoes in audiences, stage lighting is very important in stage performing. Music and light show, as a part of intelligent stage performance, make use of different key characters and elements of music, such as speed, beats, rhythms and emotions, to control the action of a series of light, including the on and off, color, and rotation [27]. On CCTV's Chinese New Year Gala in 2011, the interactively performance of actors and lasers amazed the audience. However, because of some uncertainty problem, for example, the temporary replacement of shows or actors during the performance, or the location or direction of the actor is uncertain, the current stage lighting performance cannot yet achieve complete automation. Now, technologies related to intelligent computing is penetrating into every aspects of stage performance. For example, by applying computer simulation technology, people can build three-dimensional (3D) dynamic holographic model, and provide feasible and precise guarantee for the time-space architecture of the stage [28]. Besides, as technologies such as facial expression recognition, sound recognition and electroencephalography (EEG) monitoring and analysis are getting mature, intelligent performance evaluation will become a new research direction.

In the modern movie and TV products, a large amount of advanced technologies are applied, such as computer simulation technology and motion capture technology, to make the animation and special effects more real to audiences, and also speed up the production process. Motion capture technology starts in Hollywood. Movies such as The Lord of the Rings, Rise of the Planet of the Apes and Les Aventures de Tintin et al. applied a lot of motion capture technology, and therefore created many animation characters that does not exist in the real world [29]. In motion capture technology, trackers are set on actors key body parts or facial parts, then the system capture the location of the trackers, and map the actor's expression onto the computer screen, as a reference for animation production. Movies can capture the action of real actors using motion caption technology, and merge interesting animation characters and real actions together, therefore achieve "acting" animation.

3.5 Intelligent Environment Monitoring System

Intelligent computing is also widely applied in environment monitoring.

Air monitoring is the eye of environment protection, and can inspect the source, distribution, amount, trend, transformation and occurrence regularity of harmful substances, therefore provide resources for improving environment and people's health [30, 31]. By analyzing the data of air component collected from wireless sensor network, intelligent environment monitoring system can provide the current air quality. Also, by applying technologies such as big data analysis and time series prediction, intelligent environment monitoring system can give warning to potential air quality problems in the future.

Intelligent environment monitoring system not only improve our understanding and adaptation to the environment, it also promote the development of global agriculture, fisheries and animal husbandry. The intelligent temperature and moisture control system in greenhouses can make sure that the most suitable temperature, moisture and light for plants [32]. The key for environmental monitoring for aquaculture is examination of water quality. With good water quality, plants can grow fast, therefore increase the economic benefits. Intelligent environment monitoring system for aquaculture can conduct consistent monitoring, analysis and computation on the water under different and complex water environment, to achieve accuracy, timeliness and convenience for data monitoring [33, 34].

4 Future of Intelligent Computing

Intelligent computing cannot make progress without the development of the advanced technologies including artificial intelligence, big data, wireless communication and cloud computing. Nowadays, intelligent computing is already applied in many fields. However, because of the high cost of hardware, intelligent computing is not yet prevalent in everyday use. Also, intelligent computing system requires multiple sensors to build an intelligent space, therefore, it is not very portable or inexpensive. With the development of technology, how to reduce the cost of intelligent computing system and make it more portable will be the trend of the future for intelligent computing technology. Besides, intelligent computing system is required to effectively support the scalability of intelligent space, the coexistence of heterogeneous structures of functional morphology, the concurrency of the interaction between multiple users, the dynamic correlation of information devices, and the seamless mobility of user's tasks. Therefore, the combination of intelligent computing with other advanced technologies for further improvement on the performance will be another development direction.

In the future, the research of intelligent computing will be mainly aimed at the improvements of its structure and fusion algorithms of multi-mode information. And with the progress of its theoretical research, the cost and complexity of intelligent computing systems could be reduced, so that it can be used in more areas and serve the people better.

5 Conclusions

Intelligent computing is successfully applied in different areas including transportation, healthcare, home, performance and environmental monitoring. It gathers multiple distributed intelligent network devices with the ability of sensing, communicating and information processing to a space. It conducts computing and analysis on various heterogeneous information collected, applying technologies such as neural network and big data analysis, and generates evaluation results of the space. Then it gives response to emergencies adaptively according to different requirements and needs. However, because of the complexity and large number of sensors, intelligent computing systems are not very portable or inexpensive. In the future, intelligent computing will have further development on its structure and fusion algorithms of multi-mode information to reduce cost and improve the portability, scalability and reliability. Also, combining intelligent computing with other advanced technologies to further improvement will be another development direction.

Acknowledgments. The authors are very grateful to Information Rapid Production Line Construction (I) (Project NO. PXM2017_178214_000005) of Beijing Finance Project, and also would like to thank Special Projects for Reform and Development of Beijing Institute of Science and Technology Information (2017) (Collection and Analysis key technologies research and development of Scientific and Technological Information Production Line) for the supports and assistance. The work of Fuquan Zhang was supported by the Scientific Research Foundation of Minjiang University (Project No. MYK17021).

References

1. 2016 China Big Data Market Size and Trend Forecast. http://www.chyxx.com
2. Zhou, B.: Study on the Emotional Design of Intelligent Home. Beijing Institute of Graphic Communication, Beijing (2016)
3. Lee, J.Y., Hwang, S.Y.: Applying the SPLE to develop smart home resource management systems. Int. J. Smart Home **9**(7), 135–150 (2015)

4. Huang, J.H., Xue, Z.Y., Xu, X., Tian, S.Y.: Zigbee-based intelligent home furnishing. Int. J. Smart Home **9**(1), 61–68 (2015)
5. Xiao, Z.Q., Chen, J.Y., Fu, S.: Overview on the development of intelligent transportation system in the context of big data. Softw. Guide **16**(1), 182–184 (2017)
6. Ansgar, M.M., Frank, T., Bastian, F.R., et al.: Functional safety and development process capability for intelligent transportation systems. IEEE Intell. Transp. Syst. Mag. **7**(4), 12–23 (2015)
7. Guo, Y.: Design and Implementation on Virtual Scheduling System of Great Square Art Performance. Harbin Institute of Technology, Harbin (2014)
8. Wang, K.X.: The Study of Intelligent Stage Technology Based on Data-driven. Beijing Institute of Technology, Beijing (2015)
9. Sallabi, F., Shuaib, K.: Internet of things network management system architecture for smart healthcare. In: 2016 6th International Conference on Digital Information and Communication Technology and Its Applications, pp. 165–170. Institute of Electrical and Electronics Engineers Inc. Press, Chiang Mai (2016)
10. Khan, M., Jilani, M.T., Khan, M.K., et al.: A security framework for wireless body area network based smart healthcare system. In: ICYRIME 2017 - Proceedings of the International Conference for Young Researchers in Informatics, Mathematics and Engineering, pp. 80–85. CEUR-WS Press (2017)
11. Laplante, N., Laplante, P.A., Voas, J.: Caring: an undiscovered "super-ility" of smart healthcare. IEEE Softw. **33**(6), 16–19 (2016)
12. Gong, J.J.: Philosophical Research on the Intelligence Space in the Pervasive Computing Environment. Dong Hua University, Shanghai (2012)
13. Lee, J.H., Ando, N., Hashimoto, H.: Design policy of intelligent space. In: Proceeding of IEEE SMC 1999 International Conference on Systems, Man and Cybernetics, pp. 1077–1082. IEEE Press, Tokyo (1999)
14. Jia, H.Q.: Research on the optimization of the intelligent traffic management system for the vehicle networking system based on the LTE network. Shandong Jianzhu University, Shandong (2015)
15. Jin, M.J.: Analysis on the development course of intelligent transportation systems in China. Transp. Sci. Technol. **2**, 140–142 (2013)
16. Lu, H.P., Li, R.M.: Developing trend of ITS and strategy suggestings. J. Eng. Stud. **3**(1), 6–19 (2014)
17. I-95 Corridor Coalition Beyond Boundaries. Working Together to Accelerate Improvements in Long-Distance Passenger Travel and Freight Movement. http://www.i95coalition.org/i95/Default.aspx
18. Vehicle Information and Communication System. How VICS Works. http://www.vics.or.jp/english/vics/index.html
19. INRIX. RDS-TMC. http://www.rds-tmc.co.uk
20. Wu, Y., Ding, H., Zhang, S.P.: Design and implementation of intelligent health care sofeware in community based on C#. Electron. Des. Eng. **24**(22), 83–85 (2016)
21. Demirkan, H.: A smart healthcare system framework. IT Prof. **15**(5), 38–45 (2015)
22. Chen, Z.J.: Design of Monitoring Data Transmission Network in Remote Intelligent Healthcare Based on Internet of Things. Hangzhou Normal University, Hangzhou (2012)
23. Huang, P., Ye, L., Huang, Y.H., et al.: Research on current situation of smart wearables. Sci. Technol. Vis. **18**, 106–107 (2015)
24. Wang, D.S.: Study on global wearable devices industry. Compet. Intell. **11**(5), 52–59 (2015)
25. Hu, M.D.: Research on Key Technology of Smart Home Based on Ontology. Ocean University of China, Shandong (2014)

26. Wang, X.J.: Research on the design and realization of software of intelligent stage control system. Innov. Appl. Sci. Technol. **14**, 5–75 (2016)
27. Quan, H.: Research on Expert System for Music & Light Show Scheme Auxiliary Desing. Chongqing University, Chongqing (2010)
28. Chen, Y., Din, G.Y., Huang, T.Y.: A new presentation of the digital performance age. Arts Technol. **5**, 104–105 (2010)
29. Liu, C.: The development and significance of motion capture technology in film and television programs. Contemp. TV **4**, 73–74 (2016)
30. Wolkoff, P., Nielsen, G.D.: Organic compounds in indoor air: their relevance for perceived indoor air quality? Atmos. Environ. **35**, 4407–4417 (2001)
31. Wang, Y.O., Ding, G.Y., Liu, L.Y.: A regression forecasting model of carbon dioxide concentrations based-on principal component analysis-support vector machine. Commun. Comput. Inf. Sci. **482**, 447–457 (2015)
32. Guo, P., Ma, J.H.: Design on intelligent environment monitoring system of agricultural greenhouse. J. Chin. Agric. Mech. **37**(4), 71–73 (2016)
33. Xu, D., Li, D., Fei, B., et al.: A GPRS-based low energy consumption remote terminal unit for aquaculture water quality monitoring. In: IFIP Advances in Information and Communication Technology, vol. 420, pp. 492–503 (2014)
34. Ye, T.D., Cheng, T.B., Zhou, S.B., et al.: Modeling and design research of network intelligent monitoring system about marine water environment. Comput. Meas. Control **22**(6), 1697–1699 (2014)

Directional Shuffled Frog Leaping Algorithm

Lingping Kong[1], Jeng-Shyang Pan[1,2(✉)], Shu-Chuan Chu[3], and John F. Roddick[3]

[1] Innovative Information Industry Research Center, Harbin Institute of Technology Shenzhen Graduate School, Shenzhen, China
konglingping2007@163.com

[2] Fujian Provincial Key Laboratory of Big Data Mining and Applications, College of Information Science and Engineering, Fujian University of Technology, Fuzhou, Fujian, China
jengshyangpan@fjut.edu.cn

[3] School of Computer Science, Engineering and Mathematics, Flinders University, Adelaide, Australia
{jan,roddick}@cs.flinders.edu.au

Abstract. Shuffled frog leaping algorithm is one of the popular used optimization algorithms. This algorithm includes the local search and global search two solving modes, but in this method only the worst frog from divided group is considered for improving location. In this paper, we propose a directional shuffled frog leaping algorithm (DSFLA) by introducing the directional updating and real-time interacting concepts. A direction flag is set for a frog before moving, if the frog goes better in a certain direction, it will get better in a big probability by moving a little further along that direction. The movement counter is set for preventing the frog move forward infinite. Real-time interacting works by sharing the currently optimal positions from the other groups. There should have some similarities among the best ones, and the worst individual could be improved by using those similarities. The experimental results show that the proposed approach is a very effective method for solving test functions.

Keywords: Swarm intelligence · Shuffled frog leaping algorithm · Optimization algorithm

1 Introduction

Swarm intelligence (SI) refers to the collective of natural or artificial behaviors with problem-solving ability [1,2]. SI is a system that consists of a population of simple agents or individuals interacting with each other and communicating with the environment. The inspiration sources mostly are the nature. In the past few years, the application fields based on swarm intelligence techniques reach from data mining, controlling unmanned vehicles and planetary mapping to medicine science, and also include other important areas [3]. The agents or individuals

J.-S. Pan et al. (eds.), *Advances in Smart Vehicular Technology, Transportation, Communication and Applications*, Smart Innovation, Systems and Technologies 86,
https://doi.org/10.1007/978-3-319-70730-3_31

usually can either be real or imaginary, and they can also be computational and mathematical. Researchers have proposed a lot of well-performed optimization algorithms based on swarm intelligence. Examples of commonly used swarm optimization algorithms [4,5] are briefly described below. Cat Swarm Optimization (CSO) [6,7], introduced by Chu, is inspired by the behaviors of cats, and the authors model the two major behaviors of cats into sub-models: seeking mode and tracking mode. Firefly algorithm is proposed by Xin-She Yang [8,9], which is a meta-heuristic. In this algorithm, a firefly's flash works as a signal system to attract other fireflies, then find a food source. Genetic algorithm (GA) [10], which is based on the Darwinian evolution and survival of the fittest theory, is introduced by Holland. GA is commonly used to generate optimization solutions by relying on operators such as mutation, crossover and selection. Particle swarm optimization (PSO) [1] is a computational method, which simulates the social behavior of bird flock or fish school. PSO solves a problem by involving a population of candidate particles and moving these solutions around in a search-space according to several mathematical equations over the particles position and velocity. The bee algorithm [11] is created by Pham and his coworkers, which models on the foraging behavior of honey bees. The algorithm consists of global explorative search and local exploitative search.

The remainder of the paper is organized as follows: Sect. 2 describes the basic concepts of shuffled frog-leaping algorithm [12] and relevant applications; Sect. 3 presents our proposed algorithm DSFLA; Sect. 4 presents experimental evaluations of the DSFLA and comparison with other algorithms; Sect. 5 provides the conclusion to the paper.

2 Related Works

Shuffled frog-leaping algorithm (SFLA) [13] is a memetic meta-heuristic, which is developed by Eusuff for solving combinatorial optimization problems. The SFLA includes local search and global information exchange parts. The local search is completed by modeling the similarity of particle swarm optimization method and periodically shuffled evolution process is applied into global exploration. The author Kaur [14] proposed the augmented shuffled frog leaping algorithm (ASFLA) after analyzing the merit and dement of the original SFLA. The author pointed out that only those worst individuals are taken into consideration for enhancing their positions. That is not fair for the best ones because those best individuals may be local optima, and by updating the moment of inertia of best frogs it could facilitates the best frogs to jump out to a better location with more food in their vicinity. Author Zhao [15] proposes an algorithm shuffled frog leaping algorithm (SFLA) based on enhanced learning, which evolves the solution with the help of superior information of each individual and average center of optimal ones for mutual learning. The optimal individual from small group will move based on the information of general center of the group and the best from of the whole population. AASFLA is proposed by Anandamurugan [16] with the idea of antipredator capabilities and a hybrid shuffled frog leaping algorithm, which can avoid the local minima. The author also applies it into Wireless Sensor Network (WSN) for selecting the cluster head and improving the WSN performance and lifetime. Chandirasekaran [17,18] combines the shuffled frog leaping and firefly algorithms (FFA) two algorithms and tries to estimate the optimum placement of randomly deployed sensors in wireless sensor networks. Author Fan [19] improves the shuffled frog leaping algorithm to choose the sensor node with more residual energy as a cluster heads and partition those sensor nodes into varied size of groups. As Kaur author mentioned, only those worst individuals are taken into consideration for enhancing their positions. There is no communication between the best frogs from different sets in local search at all. In this paper, we proposed directional shuffled frog leaping algorithm, which solves the problem by improving the local locations of best individuals and bringing in more interacting between group and group.

3 Directional Shuffled Frog Leaping Algorithm

In shuffled frog leaping algorithm, it simulate the major two information exchange behaviors of frogs: local search behavior between the best and worst two frogs of general center, global search behavior between the best frogs among the whole frog population. There is no communication between the best frogs from different sets in local search at all. In our directional shuffled frog leaping algorithm, we bring out the directional moving and real-time interacting concepts. If an individual moves in a certain direction and the individual goes better, there are more chances for the individual to get better after moving a little further along that direction. So in local search process, the DSFLA adds the directional flag and the movement counter before one moving. Once the individual finds a better place than the original location, it will continue to march forward and update the movement counter until it stops getting better or the counter reaches the movement threshold. In the worst case, the individual did not locate the good location during its first move, it should not move forward. The one will update with the help of the best individuals from global search on the current time, which is called the real-time interacting. That is also interpreted as sharing information from the best ones. Usually there must be some similarities among the best individuals and those frogs bear some analogy of location to food source. It would not be wrong if one individual walk to the center of superior frogs. DSFLA adds the chances for finding food by introducing the directional moving and real-time interacting concepts. DSFLA algorithm consists of a set of virtual frog population divided into several small groups of sample. During the proceeding of local search, the worst sample will evolve based on the position of the best one from the same group. The point is that the evolving direction is defined in the beginning, and it will continue to move along that direction if the position of frog gets better in the last move. Otherwise, the frog finds the food toward the global best way that concluded by the best frogs. The following are the details of DSFLA.

3.1 Steps of the DSFLA

The DSFLA is summarized in the following steps:

Step 1. Initialize. Define the frog population P, and its individual number is n, and the virtual group set G that the population is divided into with group number value m, where each group consists of n/m individuals, here m could be divided by n.

Step 2. Sort population. Rank the P in increase order of performance value. Store the best one as *gbest*, that is P_{n-1}.

Step 3. Packing process. Partition the n frogs into m groups according to the order of the population, and make sure that each group contains the better and the worse individual at the same time. The partition method could take the interval sampling. The first individual goes to the first group, and the second individual goes to the second group, the next goes to the third group, till the m group, then start with the first group again.

Step 4. Information exchange exploration. Find the worst one individual in each group, then evolving it based on *Directional searching model* below for a threshold times.

Step 5. Global shuffle. After a certain number of information exchange exploration steps within groups, copy each group back to the population P, and sort the population as Step 2. Compare the best individual P_0 with the gbest, if P_0 is better, then replace *gbest* by P_0, otherwise ignore it.

Step 6. Terminal check. If the performance value of the *gbest* is satisfied or the iteration times reach the threshold value, then stop the process. Otherwise go to Step 3.

3.2 Directional Searching Model

Suppose that each individual $P_i = [x_1, x_2, ..., x_d]$ carries d dimension characteristic parameters. Updating process actually is to move each dimension position of the individual. The steps of directional search model process are followed as:

Step 0. Set update threshold that counter the times of updating process. Note that some certain group of frogs as $F_k = [F_{k,0}, F_{k,1}, ..., F_{k,[n/m-1]}]$, then the worst frog is $F_{k,0}$, and the best one is $F_{[n/m-1]}$. The $F_{k,0} = [k|x_{0,1}, k|x_{0,2}, ..., k|x_{0,d}]$ and $F_{k,[n/m-1]} = [k|x_{[n/m-1],1}, k|x_{[n/m-1],2}, ..., k|x_{[n/m-1],d}]$ are relevant symbols illustration. Then the set $B = \bigcup F_{k,[n/m-1]}$ and $1 \leqq k \leqq m$ are the collection of the best ones from each group. $B = [F_{1,[n/m-1]}, F_{2,[n/m-1]}, ..., F_{m,[n/m-1]}]$. Pre-define the moving step length for one individual in one update, noted as *rand*, and it is a random number $rand \in [0-1]$.

Step 1. Defined the direction Flag as $f = [f_1, f_2, ..., f_d]$, and the value of Flag comes from equation one, where $i \in [1, 2, ..., d]$. And Set moving counter $MC = 0$.

$$f_i = \begin{cases} -1, & if \quad k|x_{[n/m-1],i} - k|x_{0,i} \leq 0 \\ 1, & \quad Otherwise \end{cases} \tag{1}$$

Step 2. Check the MC value, if MC is smaller than a threshold, to move the worst frog F_0 with equation two, where $i \in [1, 2, ..., d]$. After that set the $MC = MC+1$. Otherwise go to Step 5.

$$k|x_{0,i} = k|x_{0,i} + f_i \left| k|x_{[n/m-1],i} - k|x_{0,i} \right| \times rand \tag{2}$$

Step 3. Evaluate the performance of this individual, if it gets better, than go to Step 2. If not and the MC is not one, this frog should move back one step and stop process. Otherwise go to Step 4.

Step 4. Move the F_0 based on Eq. 3.

$$k|x_{0,i} = (\sum_{k=1}^{m} k|x_{[n/m-1],i}) \div m \tag{3}$$

Step 5. Terminal check. If the updating iteration times reach the threshold, then stop, otherwise go to Step 1.

4 Experimental Results

Three different algorithms are simulated in our experiment: *SFLA*, *ASFLA* [14], our scheme. We applied them with the same factor into three test function (Rosenbrock, Rastrigrin and Schaffer) to compare their performance. The test function $f(x)$is shown as Eq. 4. The parameters used in experiment are showed in Table 1. And the final results list in Table 2, from the table we can tell that *DSFLA* shows a better performance (Tables 3 and 4).

$$f_1(x) = f(x_1, x_2, ..., x_d) = \sum_{i=1}^{d} [100 \times (x_{2i-1}^2 - x_{2i})^2 + (x_{2i} - 1)^2] \tag{4}$$

$$f_2(x) = f(x_1, x_2, ..., x_d) = Ad + \sum_{i=1}^{d} \left[x_i^2 - A \times \cos(2\pi x_i) \right] \tag{5}$$

$$f_3(x) = 0.5 + \frac{\sin^2(x_1^2 - x_2^2) - 0.5}{[1 + 0.0.1(x_1^2 + x_2^2)]^2} \tag{6}$$

Table 1. Symbols description

Symbols	Description	Values
n	The number of population	100
m	The number of groups	10
d	The boundary value of parameter	$[-100, 100]$
MC	The steps value in directional search model process	2
Iteration	The iteration times used in three algorithm	100

Table 2. Comparison results in Rosenbrock function

Iteration	Methods		
	ASFLA	SFLA	DSFLA
10	0.4371	0.0481	7.18×10^{-9}
20	0.0254	0.0047	2.07×10^{-20}
26	0.0103	2.09×10^{-6}	1.25×10^{-28}
40	1.73×10^{-7}	7.27×10^{-8}	0
60	4.27×10^{-8}	7.27×10^{-8}	0
100	2.09×10^{-8}	3.39×10^{-8}	0

Table 3. Comparison results in Rastrigrin function

Values	Methods		
	ASFLA	SFLA	DSFLA
x	8.567×10^{-8}	1.897×10^{-5}	2.084×10^{-9}
y	0.9949	3.957×10^{-5}	1.045×10^{-9}
f(x)	0.9949	3.8213×10^{-7}	0.0

Table 4. Comparison results in Schaffer function

Iteration	Methods		
	ASFLA	SFLA	DSFLA
5	0.0311	0.0354	4.29×10^{-5}
10	4.44×10^{-5}	0.0354	6.35×10^{-11}
13	4.44×10^{-5}	0.01364	2.89×10^{-15}
40	5.93×10^{-7}	0.0337	0
70	1.31×10^{-10}	1.16×10^{-6}	0
100	2.18×10^{-11}	2.29×10^{-11}	0

5 Conclusion

In this paper, we propose a directional shuffled frog leaping algorithm which solves optimization problems. *DSFLA* includes grouping optimization and global search two processes. Grouping optimization process executes moving position based on direction flag array and the optimal ones from every independent group. Global search process works for sharing the useful information from each group by shuffling operation. In the end, one recently algorithm and the original *SFLA* approach are simulated to compare the performance with our proposed scheme on mostly used test function. The experimental results show that our scheme is better than the other two algorithms.

References

1. Kennedy, J.: Swarm intelligence. In: Handbook of Nature-Inspired and Innovative Computing, pp. 187–219. Springer, US (2006)
2. Derrac, J., Salvador, G., Daniel, M., Francisco, H.: A practical tutorial on the use of nonparametric statistical tests as a methodology for comparing evolutionary and swarm intelligence algorithms. Swarm Evol. Comput. **1**(1), 3–18 (2011)
3. Mavrovouniotis, M., Changhe, L., Shengxiang, Y.: A survey of swarm intelligence for dynamic optimization: algorithms and applications. Swarm Evol. Comput. **33**, 1–17 (2017)
4. Chao, Z., Feng-ming, Z., Fei, L., Hu-sheng, W.: Improved artificial fish swarm algorithm. In: 2014 IEEE 9th Conference on Industrial Electronics and Applications, pp. 748–753 (2014)
5. Thi-Kien, D., Tien-Szu, P., Trong-The, N., Shu-Chuan, C.: A compact articial bee colony optimization for topology control scheme in wireless sensor networks. J. Inf. Hiding Multimedia Sig. Process. **6**(2), 297–310 (2015)
6. Shu-Chuan, C., Pei-Wei, T., Jeng-Shyang, P.: Cat swarm optimization. In: Pacific Rim International Conference on Artificial Intelligence. Springer, Heidelberg (2006)
7. Vaclav, S., Lingping, K., Pei-Wei, T., Jeng-Shyang, P.: Sink node placement strategies based on cat swarm optimization algorithm. J. Netw. Intell. **1**(2), 52–60 (2016)
8. Xin-She, Y.: Firefly algorithm, stochastic test functions and design optimisation. Int. J. Bio-Inspired Comput. **2**(2), 78–84 (2010)
9. Trong-The, N., Jeng-Shyang, P., Shu-Chuan, C., John, F.R., Dao, T.-K.: Optimization localization in wireless sensor network based on multi-objective firefly algorithm. J. Netw. Intell. **1**(4), 130–138 (2016)
10. Bakirtzis, A., Spyros, K.: Genetic algorithms. In: Advanced Solutions in Power Systems: HVDC, FACTS, and Artificial Intelligence: HVDC, FACTS, and Artificial Intelligence, pp. 845–902 (2016)
11. Kiran, M.S., Oguz, F.: A directed artificial bee colony algorithm. Appl. Soft Comput. **26**, 454–462 (2015)
12. Eusuff, M., Kevin, E.L.: Optimization of water distribution network design using the shuffled frog leaping algorithm. J. Water Resour. Plann. Manage. **129**(3), 210–225 (2003)
13. Eusuff, M., Kevin, L., Fayzul, P.: Shuffled frog-leaping algorithm: a memetic metaheuristic for discrete optimization. Eng. Optim. **38**(2), 129–154 (2006)

14. Kaur, P., Shikha, M.: Resource provisioning and work flow scheduling in clouds using augmented Shuffled Frog Leaping Algorithm. J. Parallel Distributed Comput. **101**, 41–50 (2017)
15. Jia, Z., Min, H., Hui, S., Li, L.: Shuffled frog leaping algorithm based on enhanced learning. Int. J. Intell. Syst. Technol. Appl. **15**(1), 63–73 (2016)
16. Anandamurugan, S., Abirami, T.: Antipredator adaptation shuffled frog leap algorithm to improve network life time in wireless sensor network. Wirel. Personal Commun. 1–12 (2017)
17. Chandirasekaran, D., Jayabarathi, T.: Wireless sensor networks node localization-a performance comparison of shuffled frog leaping and firefly algorithm in LabVIEW. Indonesian J. Electr. Eng. Comput. Sci. **14**(3), 516–524 (2015)
18. Wuling, R., Cuiwen, Z.: A localization algorithm based On SFLA and PSO for wireless sensor network. Inf. Technol. J. **12**(3), 502–505 (2013)
19. Xunli, F., Feiefi, D.: Shuffled frog leaping algorithm based unequal clustering strategy for wireless sensor networks. Appl. Math. **9**(3), 1415–26 (2015)

A Bidirectional Collaborative Filtering Recommender System Based on EM Algorithm

Yu Mao[1], Fuquan Zhang[2,3(✉)], Lin Xu[4], Dakui Zhang[1], and Hui Yang[1]

[1] School of Computer Science and Technology, Beijing Institute of Technology, Beijing, China
maoyu_bit@163.com, dakuiz@163.com, zlj-1943@163.com

[2] School of Software, Beijing Institute of Technology, Beijing 100081, China
8528750@qq.com

[3] Fujian Provincial Key Laboratory of Information Processing and Intelligent Control, Minjiang University, Fuzhou 350121, China

[4] Innovative Information Industry Research Institute, Fujian Normal University, Fuzhou 350300, China
71471418@qq.com

Abstract. Collaborative filtering is a promising recommendation technique for predicting the preferences of users in recommender systems. The date coming from recommender system is not only big but also sparse. It motivates the need for a more intelligent approach to obtain the tastes of users. In this paper we present a novel bidirectional collaborative filtering method based on EM algorithm. We combine user rating with item rating to build a new rating matrix and calculate a steady result by iterating this strategy. The empirical evaluation demonstrates that our technique have encouraging result contrast to the traditional approaches.

Keywords: Collaborative filtering · Recommender system · EM algorithm · Bidirectional recommend

1 Introduction

With the arrival of the age of Big Data, people gradually entered the era of information overload from the era of information lack. The explosive of structured data and unstructured data volume growth built a huge information wall between consumers and producers. It is difficult to find the useful interesting information when people face of the huge amounts of data. Not only that, the producers of information also becomes more and more difficult to pass the accurate message to people what they need.

To solve the problem of information overload and knowledge explosion, recommender systems are proposed and become an important means of information recommendation. Recommender systems are systems able to provide personalized recommendations to users [1]. Currently, they have been widely used in many fields, such as electronic commerce, music, movie, books or recruitment.

J.-S. Pan et al. (eds.), *Advances in Smart Vehicular Technology, Transportation, Communication and Applications*, Smart Innovation, Systems and Technologies 86,
https://doi.org/10.1007/978-3-319-70730-3_32

Recommender systems are usually classified into two types: based on contents and based on collaborative filtering.

The first recommender system is relatively traditional approach in predict programming. The main idea is [2, 3]: according to people preferences in the past to calculate what they like. When a user have bought a item, the system recommend a same item or a similar goods. This kind of systems only need to know person whether to bought something rather than if like this item. It is based on the content information of the goods to extract the rich attribute information, and calculate the similarity of items on the attribute space of goods. Obviously, this model is difficult to excavating out the deep connotation of the data.

Another recommender system is collaborative filtering recommendation model [4]. Such recommender systems find out the simulation relation between users by making analysis the users historical rating data, and then according to these relationships, the model can get the similar users by some Similarity algorithms and predict results to recommend user information of their interest.

Recommender systems based on collaborative filtering can be divided into two classes, those which are based on memory and based on model.

(1) Based on memory: These recommender systems always use the K-Nearest Neighbors algorithm (K-NN algorithm) [5] for predicting the preferences of users. The basic idea is: Firstly, find out the users most similar to user u, called the neighbors of u: k_i. Secondly, obtain the rating data of user k_i, and get the similarity with u by calculation method of similar. Finally, recommend items to user u by using weighting method. Memory-based recommender systems can provide a intuitive method to recommend items to user by analysis known rating data. However, these recommender systems heavily dependent on the simplistic similarity algorithm for predicting and recommending items.

Therefore, they are not appropriate when facing with a large number of data.

(2) Based on model: These recommender systems [6] are based on the idea of using a model to predict the ratings that users make. The techniques try to further fill out the user-item matrix. They hope to understand how much a user will like an item that they did not encounter before. In order to achieve it, they utilize several machine learning algorithms to train on the vector of items for a specific user, and build a model that can predict the user rating for a new item that has just been added to the system.

However, most research papers have focused on unidirectional recommendation domain, such as movie. These data only contain users rating to items, not have items rating to users. But in some special domains (like recruitment field), it is easy to obtain both two sides data. So, it is significant to propose a new way to construct commendation model on these domains.

In our paper, we present a novel bidirectional collaborative filtering technique for predicting user rating. The model effective to get accurate predictions by combine forward and reverse matrices based on EM algorithm.

The remainder of this paper is organized as follows: the next section introduces the related works in recommender systems. Section 3 introduces our bidirectional recommend method, and describes our EM-based collaborative filtering model. We evaluate the performance of our method in Sect. 4 and draw conclusions in Sect. 5.

2 Related Works

In this paper, we mainly focus on collaborative filtering techniques based on model. Currently, popular model-based approaches are Bayesian Networks, Singular Value Decomposition, and Probabilistic Latent Semantic Analysis. All of these methods need to use a rating matrix in which each user provides information about the items. The detail research work on recommender system is surveyed in the following:

Besides model-based system use rating matrix, the recommender system based on contents also use rating matrix to predict result. The researches [7–9] have been applied in a matrix representing description of each item: each value in matrix used to describes how many times the user looks the item. Although these approaches utilized rating matrix to a certain extent, they have not describe and use a rating matrix in which each user provides information about how much he likes some items.

Collaborative filtering techniques are successful regarding commendation. There are two classes: memory-based and model-based. The memory-based [10, 11] systems use K-NN algorithm to find out the nearest similar users or the most likely items. This method made good use of matrix information, but it is not suitably designed for a big data and a great sparse data matrix as analysed system.

The model-based systems can be get a fantastic result facing with a huge data set. The paper [12] present a technique for predicting the tastes of users in recommender systems based on collaborative filtering. It is based on factorizing the rating matrix into two non negative matrices whose components lie within the range [0, 1] with an understandable probabilistic meaning. The paper [13] has been extended for collaborative filtering using EM algorithm and considering that the ratings follow either a normal distribution or a multinomial distribution. The paper [14] introduce a matrix factorization model by artificial immune network to collaborative filtering for movie recommendation. It proposes a new formulas in calculating the affinity between an antigen and an antibody to an immune network.

All of these model-based techniques are successful for resolving the problem of data sparseness and initial rating. However, these models only discuss unidirectional recommendation, they do not consider the situation that in some domains have reverse rating data.

In this research, we propose a new idea in collaborative filtering for recruitment recommendation. Unlike the previous model-based systems, our approach design a bidirectional collaborative filtering framework based on EM algorithm. From experimental results on the recruitment dataset, we found that our system is able to achieve better precision than traditional methods.

3 Bidirectional Recommendation Model

In this section we will present the model of this paper. In Sect. 3.1 we will show the initial elements of model. In Sect. 3.2 we will describe the EM algorithm. In Sect. 3.3 we will present the bidirectional recommendation model about how users rate items.

3.1 Initialization of Model

The main idea in this research is to develop a rating model by applying bidirectional method in collaborative filtering for recruitment recommendation. The Fig. 1 shows the structure of our bidirectional recommendation technique based on big data. In order to make the calculation of matrix, we will initialization set the input, output and parameter of the algorithm.

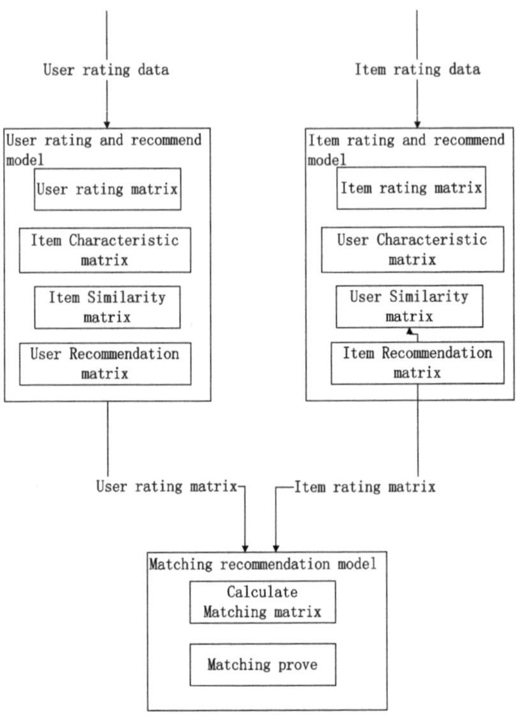

Fig. 1. Structure of bidirectional recommendation model.

The input.

n: the total numbers of users in the ratings matrix.
m: the total numbers of items in the ratings matrix.
E_u: the user rating matrix.
E_p: the item rating matrix.

T_p: the item characteristic matrix. This matrix is produced by calculating transpose the E_u matrix.

T_u: the user characteristic matrix. This matrix is produced by calculating transpose the E_p matrix.

$E(U_k/P_i)$: the value of user k rating item i. the $E(U_k/P_i)$ lies within [1, 5]. The more larger the number $E(U_k/P_i)$, the more user k like the item i.

$E(P_i/U_k)$: the value of item i rating user k

Parameters.

$\delta(0,1)$: this parameter is related to importance of each value of user rating item. The E_u need to multiply this parameter when joint calculate E_u and E_p by bidirectional recommendation algorithm. δ have to normalization processing before apply in model.

$\mu(0,1)$: this parameter is related to importance of each value of item rating user. The E_p need to multiply this parameter when joint calculate E_u and E_p by bidirectional recommendation algorithm. μ also have to normalization processing before apply in model.

$Q(0,1)$: this parameter is related to iterated algorithm

The output.

By using the input and collaborative filtering algorithm, the model outputs five matrices: item similarity matrix(D_p), user similarity matrix(D_u), user recommendation matrix(Rup), item recommendation matrix(R_{pu}) and finally recommendation matrix(RD_{up}). The D_p and D_u describe the similar of the items and the users. The R_{up} and R_{pu} represent the preliminary result about user rating item and item rating user, and then unite Parameters to calculate the finally recommendation matrix(RD_{up}).

3.2 EM Algorithm

An effective way to maximize likelihood or a posteriori estimation in problems with incomplete data is known as Expectation-Maximization (EM) algorithm [15]. We combine the EM algorithm with collaborative filtering model for recommender system. First, the initial model is trained from the R_{up} or R_{pu} set Q_l, that is, the model parameters $\hat{\theta}$ are estimated from Q_l. Then, the collaborative filtering is used to estimate the conditional probability $P(U_k|P_i; \hat{\theta})$ or $P(P_i|U_k; \hat{\theta})$ that each rating in the unknown set Q_u. Next, new parameters $\hat{\theta}$ of collaborative filtering are updated by both the originally and newly rating data. The EM algorithm looks for a local maximum likelihood by iterating these last two steps until $\hat{\theta}$ dose not change.

The EM-based collaborative filtering algorithm finds a maximum posteriority estimate of $\hat{\theta}$ by iterating E-Step and M-Step. In the E-Step, calculating the probabilistic rating of each unknown user or item by using the initial parameters $\hat{\theta}$. Then in the M-Step, new parameters of model are calculated using an estimate for R_{up} or R_{pu} until the change less than a threshold. Note that the initial parameters $\hat{\theta}$ can be achieved from just the value of rating data.

3.3 Predicting User Ratings

Once the overview of the model and algorithm framework have been generated, we then use these elements to predict a user rating for a item. The prediction process first finds R_{up} and R_{pu} using the users/items nearest neighbors based on the collaborative filtering technique, and then predicts the rating by bidirectional recommendation method and EM algorithm. We summarize the procedures in the following.

Step 1: Compute the similarity among each users by calculating Manhattan distance of the E_u row vectors.
Step 2: Construct item similarity matrix D_p by using step 1 result.
Step 3: Compute the similarity among each items by calculating Manhattan distance of the E_p row vectors.
Step 4: Construct user similarity matrix D_u by using step 3 result.
Step 5: Compute the R_{up} based on (1).

$$Ui = \sum\nolimits_{m=0}^{M} simi_{\mathrm{Im}} \times pref_{\mathrm{Im}} / \sum\nolimits_{m=0}^{\mathrm{M}} simi_{\mathrm{Im}} \tag{1}$$

The U_i is the value of user rating item, the $\mathrm{simi}_{\mathrm{Im}}$ is the similarity matrix of U_i to item Im.

Step 6: Compute the R_{pu} based on (2).

$$Pi = \sum\nolimits_{n=0}^{N} simi_{Un} \times pref_{Un} / \sum\nolimits_{n=0}^{\mathrm{N}} simi_{Un} \tag{2}$$

Step 7: Combine the R_{up} and the R_{pu} to obtain RD_{up} by using parameter δ and μ, based on (3).

$$RD_{up} = \left(R_{up} \times \delta + R_{pu} \times \mu\right) / (n + m) \tag{3}$$

Step 8: Set R_{up} to E_u and set R_{pu} set E_p, than repeat Steps 1 to 7 until parameter $\hat{\theta}$ dose not change.

4 Experiments

4.1 Experiments Corpus

In this section, we conducted several tests to evaluate the effect of our bidirectional collaborative filtering recommendation model based on EM algorithm. The test datasets we used in the experiments is a group general rating data (http://www.xiaozhaogou.com) from recruitment domain. The dataset contains more than 1000000 ratings collected from 205861 Beijing college students on 57839 recruitment informations and more than 100000 ratings collected from company on students. The rating value for this dataset is ranged from 1 to 5, the more larger the number, the more user likes item. The rating value is counted by combine click-through rate and collection rate.

4.2 Experiments Design

The experiment use 5-fold cross validation method to evaluate the result. The idea is: choose 80% to be train data set, and the rest of 20% date to be test data set. We will describe the accuracy of the predictions of our technique through Root Mean Square Error (RMSE) and Mean Absolute Error (MAE).

$$RMSE = \sqrt{\frac{\sum_{(u,i|R_{test})}\left(r_{ui} - \hat{r}_{ui}\right)^2}{|R_{test}|}}$$

$$MAE = \frac{\sum_{(u,i|R_{test})}\left|r_{ui} - \hat{r}_{ui}\right|}{|R_{test}|}$$

The $\hat{r}_{ui}$ is the prediction score, the r_{ui} is the actual score.

4.3 Experiments Result Analysis

In order to prove effect of our model, we compare the result for our approach and other techniques: Based on user collaborative filtering model, probabilistic matrix factorization, and singular value decomposition model.

From the experimental results in Table 1, it can be seen that the RMSE and MAE could be effectively improved by bidirectional collaborative filtering approach. Our result is even lower than the other recommendation algorithms.

Table 1. Accuracy of the recommendation

Technique	Relative standard	
	RMSE	MAE
Based on user	2.367	1.782
Probabilistic matrix factorization	1.832	1.265
Singular value decomposition	1.677	1.131
Our approach	1.558	1.065

For observe the learning process of the EM-based rating, the MAE at each iteration are further analysed. The Fig. 2 gives the plot of the MAE versus the learning iterations from the 0th round to the 10th round on data set.

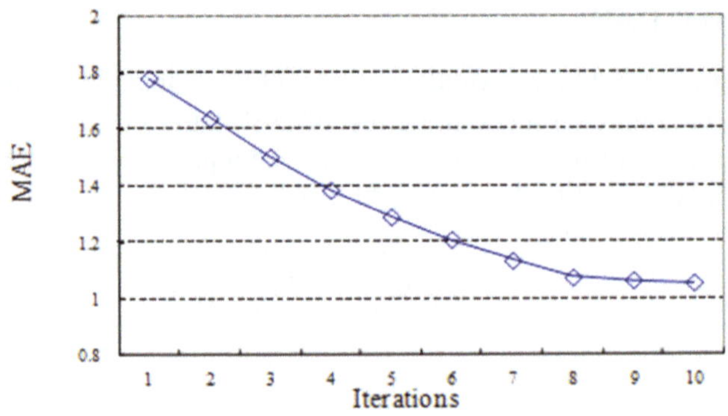

Fig. 2. The MAE at each learning iterations.

In Fig. 2, it could be seen that the MAE dis-creases with the growing of learning iterations on data set, and it is noteworthy from the figure that the MAE significantly dis-creases in several of the initially iterations and converge quickly within just a few iterations.

5 Conclusions

In order to accurately for rating taste in commendation system, it is promising to provide the bidirectional technique respond to the data set in some domain. In this paper we have presented a novel bidirectional collaborative filtering approach based on EM algorithm. Our method can combine two ways of rating data and iteration model to calculate user commendation matrix.

Although our system is very competitive with traditional commendation system in data cold starting and data scalability, our approach is not suitable to apply in some domains which have not reverse rating data. In the future research, we could try to use transfer learning mechanism to solve this problem.

Acknowledgments. The work of Fuquan Zhang was supported by the Scientific Research Foundation of Minjiang University (No. MYK17021).

References

1. Bobadilla, J., Ortega, F., Hernando, A., Gutiérrez, A.: Recommender systems survey. Knowl. Based Syst. **46**, 109–132 (2013)
2. Das, G., Mannila, H.: Context-based similarity measures for categorical databases. In: European Conference on Principles of Data Mining and Knowledge Discovery, pp. 201–210. Springer (2000)
3. Mooney, R.J., Roy, L.: Content-based book recommending using learning for text categorization. In: The Fifth ACM Conference on Digital Libraries, pp. 195–204. ACM (2000)

4. Su, X., Khoshgoftaar, T.M., et al.: A survey of collaborative filtering techniques. Adv. Artif. Intell. **2009**(12), 4 (2014)
5. Ning, X., Desrosiers, C., Karypis, G.: A comprehensive survey of neighborhood-based recommendation methods. In: Recommender Systems Handbook, pp. 107–144. Springer (2015)
6. Koren, Y.: Factorization meets the neighborhood: a multifaceted collaborative filtering model. In: ACM SIGKDD International Conference on Knowledge Discovery and Data Mining, pp. 426–434. ACM (2008)
7. Pazzani, M.J., Billsus, D.: Content-based recommendation systems. In: Adaptive Web, pp. 325–341. Springer (2007)
8. Lops, P., Gemmis, M.D., Semeraro, G.: Content-based recommender systems: state of the art and trends. In: Recommender Systems Handbook, pp. 73–105. Springer, New York (2011)
9. Lops, P., de Gemmis, M., Semeraro, G., et al.: Content-based and collaborative techniques for tag recommendation: an empirical evaluation. J. Intell. Inf. Syst. **40**(1), 41–61 (2013)
10. Deshpande, M., Karypis, G.: Item-based top- N recommendation algorithms. ACM Trans. Inf. Syst. **22**(1), 143–177 (2004)
11. Cacheda, F., Carneiro, V., Fernández, D., et al.: Comparison of collaborative filtering algorithms. ACM Trans. Web **5**(1), 1–33 (2011)
12. Hernando, A., Bobadilla, J., Ortega, F.: A non negative matrix factorization for collaborative filtering recommender systems based on a Bayesian probabilistic model. Knowl. Based Syst. **97**(C), 188–202 (2016)
13. Hofmann, T.: Latent semantic models for collaborative filtering. ACM Trans. Inf. Syst. **22**(1), 89–115 (2013)
14. Koren, Y., Bell, R., Volinsky, C.: Matrix factorization techniques for recommender systems. Computer **42**(8), 30–37 (2009)
15. Nigam, K., McCallum, A.K., Thrun, S., Mitchell, T.: Text classification from labeled and unlabeled documents using EM. Mach. Learn. **39**(2–3), 103–134 (2000)

A Novel Hybrid GWO-FPA Algorithm for Optimization Applications

Jeng-Shyang Pan[1], Thi-Kien Dao[3], Shu-Chuan Chu[2], and Trong-The Nguyen[3(✉)]

[1] Fujian Provincial Key Laboratory of Big Data Mining and Applications, College of Information Science and Engineering, Fujian University of Technology, Fuzhou, China
jengshyangpan@fjut.edu.cn

[2] School of Computer Science, Engineering and Mathematics, Flinders University, Adelaide, Australia
jan.chu@flinders.edu.au

[3] Department of Information Technology, Haiphong Private University, Haiphong, Vietnam
jvnkien@gmail.com, vnthe@hpu.edu.vn

Abstract. The recent trend of research is to hybridize two or several numbers of variants to find out the better quality of solution in practical optimization applications. In this paper, a new approach hybrid Grey Wolf Optimizer (GWO)-Flower Pollination Algorithm (FPA) is proposed based on the combination of exploitation phase in GWO and exploration stage in FPA. The hybrid proposed GWOFPA improves movement directions and speed of the grey wolves in updating positions of FPA. The simulation uses six benchmark tests for evaluating the performance of the proposed method. Compared other metaheuristics such as Particle Swarm Optimization (PSO), FPA, and GWO, the simulation results demonstrate that the proposed approach offers the better performance in solving optimization problems with or without unknown search areas.

Keywords: Optimization · Hybrid GWO-FPA algorithm · Grey Wolf Optimizer · Flower Pollination Algorithm

1 Introduction

Highly efficient technique in searching the best possible results in the benchmark and practical applications is the global optimization method [1]. In optimization, only a few results are compared the best one which is known as the goal. Classical optimization approaches have some deficiencies of finding the optimal global solutions of optimization problems [2]. These shortcomings are primarily interdependent on their original search systems. These conventional algorithms are strongly under effects of choosing proper types of variables, objectives and constraints functions. They also do not grant a universal solution method that can be applied to find the global optimal solution of the duties were several types of constrained functions, variables, and objective are used [3]. For covering these deficiencies, a new technique with the name

J.-S. Pan et al. (eds.), *Advances in Smart Vehicular Technology, Transportation, Communication and Applications*, Smart Innovation, Systems and Technologies 86,
https://doi.org/10.1007/978-3-319-70730-3_33

of metaheuristics was originated, which is mainly developed from artificial intelligence research that originated by scientists or researchers. Nature-inspired methods are developed for solving the several types of hard global optimization functions without having to the full accommodate to each function.

Recently, scientists and scholars have developed several numbers of metaheuristics in order to find the best global optimal solution of benchmark and real-life applications. The nature-inspired techniques have been originated, some of them are Particle Swarm Optimization (PSO) [4], Genetic Algorithm (GA) [5], Grey Wolf Optimization (GWO) [6], Flower Pollination Algorithm (FPA) [7], Bat Algorithm (BA) [8]. Moreover, in the case of the hybrid convergence, nature-inspired algorithm hybridizations using batch modeling are combinations amid evolutionary techniques and methods of neighborhood or course [9].

This paper introduces a new hybrid model combining Grey Wolf Optimizer (GWO) and Flower Pollination Algorithm (FPA) named GWOFPA. The proposed algorithm comprises of best characteristics of both GWO and FPA. The performance of the proposed variant is tested on six standard benchmarks. The solutions are compared relying on the metaheuristics reported in the review of the literature.

The rest of paper is organized as follows. Sections 2 and 3 review the GWO and FPA respectively. Section 4 presents the proposed GWOFPA approach. Section 5 discusses the simulation results. Section 6 gives the conclusion.

2 Grey Wolf Optimizer (GWO)

A new population-based nature-inspired algorithm called Grey Wolf Optimization (GWO) was developed by Mirjalili et al. [6]. GWO approach mimics the hunting behavior and social leadership of grey wolves in nature. Four types of grey wolves such as alpha, beta, delta, and omega are employed for simulating the command hierarchy. The first three best position (fittest) wolves are indicated as α, β and δ who guide the other wolves (ω) of the groups toward promising areas of the search space. The position of each wolf of the group is updated using the following mathematical equations:

The *encircling behavior* of each agent of the crowd is calculated by the following mathematical equations:

$$\vec{d} = \left| c \cdot \vec{x}_p^t - \vec{x}^t \right| \tag{1}$$

$$\vec{x}^{t+1} = \vec{x}_p^t - \vec{a} \cdot \vec{d} \tag{2}$$

The vectors a and c are formulate as below:

$$\vec{a} = 2l \cdot r_1 \tag{3}$$

$$\vec{c} = 2 \cdot r_2 \tag{4}$$

Hunting: In order to mathematically simulate the hunting behavior, we suppose that the alpha, beta and delta have better knowledge about the potential location of prey. The following equations are developed in this regard.

$$\vec{d}_\alpha = |\vec{c}_1 \cdot \vec{x}_\alpha - \vec{x}|,\ \vec{d}_\beta = |\vec{c}_2 \cdot \vec{x}_\beta - \vec{x}|,\ \vec{d}_\delta = |\vec{c}_3 \cdot \vec{x}_\delta - \vec{x}| \tag{5}$$

$$\vec{x}_1 = \vec{x}_\alpha - \vec{a}_1 \cdot (\vec{d}_\alpha),\ \vec{x}_2 = \vec{x}_\beta - \vec{a}_2 \cdot (\vec{d}_\beta),\quad \vec{x}_3 = \vec{x}_\delta - \vec{a}_3 \cdot (\vec{d}_\delta) \tag{6}$$

$$\vec{x}^{t+1} = \frac{\vec{x}_1 + \vec{x}_2 + \vec{x}_3}{3} \tag{7}$$

Search for prey and attacking prey:

The $\vec{a}$ is random value in the gap $[-2a, 2a]$. When random value $|\vec{a}| < 1$ the wolves are forced to attack the prey. Searching for prey is the exploration ability and attacking the prey is the exploitation ability. The arbitrary values of $\vec{a}$ are utilized to force the search to move away from the prey. When $|\vec{a}| > 1$, the members of the population are enforced to diverge from the prey.

3 Flower Pollination Algorithm (FPA)

The biological flower pollination inspires a new population-based algorithm called FPA [7]. The pollination rules state in FPA as follows. The global pollination considered cross-pollination that pollinators obey Lévy flights. The local pollination is known as self-pollination. The reproduction probability recognized flower constancy which is proportional to the resemblance of the two flowers in concerned. FPA used a switching probability p ∈ [0, 1] to control between the local and global pollination. Assumed, FPA considered as global and local pollination. Thus the local pollination is modeled as follows.

$$\vec{x}_{ij}^{t+1} = \vec{x}_{ij}^t + u \times (\vec{x}_{ih}^t - \vec{x}_{ik}^t) \tag{8}$$

where $\vec{x}_{ih}^t$, $\vec{x}_{ik}^t$ are pollen of different flowers but they are in the same plant species. u is generated from the uniform distribution [0, 1]. A random walk for local process if x_{ih}^t and x_{ik}^t come from the same species or selected from the same population of plants.

Pollens of the flowers in the global pollination are moved by pollinators e.g. insects, and pollens can be carried for a long distances. This process guarantees pollination and reproduction of the fittest solution represented as. The flower constancy is expressed mathematically as:

$$\vec{x}_{ij}^{t+1} = \vec{x}_{ij}^t + \gamma \times L(\lambda) \times \left(\vec{x}_{ij}^t - g^*\right) \tag{9}$$

where x_i is solution vector at iteration t, and γ is a scaling factor to control the step size. Lévy flight can be used to mimic the characteristic transporting of insects over a long distance with various length steps, thus, $L > 0$.

$$L = \frac{\lambda\Gamma(\lambda) \times \sin\left(\frac{\pi\lambda}{2}\right)}{\pi \times s^{i+\lambda}}, (s \gg s_0) \tag{10}$$

where $\Gamma(\lambda)$ is the standard gamma function, and this distribution is valid for large steps $s > 0$. A variable p is switching probability or the proximity probability that can be used to change the global pollination to intensive local pollination and reverse.

4 Hybrid Grey Wolf Optimizer-Flower Pollination Algorithm (GWOFPA)

This section presents the implementation of the low-level coevolutionary to hybridize GWO and FPA by merging the functionality of both approaches. Two different techniques including exploitation and exploration are involved in generating final optimal solution to the optimization problem. By this modification, we extend the performance of exploitation in GWO with the fulfillment of the exploration in FPA to produce both approaches' strength. However, FPA applies the exploration phase to updating positions.

$$\vec{d}_\alpha = \begin{cases} u \times (|\vec{c}_1 \times \vec{x}_\alpha - \vec{x}|), & rand() < 0.5 \\ \gamma \times L(\lambda) \times (|\vec{c}_1 \times \vec{x}_\alpha - \vec{x}|), & rand() \geq 0.5 \end{cases} \tag{11}$$

where $\vec{d}_\alpha$ is a modified dominance coefficient from Eq. (1) is to exploit diversity search agent for the proposed algorithm. The same done with the alpha, beta, and omega are applied in Eq. (5). New agent position can be mathematically simulated as follows.

$$\vec{x}_1 = \vec{x}_\alpha - \vec{a}_1 \times \left(\vec{d}_\alpha\right) \tag{12}$$

where $\vec{x}_1$ is done first agent position. Do the same for the second and third locations ($\vec{x}_2$ *and* $\vec{x}_3$ from Eq. (5). Figure 1 shows the pseudocode of the GWOFPA algorithm.

```
GWOFPA algorithm
Initialization the parameters such as:
population X_i(i = 1,2,..N), d,a and c
Compute the fitness of each search member X_α, X_β, X_δ
        // the first, second & third best search member
while (t <Max. iter)
    for every search member
        Get the current position by Eqs. (2) and (7)
    end for
      Get a , c, d by Eqs. (3)(4)and (11)
      Calculate the fitness of all search member
      Update X_α, X_β, X_δ by Eq.(12)
      t = t+1
end while
return X_α
```

Fig. 1. Pseudo code of the GWOFPA algorithm

5 Experimental Results

In this section, we evaluate the proposed GWOFPA approach performance quality by executing a set of benchmark problems to test the solution quality, solution stability, convergence speed and ability to find the global optimum. Twenty runs of the testing functions in the experiments outcome values are averaged with different random seeds.

Table 1 lists the initialization for a set of the trial functions [10]. Maxgen column in Table 1 is a maximum number of iterations (it can be set to 500, 1000, 1500,.., 10000).

Table 1. Initialization for dimensions, max generations, and boundaries of the testing functions

Testing problems	Bounds	Dims	Maxgen
$f_1(x) = \sum_{i=1}^{n} \sin(x_i) \cdot (\sin(\frac{ix_i^2}{\pi}))^{2m},\ m = 10$	±500	30	2000
$f_2(x) = \sum_{i=1}^{n} [x_i^2 - 10\cos(2\pi x_i) + 10]$	±5.12	30	2000
$f_3(x) = \sum_{i=1}^{n} -x_i \sin(\sqrt{\lvert x_i \rvert})$	0, π	30	2000
$f_4(x) = [e^{-\sum_{i=1}^{n} (x_i/\beta)^{2m}} - 2e^{-\sum_{i=1}^{n} x_i^2}] \prod_{i=1}^{n} \cos^2 x_i,\ m = 5$	±20	30	2000
$f_5(x) = -\sum_{i=1}^{4} c_i \exp(-\sum_{j=1}^{6} a_{ij}(x_j - p_{ij})^2)$	0, 10	4	2000
$f_6(x) = -\sum_{i=1}^{5} [(X - a_i)(X - a_i)^T + c_i]^{-1}$	0, 10	4	2000

Let $X = \{x_{i1}, x_{i2}, \ldots, x_{im}\}$ be the real value vectors of m-dimensional for GWOFPA. We set the population size N (N can be set to 10, 20,.., 100) for the algorithms of GWOFPA, GWO, FPA, and PSO to 40 for all runs in the experiments. Table 2 displays the outcome of implement for testing problems in Table 1 for the optimization is to maximize and minimize the results.

Table 2. The obtained results of proposed GWOFPA, GWO, and FPA for the benchmark functions

$f_i(x)$	PSO		GWO		FPA		**GWOFPA**	
	f_{min}	f_{max}	f_{min}	f_{max}	f_{min}	f_{max}	f_{min}	f_{max}
1	1.61E+01	6.87E+04	8.80E−06	4.87E+04	7.96E−06	8.06E+04	**6.39E−06**	**7.89E+04**
2	1.39E+01	7.41E+14	7.50E−02	1.23E+12	7.27E−02	2.13E+11	**6.10E−02**	**5.71E+12**
3	3.25E+03	1.09E+05	3.96E−01	1.25E+05	4.12E−01	2.76E+05	**3.48E−01**	**1.12E+05**
4	1.79E+01	8.67E+01	2.02E+00	7.97E+01	2.00E+00	8.62E+01	**1.71E+00**	**8.65E+01**
5	8.88E+02	3.09E+08	3.23E+01	1.08E+08	3.88E+01	2.53E+08	**2.72E+01**	**2.54E+08**
6	9.09E+01	7.38E+04	4.36E+00	4.62E+04	4.79E+00	7.55E+04	**3.96E+00**	**6.63E+04**
AVG	1.08E+01	1.30E+02	6.53E+00	1.38E+02	7.68E+00	1.16E+02	**5.55E+00**	**1.29E+02**

We tested each benchmark function with 2000 iterations per a run. We also compare the simulation results of the proposed method with those obtained results of the previous algorithms such as the GWO, FPA, and PSO as shown in Tables 3 and 4. For further parameters setting could be found in [6, 7] with the initial range, the dimension and total iterations for all test functions in Table 1 for GWO, FPA, and PSO.

Table 3. Comparison of the the proposed GWOFPA with GWO, and FPA, with for solving the testing problems in term of the quality performance evaluation

Testing functions	Obtained results			Comparison	
	GWO	FPA	**GWOFPA**	with GWO	with FPA
1	8.80E−06	7.96E−06	6.39E−06	27%	20%
2	7.50E−02	7.27E−02	6.10E−02	19%	16%
3	3.96E−01	4.12E−01	3.48E−01	12%	16%
4	2.02E+00	2.00E+00	1.71E+00	15%	14%
5	3.23E+01	3.88E+01	2.72E+01	16%	30%
6	4.36E+00	4.79E+00	3.96E+00	9%	17%
Avge	**6.53E+00**	**7.68E+00**	**5.55E+00**	**16%**	**19%**

Table 4. Comparison of the proposed algorithm quality performance with PSO for testing problems

Testing functions	Execution time		Comp. times	Performances		Comp. qualities
	PSO	GWOFPA		PSO	GWOFPA	
1	1.4793	1.4371	2%	1.51E+00	1.22E+00	23%
2	1.8672	1.8686	3%	−4.17E+03	−5.09E+03	18%
3	1.9871	1.9503	4%	1.69E+02	1.37E+02	23%
4	0.7776	0.7834	1%	2.70E-03	2.50E-03	8%
5	1.8918	1.9072	2%	−2.34E+00	−3.09E+00	24%
6	1.9891	1.9792	1%	−8.15E+00	−9.82E+00	17%
Avg	**1.1703**	**1.1543**	**2%**	**−6.68E+02**	**−8.27E+02**	18%

Figures 2 and 3 show the experimental results for the first two benchmark functions over 20 times output obtained from the proposed GWOFPA, FPA, GWO, and PSO methods with the same iteration of 2000. Appearly, these figures show all of the cases of testing functions in the GWOFPA have performance quality higher the other algorithm regarding the accuracy and convergence rate.

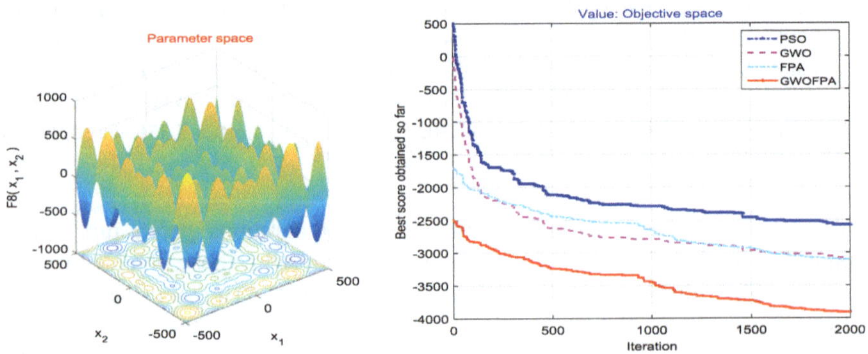

Fig. 2. Comparison the experimental result curves of the proposed GWOFPA with PSO, FPA, and GWO for the testing function f1

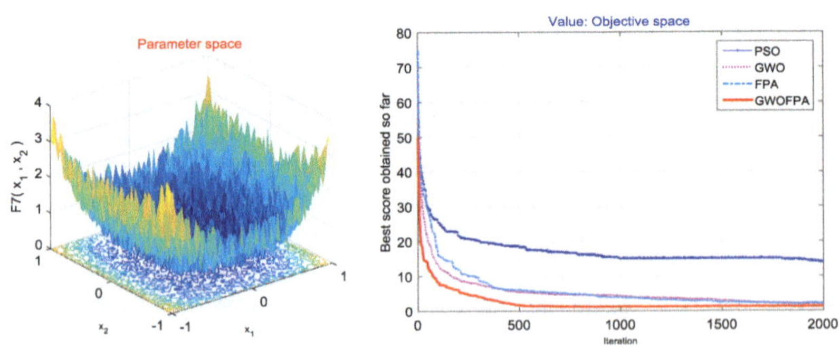

Fig. 3. Comparison the experimental result curves of the proposed GWOFPA with PSO, FPA, and GWO for the testing function f2

Table 3 displays the performance of the proposed GWOFPA in comparison with the algorithms of GWO, FPA for the testing functions. The result of the proposed algorithm on all of these cases of testing shows that the proposed algorithm provides 27% and 30% higher than those obtained from primary methods of GWO and FPA respectively. However, the figure for the minimum cases is only the increase 9% and 14% than the GWO and FPA respectively for a set of testing functions. In general, the proposed GWOFPA increases the average values of the cases 16% and 19% than obtained from the GWO and FPA methods are respectively for testing problems in terms of the convergence rates.

Table 4 displays the performing quality and running time comparison of the proposed GWOFPA with PSO method for the testing problems. The columns of correlation times and conditions are calculated as absolute of the obtained from GWOFPA minus that got from PSO then divided the received value of the GWOFPA method. The results of the proposed method on all of these cases of testing multimodal benchmark problems show that GWOFPA process almost increases higher quality and shorter running time than those obtained from PSO method. In general, the proposed algorithm achieved the standard cases of various tests for the convergence, and accuracy increased more than those obtained from the PSO method is 18%, and for the speed faster than that got from PSO method is 2% average respectively.

6 Conclusion

In this paper, we presented a novel hybrid approach for the optimization applications based on a combination of Grey Wolf Optimizer (GWO) and Flower Pollination Algorithm (PFA), namely GWOFPA. We use the location update equation of FPA for updating the positions of the grey wolves in GWO to explore and exploit the diversity of the algorithm efficiently. In the simulation, a set of functions are applied to verify the accuracy, convergent behavior, best global optimal solution of the newly developed approach. Results reveal that the proposed approach provides highly competitive solutions as compared to other algorithms.

References

1. Nguyen, T.-T., Pan, J.-S., Chu, S.-C., Roddick, J.F., Dao, T.-K.: Optimization localization in wireless sensor network based on multi-objective firefly algorithm. J. Netw. Intell. **1**, 130–138 (2016)
2. Dao, T., Pan, T., Nguyen, T.: A compact artificial bee colony optimization for topology control scheme in wireless sensor networks. J. Inf. Hiding Multimed. Sig. Process. **6**, 297–310 (2015)
3. Boussaïd, I., Lepagnot, J., Siarry, P.: A survey on optimization metaheuristics. Inform. Sci. **237**, 82–117 (2013)
4. Eberhart, R., Kennedy, J.: A new optimizer using particle swarm theory. In: Proceedings of the Sixth International Symposium on Micro Machine and Human Science, MHS 1995, pp. 39–43. IEEE, New York (1995)
5. Srinivas, M., Patnaik, L.M.: Genetic algorithms: a survey. Computer (Long. Beach. Calif). **27**, 17–26 (1994)
6. Mirjalili, S., Mirjalili, S.M., Lewis, A.: Grey Wolf Optimizer. Adv. Eng. Softw. **69**, 46–61 (2014)
7. Yang, X.S.: Flower pollination algorithm for global optimization. In: Lecture Notes in Computer Science (including subseries Lecture Notes in Artificial Intelligence and Lecture Notes in Bioinformatics), pp. 240–249 (2012)
8. Yang, X.S.: A new metaheuristic bat-inspired algorithm. In: González, J., Pelta, D., Cruz, C., Terrazas, G., Krasnogor, N. (eds.) Studies in Computational Intelligence, pp. 65–74. Springer, Heidelberg (2010)
9. Pan, T.-S., Dao, T.-K., Nguyen, T.-T., Chu, S.-C.: Hybrid particle swarm optimization with bat algorithm. In: Advances in Intelligent Systems and Computing, pp. 37–47 (2015)

Seam Carving Algorithm Based on Saliency

Wen Lin[1], Fuquan Zhang[2,3](✉), Renbao Lian[1], Lin Xu[4], Xueyun Chen[5], and Linghong Kuang[6]

[1] School of Electronic Information Science, Fujian Jiangxia University, Fuzhou University Town, Fuzhou 350108, China
2826550@qq.com, 47872247@qq.com

[2] School of Software, Beijing Institute of Technology, Beijing 100081, China
8528750@qq.com

[3] Fujian Provincial Key Laboratory of Information Processing and Intelligent Control, Minjiang University, Fuzhou 350121, China

[4] Innovative Information Industry Research Institute, Fujian Normal University, Fuzhou 350300, China
xulin@fjnu.edu.cn

[5] Institute of Information Engineering, Longyan University, Town DongXiao, XinLuo District, Longyan 364012, China
6288024@qq.com

[6] School of Information Science and Engineering, Fujian University of Technology, Fuzhou University Town, Fuzhou 350118, China
2852183@qq.com

Abstract. This paper proposes a seam carving algorithm based on saliency. This algorithm makes saliency detection to the source image. Images are classified according to the gray-scale of saliency detection. Adding different energy protection methods, it can protect the foreground area of a subject and detailed areas of the edge. Then the cumulative energy map is calculated. According to the principle of deleting the minimum energy pixel, single pixel-wide carves chosen to be deleted or copied. The experimental result shows that the algorithm avoids image restoration caused by extracting too many pixels in saliency area and non equal-ratio scaling. It reduces the deformed structures caused by displacement of edge pixel and also improves the integrity of algorithm.

Keywords: Image resizing · Seam carving · Saliency · Classification · Pixel energy

1 Introduction

1.1 Background

Numerous digital images are frequently used in daily life with the rapid development of multimedia technology and display devices. Different display devices have distinct sizes and aspect ratios. The images will be resized and image resizing should keep the most important content as far as possible. Ordinary image resizing methods are nearest

J.-S. Pan et al. (eds.), *Advances in Smart Vehicular Technology, Transportation, Communication and Applications*, Smart Innovation, Systems and Technologies 86,
https://doi.org/10.1007/978-3-319-70730-3_34

neighbor interpolation and linear interpolation [1]. They are more suitable for the images which are same aspect ratios. Hence seam carving for content-aware image resizing is presented by Avidan et al.

1.2 Related Work

A seam is an optimal path of pixels on a single image from left to right, or top to bottom. The optimality of the seam is defined by energy function of the image. We change the aspect ratio of the image by carving out or inserting seams repeatedly. Seam caving supports both reduction and expansion for image [2]. After that, more research works about image resizing are conducted. Optimized image resizing using seam carving and scaling is presented by Dong et al. The method operates by joint use of seam carving and image scaling. It preserves both the important regions and the global visual effect [3]. Huang et al. present a real-time resizing method for content-aware image [4]. It reduces the time of looking for the seam. Liu et al. using saliency-based continuous seam carving to efficiently display images on small screens [5]. Frankovich and Wong via integration of energy gradient function to retain high image detail concentration [6]. Noh and Han proposes a new energy function for seam carving based on forward gradient differences to preserve regular structures in images [7]. Cao et al. improve the image quality and avoid removing the seams adjacently via strip constraints [8]. Wang and Akleman simply uses the original energy function of seam carving. They resize image with cuts in continuous domain [9]. Shen et al. presents a method which achieves a better seam carving performance by cutting the near objects less seams while removing distant objects more seams [10]. Yue et al. proposes improved seam carving for stereo image resizing. The method protects the important content, reduces the visual distortion and guaranteed the geometric consistency of resized stereo images [11]. Dekel et al. proposes a novel method for retargeting a pair of stereo images [12]. Wang and Yuan presents a high quality image resizing. It retains the details when stretching an image and a further extension on video enlargement is also presented as an example [13]. Seam carving is also used for media retargeting by Shamir and Avidan [14]. Zhang et al. proposes a method allows distortion to be diffused better in all directions, and important image edges are well-preserved [15]. Geometry seam carving is presented by Dekkers et al. The method enables the deformation of a given mesh in a way that causes stronger distortion in homogeneous mesh regions while salient features are preserved much better [16].

2 Overview

Biologic visual system research indicates that humans have a visual attention mechanism and it is selective. It is caused by the saliency of vision which is great different from the background and then selected according to the subjective consciousness of the observer [17].

Experimental results show that seam carving is effective to the images without visual focus. It uses the same energy map for all images. Hence the resizing result loses much important information about the visual focus if the image energy is concentrated and the seam goes through the salient area. It also loses much important information about edge if the image energy is scattered and it includes much slender vertical or horizontal region [18].

This paper proposes a seam carving algorithm based on saliency and classification. Images can be classified into two types according to visual sense. One is that having obvious visual focus and the image energy is concentrated, the other is that having no obvious visual focus and the image energy is scattered, using different methods to protect the image energy. When the image energy is concentrated, the salient area should be protected. When the image has no visual focus or there is something long or narrow, the information about edge should be protected. Hence visual focus and details can be obtained more. Then, seam carving is applied to resized the image.

3 Seam Carving Algorithm Based on Saliency

3.1 Saliency Detection

Image saliency refers to the difference between the salient object and the background in the image. It describes the ability of object to be prominent from the background. The purpose of saliency detection is to generate a saliency map as gray scale map. It shows the importance level by different gray scale. The gray scale is higher, the saliency is more obvious. And it has more possibility to belong to the saliency area [19].

By analyzing the log-spectrum of an input image, Hou and Zhang extracts the spectral residual of an image in spectral domain, and proposes a fast method to construct the corresponding saliency map in spatial domain [20]. Achanta et al. presents a method for salient region detection that outputs full resolution saliency maps with well-defined boundaries of salient objects [21]. Zhang et al. presents a bayesian framework for saliency using natural statistics [22]. Cheng et al. proposes regional contrast based saliency extraction algorithm, which simultaneously evaluates global contrast differences and spatial coherence. The proposed algorithm is simple, efficient, and yields full resolution saliency maps [23].

In this paper, the salient region is checked by the algorithm based on global contrast. Figure 1(b) is the saliency map of Fig. 1(a).

(a) Original Image (b) Saliency Map (c) Blocks

Fig. 1. Examples of saliency map and block partition

3.2 Image Classification

Images are classified into two types according to the saliency maps. Each of them uses different method to protect the image energy.

Global Objective Function. Suppose that the size of original image is n × m. Either m or n is divided by 10, the remainder is 0. The image S is divided into i × j blocks. The width of each block is parameter a. The blocks compose a set Z. As shown in the Fig. 1(c). The calculation procedure of target parameter λ will be shown.

1. Calculate the percentage of energy for each block.

$$EN(i,j) = Z(i,j) / \sum_{i=1}^{n} \sum_{j=1}^{m} S(n,m) \tag{1}$$

   Z (i, j) is the sum of energy of all the pixel in each block.
2. Calculate the average energy of each block.

$$AE = \sum_{i=1}^{n} \sum_{j=1}^{m} S(n,m)/(i \times j) \tag{2}$$

3. Calculate P. Iterate through EN, get the number of the block which energy is higher than verified ratio k × AE.
4. Calculate λ, which is the percentage of P.

$$\lambda = P/(i \times j) \tag{3}$$

5. The image is classified according to the weight W. W is got through a large number of experiments with different parameters. Analyze the results of λ and the subjective consciousness of vision.

Training Data. Some experimental data are listed below. NO. 1 and NO. 2 are the images without visual focus, NO. 3 and NO. 4 are the images with visual focus (Tables 1, 2 and 3).

Table 1. The result of k and λ when a = 10

λ \ k	1	1.5	1.6	1.7	1.8	1.9	2
1	0.383	0.229	0.220	0.162	0.025	0.018	0.009
2	0.423	0.114	0.097	0.073	0.066	0.064	0.061
3	0.228	0.183	0.178	0.173	0.169	0.167	0.163
4	0.275	0.215	0.211	0.207	0.196	0.186	0.181

Table 2. The result of k and λ when a = 15

λ \ k	1	1.5	1.6	1.7	1.8	1.9	2
1	0.217	0.190	0.185	0.179	0.176	0.173	0.173
2	0.231	0.201	0.201	0.190	0.190	0.183	0.176
3	0.442	0.238	0.210	0.183	0.165	0.153	0.139
4	0.449	0.217	0.186	0.163	0.123	0.080	0.069

Table 3. The result of k and λ when a = 20

λ \ k	1	1.5	1.6	1.7	1.8	1.9	2
1	0.224	0.188	0.182	0.182	0.172	0.167	0.167
2	0.250	0.188	0.188	0.181	0.181	0.169	0.169
3	0.436	0.158	0.127	0.115	0.103	0.079	0.073
4	0.439	0.139	0.114	0.075	0.068	0.057	0.036

Compare the experimental data and the results of the subjective consciousness of vision, the accuracy of image classification is highest when a = 10, k = 1.9 and w = 1.5 are set.

3.3 Energy Protection

Protect the area which has great influence on human vision by increasing the energy of it in gradient map.

When λ > w, there is visual focus in the image, people pay more attention to the focus. The gradient map was got from the operator of Canny. Superposed the gradient map and saliency map, an updated gradient map is formed. It can protect the visual focus region, but it will pay a price of losing some pixel which is less concerned. After that, energy map with the updated gradient map is got.

Figure 2 shows the comparison before and after the saliency map is superposed. Figure 2(a) and (b) are the gradient maps and energy maps of seam carving. Figure 2(c) and (d) are the updated gradient maps and energy maps of new algorithm.

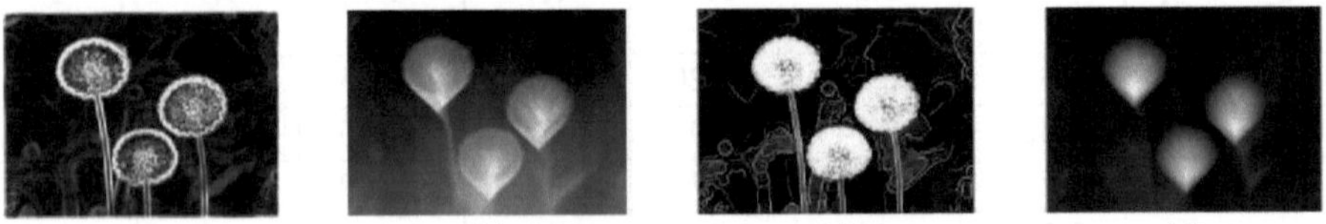

(a) (b) (c) (d)

Fig. 2. The comparison chart before and after protection

When $\lambda \le w$, seam carving is effective. But it is easy to loss details if there is high frequency and vertical region. We get the gradient map by the operator of Sobel. Get an updated gradient map by protecting the details of edge through dilation. After that, an energy map with the updated gradient map is got.

Linear structuring element is selected in this paper and the parameter is 3.

Figure 3 shows the comparison before and after dilation. Figure 3(a) and (b) are the gradient maps and energy maps of seam carving, Fig. 3(c) and (d) are the updated gradient maps and energy maps of new algorithm.

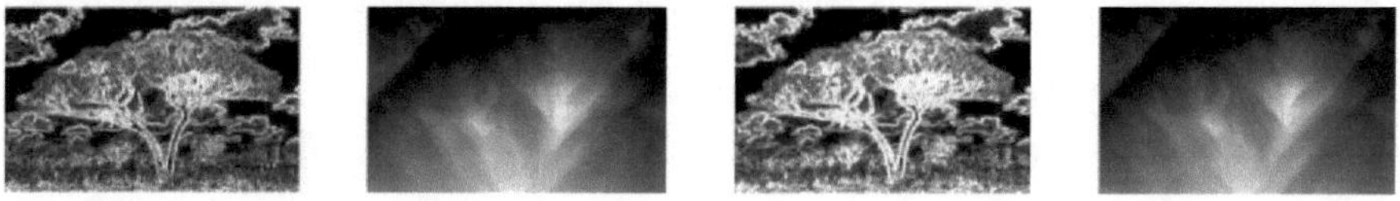

(a) (b) (c) (d)

Fig. 3. The comparison chart before and after dilation

3.4 Image Resizing

Let I (original image) be an n × m image and define a vertical seam to be:

$$s^x = \{s_j^x\}_{j=1}^n = \{(x(i), i)\}_{j=1}^n, s.t.\forall i, |x(i) - x(i-1)| \le 1 \quad (4)$$

The horizontal seam to be:

$$s^y = \{s_j^y\}_{j=1}^m = \{(j, y(j))\}_{j=1}^m, s.t.\forall j, |y(j) - y(j-1)| \le 1 \quad (5)$$

In order to resize the original image to n × k, it will delete m-k seams or insert k-m seams.

At first, get the saliency map S of original image then calculate the target parameter λ according to S. The gradient map is got through canny operator or Sobel operator according to λ and w. When the updated gradient map is formed, superposed saliency map or dilation according to λ and w. Finally, obtain energy map, removes low energy pixels and keeps the high energy ones by seam carving for content-aware image resizing.

4 Experiment

After comparing the result with the seam carving algorithm, the newly applied algorithm can be further validated. Parts of the result are shown as following.

4.1 Comparison of the Images with Visual Focus and the Energy Is Concentrated

Figure 4 is the original and resized images when $\lambda > 0.15$. Figure 4(a) is original images; Fig. 4(b) and (d) are the resized images by single seam carving; Fig. 4(c) and (e) are the resized images by the updated algorithm.

Fig. 4. Resized images by single seam carving and the updated algorithm

Mushrooms and lotus leaves are distorted and there are zigzag cracks on the stalks of dandelions and the necks of flamingos in the resized image by single seam carving because of the lost details. The updated algorithm protects the details and avoids the distortion effectively. Figure 5(a) is the details by single seam carving; Fig. 5(b) is the resized image by the updated algorithm.

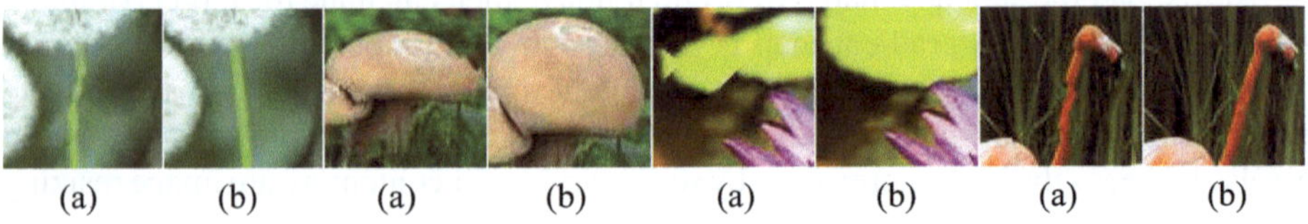

Fig. 5. Magnified image of details

4.2 Comparison of the Images Which the Energy Is Scattered

Figure 6 is the original and resized images when $\lambda \leq 0.15$. Figure 6(a) is original images; Fig. 6(b) and (d) are the resized images by single seam carving; Fig. 6(c) and (e) are the resized images by the updated algorithm.

The results of the updated algorithm retain more details about foreground and edge like the trunk and the bridge. It also proves that the updated algorithm retains more

Fig. 6. Resized image by single seam carving and the updated algorithm

details and avoids the distortion effectively. Figure 7(a) is the details by single seam carving; Fig. 7(b) is the resized image by the updated algorithm.

Fig. 7. Magnified image of details

4.3 Experimental Data

Different size of images are chosen randomly. Resize the images by using single seam carving and the updated algorithm. Compare the results to the subjective judgment of human mind. 7% of them do not meet the demand of human visual perception, and the effect of 16% improved ones is not obvious, 77% is effective.

4.4 Comparison of Running Time

Running time of image resizing by single seam carving and the updated algorithm is compared. The result shows that the running time of the updated algorithm can satisfy human's higher visual requirement. However, the average running time is 17.97% more than the original one. Figure 8 is the chart of the percentage of running time difference.

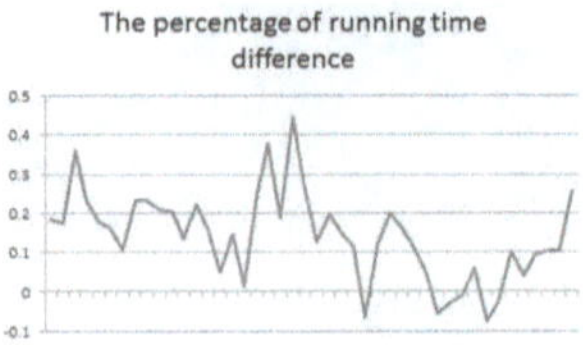

Fig. 8. The percentage of running time difference

4.5 Limitations

Figure 9 shows the images which improvement effect is not obvious.

(a1) (b1) (c1) (a2) (b2) (c2)

Fig. 9. The images which improvement effect is not obvious

The limitations of the new algorithm are caused by two reasons: (1) Classification error. Figure 9(a1) is an image with visual focus, but it is distinguished that the energy is scattered because the focus is too big. Figure 9(b1) is the result. If the classification is correct, Fig. 9(c1) will be got. (2) When the energy of the image concentrate in a narrow region, dilation is more suitable for the image. But if $\lambda > 0.15$ in the experiment, it superposed the gradient map and saliency map. Figure 9(b2) is the result. It loses many details about edge. If dilation is used to protect the image energy, more details of edge will be retained and get Fig. 9(c2).

5 Conclusions and Future Work

This paper presets a seam carving algorithm based on saliency. This algorithm makes saliency detection to the source image. Images are classified according to the gray-scale of saliency detection for the first time. Adding different energy protection methods, it can protect the foreground area of a subject and detailed areas of the edge. The results shows that the updated algorithm can avoid distortion and retain more details. But the algorithm still has limitations and the running time is longer than single seam carving.

In the future, the method of image classification will be improved in order to improve the classification accuracy, search for better energy protection methods and reduce the complexity of the algorithm to reduce the running time.

Acknowledgment. The paper is supported by the foundation of Technologies Research of the Education Department of Fujian Province No. JA15535. The work of Fuquan Zhang was supported by the Scientific Research Foundation of Minjiang University (No. MYK17021). Thanks to Pro. Shiming Hu and Dr. Mingming Chen.

References

1. Gonzalez, R.C., Woods, R.E.: Digital Image Processing. Publishing House of Electronic Industry, Beijing (2017)
2. Avidan, S., Shamir, A.: Seam carving for content-aware image resizing. ACM Trans. Graph. **26**(3) (2007). Article No. 10
3. Dong, W., Zhou, N., Paul, J., et al.: Optimized image resizing using seam carving and scaling. ACM Trans. Graph. **28**(5), 1–10 (2009)
4. Huang, H., Fu, T.N., Rosin, P.L., et al.: Real-time content-aware image resizing. Sci. China Ser. F Inform. Sci. **52**(2), 172–182 (2009)
5. Liu, Z., Yan, H., Ngan, K.N., et al.: Adaptive image retargeting using saliency-based continuous seam carving. Opt. Eng. **49**(1), 017002 (2010)
6. Frankovich, M., Wong, A.: Enhanced seam carving via integration of energy gradient functionals. IEEE Sig. Process. Lett. **18**(6), 375–378 (2011)
7. Noh, H., Han, B.: Seam carving with forward gradient difference maps. In: Multimedia: Proceedings of the 20th ACM International Conference, pp. 709–712 (2012)
8. Cao, L., Wu, L., Wang, J.: Fast seam carving with strip constraints. In: Internet Multimedia Computing and Service: Proceedings of the 4th International Conference (ICIMCS 2012), pp. 148–152 (2012)
9. Wang, Y., Akleman, E.: Calligraphic cutting: extreme image resizing with cuts in continuous domain. In: ACM SIGGRAPH 2012 Talks (SIGGRAPH 2012), p. 1 (2012)
10. Shen, J., Wang, D., Li, X.: Depth-aware image seam carving. IEEE Trans. Cybern. **43**(5), 1453–1461 (2013)
11. Yue, B., Hou, C., Zhou, Y.: Improved seam carving for stereo image resizing. EURASIP J. Wirel. Commun. Netw. **2013**(1), 1–6 (2013)
12. Dekel, T., Moses, Y., Avidan, S.: Stereo seam carving a geometrically consistent approach. IEEE Trans. Pattern Anal. Mach. Intell. **35**(10), 2513–2525 (2013)
13. Wang, Q., Yuan, Y.: High quality image resizing. Neurocomputing **131**, 348–356 (2014)
14. Shamir, A., Avidan, S.: Seam carving for media retargeting. Commun. ACM **52**(1), 77–85 (2009)
15. Zhang, G., Cheng, M., Hu, S., et al.: A shape-preserving approach to image resizing. Comput. Graph. Forum **28**(7), 1897–1906 (2009)
16. Dekkers, E., Kobbelt, L.: Geometry seam carving. Comput. Aid. Des. **46**, 120–128 (2014)
17. Pashler, H.E.: Attention. Psychology Press, Philadelphia (1998)
18. Nie, D.-D., Ma, Q.-Y.: Seam carving algorithm based on global image structure. Appl. Res. Comput. **28**(3), 1198–1200 (2011)
19. Xie, Y.-L.: Saliency detection based on bayesian framework. DaLian University of Technology (2011)
20. Hou, X., Zhang, L.: Saliency detection: a spectral residual approach. In: IEEE Conference on Computer Vision and Pattern Recognition (2007)
21. Achanta, R., Hemami, S., Estrada, F., et al.: Frequency-tuned salient region detection. In: IEEE Conference on Computer Vision and Pattern Recognition (2009)
22. Zhang, L., Tong, M., Marks, T., et al.: Sun: a bayesian framework for saliency using natural statistics. J. Vis. **8**(7), 1–20 (2008)
23. Cheng, M., Zhang, G., Mitra, N.J., et al.: Global contrast based salient region detection. In: IEEE Conference on Computer Vision and Pattern Recognition (2011)

Applied Mathematics and Its Applications

Time Domain Acoustic Scattering from Locally Perturbed Flat Substrates

Bo Chen[1(✉)], Fuquan Zhang[2,3], Lin Xu[4], and Minghui Liu[5]

[1] College of Science, Civil Aviation University of China, Tianjin, China
charliecb@163.com
[2] Fujian Provincial Key Laboratory of Information Processing and Intelligent Control, Minjiang University, Fuzhou, China
8528750@qq.com
[3] School of Software, Beijing Institute of Technology, Beijing, China
[4] Innovative Information Industry Research Institute, Fujian Normal University, Fuzhou, China
xulin@fjnu.edu.cn
[5] School of Criminal Investigation and Counter Terrorism, People's Public Security University of China, Beijing, China
liuminghuime@163.com

Abstract. This paper is concerned with the analysis of the time domain acoustic scattering from locally perturbed flat substrates. For this three dimensional scattering problem with unbounded scatterer, boundary integral equation is directly established on the boundary of the unbounded scatterer without application of the symmetric continuation. The well-posedness of the time domain boundary integral equation is proved.

Keywords: Time domain · Acoustic scattering · Locally perturbation · Well-posedness · Boundary integral equation

1 Introduction

We consider the time domain acoustic scattering problem with a kind of unbounded scatterer, that is, the locally perturbed substrate. The scattering problem have got extensive attention due to its practical applications such as geophysical exploration and medical imaging. For decades, people have their attention on the study of the frequency domain problems [1,2], which cares about the time-harmonic case and is easier to compute. In recent years, with the development of the computer and numerical computation, some analysis of the time domain scattering problems have been published [3,4], which is more close to the practical applications.

For the acoustic scattering problem with locally perturbed substrates, the difficulty caused by the unbounded scatterer must be considered. The symmetric continuation is the most popular strategy to deal with this kind of scatterers,

J.-S. Pan et al. (eds.), *Advances in Smart Vehicular Technology, Transportation, Communication and Applications*, Smart Innovation, Systems and Technologies 86,
https://doi.org/10.1007/978-3-319-70730-3_35

which turns the unbounded scatterer into a bounded symmetric scatterer. The corresponding frequency domain problem has been well-studied [5,6] and the two dimensional time domain problem has been solved using the symmetric continuation strategy [7]. However, in this paper, we try to give a direct analysis of the three dimensional problem in time domain without symmetric continuation, which is more convenient and intuitional. A new definition of the retarded potential is given for the analysis.

The scattering problem is solved using the retarded potential boundary integral equation (RPBIE) method. The boundary integral equation method is a useful tool to solve the scattering problem [8] and the time domain boundary integral equation is called "retarded" since it is defined on the bases of a retarded time [9]. We use a first kind RPBIE to solve the scattering problem and the well-posedness analysis of the RPBIE is given in this paper, that is, the unique solvability and the stability.

2 The Scattering Problem

The scatterer is considered as the flat substrate with a locally raised area, which we call the locally perturbed substrate. Consider the scattering of acoustic in the upper space of the unbounded scatterer. The Cartesian coordinate (x_1, x_2, x_3) is adopted. The upper boundary of the scatterer is considered to be sound-soft and the scatterer may have several disjoint components. The upper boundary of the locally raised area is defined as Γ_0 and the rest of the boundary is Γ_c. The unbounded domain upon the boundary $\Gamma := \Gamma_0 \cup \Gamma_c$ is Ω^+.

We consider a point source y in the upper space separated from the unbounded scatterer. A causal signal $\lambda(t)$ is emitted from the source point. The incident field is

$$u^i(t, x; y) = k(t, x; y) * \lambda(t), \quad t \in \mathbb{R}, x \in \mathbb{R}^3 \backslash \{y\},$$

where $k(t, x; y) * \lambda(t)$ is the time convolution of $k(t, x; y)$ and $\lambda(t)$ and $k(t, x; y)$ is the Green's function of the operator $c^{-2}\partial_{tt} - \Delta$ in the three dimensional space

$$k(t, x; y) := \frac{\delta(t - c^{-1}|x - y|)}{4\pi|x - y|},$$

in which δ is the Dirac delta function, $c > 0$ is the sound speed of the homogeneous background medium and Δ is the Laplace operator in three dimensional space.

For $x = (x_1, x_2, x_3) \in \mathbb{R}^3$, denote by $x^r = (x_1, x_2, -x_3)$ the symmetry point of x with respect to the $x_1 - x_2$ plane. Then the reflected field corresponding to $u^i(t, x; y)$ is defined as

$$u^r(t, x; y) := -k(t, x; y^r) * \lambda(t), \quad t \in \mathbb{R}, x \in \mathbb{R}^3 \setminus \{y^r\}.$$

Define a new incident field $u^{i,r} := u^i + u^r$ in $\mathbb{R} \times (\mathbb{R}^3 \backslash \{y, y^r\})$. Then $u^{i,r} = G(t, x; y) * \lambda(t)$, where G is the Green's function of the operator $c^{-2}\partial_{tt} - \Delta$ in the half space $\mathbb{R} \times \mathbb{R}^3_+$ corresponding to the Dirichlet boundary condition which satisfies

$$G(t, x; y) = k(t, x; y) - k(t, x; y^r).$$

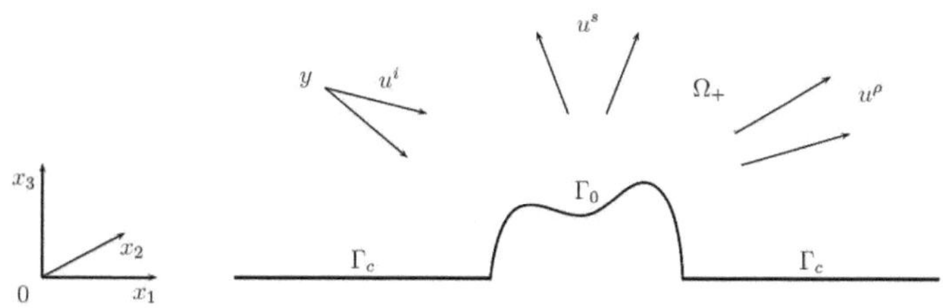

Fig. 1. Sketch of the geometry of the scattering problem.

The sketch of the geometry of the scattering problem can be seen in Fig. 1. The scattering problem related to the scattered field $u^s := u - u^{i,r}$ is considered. The scattered field u^s is the solution of the following scattering problem: Given a locally perturbed flat substrate as the scatterer and the incident field u^i, find a scattered field u^s which satisfies the homogeneous wave equation, Dirichlet boundary condition and the homogeneous initial conditions

$$c^{-2}\partial_{tt}u^s - \Delta u^s = 0 \quad \text{in } \mathbb{R} \times \Omega_+, \tag{1}$$

$$u^s = -u^{i,r} \quad \text{on } \mathbb{R} \times \Gamma, \tag{2}$$

$$u^s(0, \cdot) = \partial_t u^s(0, \cdot) = 0 \quad \text{in } \Omega_+. \tag{3}$$

3 The Retarded Single Layer Potential

Special retarded potentials are needed for the analysis of the time domain scattering problem. For simplicity, only the retarded single layer potential will be mentioned in this paper. For the scattering problem with the scatterer D, the retarded single layer potential on ∂D is defined as (see [10])

$$(SL_{\partial D}\psi)(t, x) := \int_{\mathbb{R}} \int_{\partial D} k(t - \tau, x; y)\psi(\tau, y)\mathrm{d}s_y \mathrm{d}\tau, \quad t \in \mathbb{R},\ x \in \mathbb{R}^3 \backslash \partial D.$$

In the following analysis, we attempt to define the potential on the unbounded surface Γ of the scatterer.

Consider the wave field $u^{i,r}$ we have defined. In fact, $u^{i,r}$ can be regarded as the scattering field emitted from two symmetric source points y and y^r. The signal λ is emitted from the point source y and the signal $-\lambda$ is emitted from the point source y^r at the same time.

Then we consider the retarded single layer potential based on the concept of $u^{i,r}$. Define $\Gamma^r := \{y : y^r \in \Gamma\}$. Assume that the signal ψ is simultaneously emitted from all the points $y \in \Gamma$, and the signal $-\psi$ is simultaneously emitted from all the points $y^r \in \Gamma^r$. Then the superposition of the wave field is defined as the retarded single layer potential on Γ, which can be written as

$$(SL_{\Gamma}^{half}\psi)(t,x) := \int_{\mathbb{R}}\int_{\Gamma} G(t-\tau,x;y)\psi(\tau,y)\mathrm{d}s_y\mathrm{d}\tau, \quad t \in \mathbb{R},\ x \in \mathbb{R}^3\backslash\Gamma. \tag{4}$$

The retarded single layer potential is defined in the space $\mathbb{R}^3 \setminus \Gamma$. We also need the definition of the retarded single layer operator on Γ:

$$(S_{\Gamma}^{half}\psi)(t,x) := \int_{\mathbb{R}}\int_{\Gamma} G(t-\tau,x;y)\psi(\tau,y)\mathrm{d}s_y\mathrm{d}\tau, \quad t \in \mathbb{R},\ x \in \Gamma.$$

Note that we define the potential with the Green's function in half space, which is consistent with the potential we use in two dimensional case [7]. However, other than the analysis of the two dimensional case, we attempt to establish the boundary integral equation directly on the upper boundary Γ of the unbounded scatterer instead of turning the unbounded scatterer into a bounded symmetric scatterer with symmetric continuation.

For the unbounded scatterer with $\Gamma = \Gamma_0 \cup \Gamma_c$ as the upper boundary, define by Ω_- the space underneath Γ and Ω_- is regarded as the interior of the scatterer. Then we can define the unit outward normal vector to Γ as ν. The restrictions of the function u to the boundary Γ from exterior and interior are defined as $\gamma^+ u$ and $\gamma^- u$, respectively. The normal derivative on Γ from the exterior and interior are defined as $\partial_\nu^+ u$ and $\partial_\nu^- u$, respectively. The definitions of $\partial_\nu^\pm u$ are given using the Green's formula:

$$\langle \partial_\nu^- v, \gamma^- u\rangle_\Gamma := (\Delta u, v)_{\Omega_-} + (\nabla u, \nabla v)_{\Omega_-}, \quad \forall v \in H^1(\Omega_-),$$
$$\langle \partial_\nu^+ v, \gamma^+ u\rangle_\Gamma := -(\Delta u, v)_{\Omega_+} - (\nabla u, \nabla v)_{\Omega_+}, \quad \forall v \in H^1(\Omega_+).$$

The jump relation is the key definition to analyse the layer potentials. Define by $[\![\gamma u]\!] = \gamma^- u - \gamma^+ u$ and $[\![\partial_\nu u]\!] = \partial_\nu^- u - \partial_\nu^+ u$ the jumps on Γ. Similar with the analysis of Sayas in Sect. 1.3 of [11], we have the following conclusion for the jump relation of the single layer potential $(SL_\Gamma\psi)(t,x)$:

Lemma 1 [11]. *Assume that Γ_0 is C^2 continuous surface, Γ is* Lipschitz *continuous surface and ψ is a continuous function. Then $(SL_\Gamma\psi)(t,x)$ is continuous for all $x \in \mathbb{R}^3$, which means the jump of $(SL_\Gamma\psi)(t,x)$ on Γ is zero:*

$$[\![\gamma(SL_\Gamma\psi)]\!] = 0 \quad \text{on } \mathbb{R}\times\Gamma.$$

The jump of the normal derivative of $(SL_\Gamma\psi)(t,x)$ on Γ is

$$[\![\partial_\nu(SL_\Gamma\psi)]\!] = \psi \quad \text{on } \mathbb{R}\times\Gamma.$$

Notice that the retarded single layer potential we define is in fact the difference of the classical retarded single layer potentials on Γ and $\Gamma^r := \{x : x^r \in \Gamma\}$, that is,

$$(SL_{\Gamma}^{half}\psi)(t,x) = (SL_\Gamma\psi)(t,x) - (SL_{\Gamma^r}\psi)(t,x). \tag{5}$$

Then the continuity of the potential $(SL_{\Gamma}^{half}\psi)(t,x)$ can be directly inferred from the continuity of $(SL_{\Gamma}\psi)(t,x)$ and the jump of the normal derivative of $(SL_{\Gamma}^{half}\psi)(t,x)$ can be computed with the normal derivative of $(SL_{\Gamma}\psi)(t,x)$. Then we get the following proposition.

Proposition 1. *Assume that Γ_0 is C^2 continuous surface, Γ is* Lipschitz *continuous surface and ψ is continuous function. Then $(SL_{\Gamma}^{half}\psi)(t,x)$ is continuous for all $x \in \mathbb{R}^3$:*

$$[\![\gamma(SL_{\Gamma}^{half}\psi)]\!] = 0 \quad \text{on } \mathbb{R}\times\Gamma.$$

The normal derivative of $(SL_{\Gamma}^{half}\psi)(t,x)$ on Γ_c is continuous, and the jumps of the normal derivative of $(SL_{\Gamma}^{half}\psi)(t,x)$ on Γ_0 and Γ_0^r are

$$[\![\partial_\nu(SL_{\Gamma}^{half}\psi)]\!] = \psi \quad \text{on } \mathbb{R}\times\Gamma_0,$$
$$[\![\partial_\nu(SL_{\Gamma}^{half}\psi)]\!] = -\psi \quad \text{on } \mathbb{R}\times\Gamma_0^r,$$

where the unit outward normal vector on Γ_0^r is considered as part of the unit outward normal vector on the closed surface $\Gamma_0 \cup \Gamma_0^r$.

Proof. The continuity of $(SL_{\Gamma}^{half}\psi)(t,x)$ for all $x \in \mathbb{R}^3$ can be inferred from (5) and Lemma 1, then $(SL_{\Gamma}^{half}\psi)(t,x)$ is continuous on Γ.

For the normal derivative of the retarded single layer potential, consider the upper space of $\Gamma = \Gamma_0 \cup \Gamma_c$ as the exterior, then Lemma 1 implies

$$[\![\partial_\nu(SL_{\Gamma}\psi)]\!] = \psi,\ [\![\partial_\nu(SL_{\Gamma^r}\psi)]\!] = \psi \quad \text{on } \mathbb{R}\times\Gamma_c.$$

Then the continuity of $(SL_{\Gamma}^{half}\psi)(t,x)$ on Γ_c can be got from (5).

Take the outward normal derivative on the closed surface $\Gamma_0 \cup \Gamma_0^r$ as the outward normal derivative on Γ_0 and Γ_0^r. Lemma 1 implies

$$[\![\partial_\nu(SL_{\Gamma}\psi)]\!] = \psi \quad \text{on } \mathbb{R}\times\Gamma_0,$$
$$[\![\partial_\nu(SL_{\Gamma^r}\psi)]\!] = \psi \quad \text{on } \mathbb{R}\times\Gamma_0^r.$$

Obviously, $\partial_\nu(SL_{\Gamma}\psi)$ and $\partial_\nu(SL_{\Gamma^r}\psi)$, respectively, are continuous on Γ_0^r and Γ_0. Then the jump of the normal derivative of $(SL_{\Gamma}^{half}\psi)(t,x)$ on Γ_0 and Γ_0^r can be inferred from (5).

4 The Retarded Potential Boundary Integral Equation Method

Referring to the analysis of Sect. 2.4 in [11] by Sayas, we quote the following uniqueness result.

Lemma 2 [11]. *Assume that $u \in \mathcal{L}'_\sigma(\mathbb{R}, H^1_\Delta(\mathbb{R}^3 \setminus \Gamma))$ and Γ is* Lipschitz *continuous surface. If u satisfies*

$$c^{-2}\ddot{u} - \Delta u = 0 \quad \text{in } \mathbb{R}\times(\mathbb{R}^3\backslash\Gamma)$$

in the sense of the causal tempered $L^2(\mathbb{R}^3 \setminus \Gamma)$-value generalized function, where $\ddot{u}$ is the generalized second order time derivative of u. Moreover, assume that the statements

$$[\![\gamma u]\!] = 0,\ [\![\partial_\nu u]\!] = 0 \quad \text{on } \mathbb{R} \times \Gamma$$

are true in the sense of $H^{1/2}(\Gamma)$-value and $H^{-1/2}(\Gamma)$-value, respectively. Then $u = 0$.

We attempt to establish the boundary integral equation on the bases of the uniqueness result in Lemma 2 and the property of the retarded single layer potential. Consider the scattering problem

$$c^{-2}\partial_{tt}u^s - \Delta u^s = 0 \quad \text{in } \mathbb{R} \times (\mathbb{R}^3 \backslash \Gamma), \tag{6}$$

$$\gamma^{\pm} u^s = -u^{i,r} \quad \text{on } \mathbb{R} \times \Gamma, \tag{7}$$

$$u^s(0,\cdot) = \partial_t u^s(0,\cdot) = 0 \quad \text{in } \mathbb{R}^3 \backslash \Gamma. \tag{8}$$

Obviously, if u^s is the solution of the scattering problem (6)–(8), then $u^s|_{\mathbb{R}\times(\mathbb{R}^2\backslash\overline{D})}$ is the solution of the scattering problem (1)–(3).

Proposition 2. *The solution u^s of the scattering problem* (6)–(8) *can be represented as*

$$u^s = SL_\Gamma^{half}\psi \quad \text{in } \mathbb{R} \times (\mathbb{R}^3 \backslash \Gamma), \tag{9}$$

where the causal density function ψ is determined by the jump of the normal derivative of u^s on Γ:

$$\psi := [\![\partial_\nu u^s]\!] \quad \text{on } \mathbb{R} \times \Gamma.$$

Proof. It is easy to see that the retarded single layer potential $u^s = SL_\Gamma^{half}\psi$ satisfies the wave Eq. (6). Proposition 1 implies that the retarded single layer potential $(SL_\Gamma^{half}\psi)(t,x)$ is continuous on Γ. Thus $u^s = SL_\Gamma^{half}\psi$ satisfies the continuity condition (7). Taking into consideration the causality of the density function ψ, $u^s = SL_\Gamma^{half}\psi$ is also a causal function and satisfies (8).

Then $u^s = SL_\Gamma^{half}\psi$ is the solution of the scattering problem (6)–(8). The uniqueness result in Lemma 2 implies the unique solvability of the scattering problem (6)–(8). Thus the unique solution of the scattering problem (6)–(8) is $u^s = SL_\Gamma^{half}\psi$.

The relationship of the jumps of the normal derivatives of ψ and u^s can be inferred from the jump of the retarded single layer potential described in Proposition 1. This completes the proof.

Using the boundary condition (7), we can get the following RPBIE:

$$S_\Gamma^{half}\psi = -u^{i,r} \quad \text{on } \mathbb{R} \times \Gamma. \tag{10}$$

We have mentioned that if u^s is the solution of the scattering problem (6)–(8), then $u^s|_{\mathbb{R}\times(\mathbb{R}^2\backslash\overline{D})}$ is the solution of the scattering problem (1)–(3). Then Lemma 2 and Proposition 2 imply that the unique solution of the scattering problem (1)–(3) can be represented by (9). Moreover, the boundary condition (2) implies that the density function ψ in (9) satisfies the boundary integral Eq. (10). Then the RPBIE method to solve the scattering problem (1)–(3) is:

(i) Get the value of the density function ψ on $\mathbb{R} \times \Gamma$ by solving the RPBIE (10).
(ii) The solution u^s of (1)–(3) is given by the formula (9).

5 The Well-Posedness Analysis

Firstly, we recall some definition of the spaces and the integration transform. For given Lipschitz domain $\Omega \in \mathbb{R}^3$, we use the definition of the L^2-inner product $(u, v)_\Omega$, the L^2-norm $\|u\|_\Omega$ and the $H^1(\Omega)$-norm $\|u\|_{H^1(\Omega)}$ (for details, see [11]). For $c > 0$, define $\underline{c} := min\{1, c\}$ and the norm

$$|||u|||_{c,\Omega}^2 := \|\nabla u\|_\Omega^2 + c^2\|u\|_\Omega^2.$$

Let Γ be the boundary of the Lipschitz domain Ω. For Γ and the open surface $\Gamma_0 \subset \Gamma$, we recall the $L^2(\Gamma)$-inner product $\langle \xi, \eta \rangle_\Gamma$. Moreover, we use the definition of the trace spaces $H^{1/2}(\Gamma)$ and $H^{-1/2}(\Gamma)$ with the norms $\|\cdot\|_{H^{1/2}(\Gamma)}$ and $\|\cdot\|_{H^{-1/2}(\Gamma)}$, respectively (see [12] for details).

For a Hilbert space X, denote by $\mathcal{D}'(X)$ and $\mathcal{S}'(X)$ the X-valued distribution and the tempered distributions on the real line, respectively. For $\sigma > 0$, define $\mathbb{C}_\sigma := \{\omega \in \mathbb{C} : \text{Im}(\omega) > \sigma\}$ and $\mathbb{C}_+ := \{\omega \in \mathbb{C} : \text{Im}(\omega) > 0\}$. The Fourier-Laplace transform is defined as

$$\mathcal{L}[f](\omega) := \int_{-\infty}^{\infty} \mathrm{e}^{\mathrm{i}\omega t} f(t)\mathrm{d}t, \quad \omega \in \mathbb{C}_\sigma.$$

For $\sigma \in \mathbb{R}$, define

$$\mathcal{L}'_\sigma(\mathbb{R}_+, X) := \left\{f \in \mathcal{D}'(X) : \mathrm{e}^{-\sigma t} f \in \mathcal{S}'(X),\ f(t) = 0,\ \forall t < 0\right\}.$$

For $\sigma \in \mathbb{R}$ and $p \in \mathbb{R}$, define the space

$$H_\sigma^p(\mathbb{R}_+, X) := \left\{ f \in \mathcal{L}'_\sigma(\mathbb{R}_+, X) : \int_{-\infty+\mathrm{i}\sigma}^{\infty+\mathrm{i}\sigma} |\omega|^{2p} \|\mathcal{L}[f](\omega)\|_X^2 \mathrm{d}\omega < \infty \right\},$$

with the norm (see [7] for details)

$$\|f\|_{H_\sigma^p(\mathbb{R}_+, X)} := \left(\int_{-\infty+\mathrm{i}\sigma}^{\infty+\mathrm{i}\sigma} |\omega|^{2p} \|\mathcal{L}[f](\omega)\|_X^2 \mathrm{d}\omega \right)^{1/2}.$$

As one of the focuses of this paper, we give the well-posedness analysis of the Eq. (10). For simplification, we set the sound speed of the background medium to be $c = 1$ in the rest of this paper. We claim that the assumption does not affect the validity of our analysis, that is because simple transform implies the validity of the analysis for $c \neq 1$. Taking Fourier-Laplace transform of the Eq. (1) with respect to t implies

$$\Delta U(\omega) + \omega^2 U(\omega) = 0 \quad \text{in } \Omega_+.$$

Define $SL_\Gamma^{F,half}(\omega) := \mathcal{L}[SL_\Gamma^{half}]$ and $S_\Gamma^{F,half}(\omega) := \mathcal{L}[S_\Gamma^{half}]$. Then we have

$$SL_\Gamma^{F,half}(\omega)\psi = \int_\Gamma \Phi_{3,\omega}^{half}(\cdot;y)\psi(y)\mathrm{d}s_y \quad \text{in } \mathbb{R}^3 \setminus \Gamma,$$
$$S_\Gamma^{F,half}(\omega)\psi = \int_\Gamma \Phi_{3,\omega}^{half}(\cdot;y)\psi(y)\mathrm{d}s_y \quad \text{on } \Gamma,$$

in which $\Phi_{3,\omega}^{half}(x;y)$ is defined as

$$\Phi_{3,\omega}^{half}(x;y) := \frac{\mathrm{e}^{\mathrm{i}\omega|x-y|}}{4\pi|x-y|} - \frac{\mathrm{e}^{\mathrm{i}\omega|x-y^r|}}{4\pi|x-y^r|}.$$

Thus the scattering problem (1)–(3) is transformed into a corresponding Laplace domain problem. The solution $U(\omega)$ of the Laplace domain problem has the representation

$$U(\omega) = SL_\Gamma^{F,half}(\omega)\Lambda(\omega),$$

where $\Lambda : \mathbb{C}_+ \to H^{-1/2}(\Gamma)$ is the solution of the boundary integral equation

$$S_\Gamma^{F,half}(\omega)\Lambda(\omega) = \mathcal{L}[-u^{i,r}].$$

To prove the well-posedness of the Eq. (10), we give the analysis of the Laplace domain problem firstly and come back to time domain after that. As a base, we quote the result of Proposition 2.5.2 of [11].

Lemma 3 [11]. *Assume that Ω is* Lipschitz *continuous domain, $C_\Omega > 0$ is a constant depends on Ω. Moreover, $\omega \in \mathbb{C}_+$ and $U(\omega) \in H^1_\Delta(\Omega)$ satisfy $\Delta U(\omega) + \omega^2 U(\omega) = 0$ on Ω. Then*

$$\|\partial_\nu U(\omega)\|_{H^{-1/2}(\partial\Omega)} \le C_\Omega \left(\frac{|\omega|}{\underline{\sigma}}\right)^{1/2} |||U(\omega)|||_{|\omega|,\Omega}.$$

We have the following consequence in Laplace domain for the scattering problem with locally perturbed flat substrates.

Proposition 3. *Assume that Γ is* Lipschitz *continuous surface. Then there exists a constant $C_\Gamma > 0$ depends on Γ such that for any given $\omega \in \mathbb{C}_+$, the operator $S_\Gamma^{F,half}(\omega) : H^{-1/2}(\Gamma) \to H^{1/2}(\Gamma)$ is reversible and the inverse operator $S_\Gamma^{F,half}(\omega)^{-1}$ satisfies*

$$\|S_\Gamma^{F,half}(\omega)^{-1}\|_{H^{1/2}(\Gamma)\to H^{-1/2}(\Gamma)} \le C_\Gamma^{-1}\frac{|\omega|^2}{\sigma\underline{\sigma}}.$$

Proof. Set the density function $\Lambda(\omega) \in H^{-1/2}(\Gamma)$. The function $U(\omega) \in H^1_\Delta(\mathbb{R}^3 \setminus \Gamma)$ has the representation $U(\omega) = SL_\Gamma^{F,half}(\omega)\Lambda(\omega)$. Then $U(\omega)$ is the unique solution of the following problem:

$$\Delta U(\omega) + \omega^2 U(\omega) = 0 \quad \text{in } \mathbb{R}^3 \setminus \Gamma,$$
$$[\![\gamma U(\omega)]\!] = 0 \quad \text{on } \Gamma,$$
$$[\![\partial_\nu U(\omega)]\!] = \Lambda(\omega) \quad \text{on } \Gamma.$$

Combining with the previous analysis, we can infer that $\Lambda(\omega)$ satisfies the integral equation

$$S_{\Gamma}^{F,half}(\omega)\Lambda(\omega) = \gamma^{\pm}U(\omega).$$

Thus

$$\begin{aligned}
\left\langle\overline{\Lambda(\omega)}, S_{\Gamma}^{F,half}(\omega)\Lambda(\omega)\right\rangle_{\Gamma} &= \langle\partial_{\nu}^{-}v, \gamma^{-}U(\omega)\rangle_{\Gamma} - \langle\partial_{\nu}^{+}v, \gamma^{+}U(\omega)\rangle_{\Gamma}\\
&= \left(\Delta\overline{U(\omega)}, U(\omega)\right)_{\mathbb{R}^3\setminus\Gamma} + \left(\nabla\overline{U(\omega)}, \nabla U(\omega)\right)_{\mathbb{R}^3\setminus\Gamma}\\
&= -\overline{\omega}^2\left(\overline{U(\omega)}, U(\omega)\right)_{\mathbb{R}^3\setminus\Gamma} + \left(\nabla\overline{U(\omega)}, \nabla U(\omega)\right)_{\mathbb{R}^3\setminus\Gamma}.
\end{aligned}$$

Define $\sigma := \mathrm{Im}(\omega)$. Lemma 3 and $\Lambda(\omega) = [\![\partial_{\nu}U(\omega)]\!]$ imply

$$\begin{aligned}
&\mathrm{Im}\left(\mathrm{e}^{\mathrm{iArg}\omega}\left\langle\overline{\Lambda(\omega)}, S_{\Gamma}^{F,half}(\omega)\Lambda(\omega)\right\rangle_{\Gamma}\right)\\
=&\mathrm{Im}\left(-\frac{\omega}{|\overline{\omega}|}\overline{\omega}^2\left(\overline{U(\omega)}, U(\omega)\right)_{\mathbb{R}^3\setminus\Gamma} + \frac{\omega}{|\overline{\omega}|}\left(\nabla\overline{U(\omega)}, \nabla U(\omega)\right)_{\mathbb{R}^3\setminus\Gamma}\right)\\
=&\frac{\sigma}{|\omega|}|||U(\omega)|||_{|\omega|,\mathbb{R}^3\setminus\Gamma}^2 \geq \frac{\sigma\underline{\sigma}}{|\omega|^2}\left(C_{\Omega_-}^2\|\partial_{\nu}^{-}U(\omega)\|_{H^{-1/2}(\Gamma)}^2 + C_{\Omega_+}^2\|\partial_{\nu}^{+}U(\omega)\|_{H^{-1/2}(\Gamma)}^2\right)\\
\geq&C_{\Gamma}\frac{\sigma\underline{\sigma}}{|\omega|^2}\|\Lambda(\omega)\|_{H^{-1/2}(\Gamma)}^2.
\end{aligned}$$

where C_{Ω_-} and C_{Ω_+} are, respectively, positive constant depends on Ω_- and Ω_+. It can be seen from the sketch of the scattering problem, the two constants can be unified by the constant $C_{\Gamma} > 0$ depends on Γ in the inequality.

Then Lax-Milgram theorem implies that $S_{\Gamma}^{F,half}(\omega)$ is reversible and we also get the estimation of the inverse operator $S_{\Gamma}^{F,half}(\omega)^{-1}$.

Then, coming back to time domain, we have the following result:

Theorem 1. *Let $\sigma > 0$, $p \in \mathbb{R}$ and $u^{i,r} \in H_{\sigma}^{p}(\mathbb{R}_+, H^{1/2}(\Gamma))$. Then the boundary integral Eq. (10) is unique solvable for $\psi \in H_{\sigma}^{p-2}(\mathbb{R}_+, H^{-1/2}(\Gamma))$. Moreover, there exists a constant $C_{\sigma,\Gamma}$ depends only on σ and Γ such that*

$$\|\psi\|_{H_{\sigma}^{p-2}(\mathbb{R}_+,H^{-1/2}(\Gamma))} \leq C_{\sigma,\Gamma}\|u^{i,r}\|_{H_{\sigma}^{p}(\mathbb{R}_+,H^{1/2}(\Gamma))}.$$

Proof. Combining the result of Proposition 3 with Lemma 2 of [10] and a Fourier argument, the theorem is proved.

6 Conclusion

We have analysed the time domain acoustic scattering problem with locally perturbed flat substrates, a kind of unbounded scatterers in three dimensional space. For this time domain scattering problem, retarded potential boundary integral equation has been directly established on the boundary of the unbounded scatterer. The unique solvability and stability of the time domain boundary integral equation have been proved.

Acknowledgements. The work of Bo Chen was supported by the National Natural Science Foundation of China (NSFC) grants 11371172 and 11671170 and the Scientific Research Foundation of Civil Aviation University of China (No. 2017QD04S). The work of Minghui Liu was supported by the NSFC Grant 11626230 and the Fundamental Research Funds for the Central Universities Grant 2016JKF01101. The work of Fuquan Zhang was supported by the Scientific Research Foundation of Minjiang University (No. MYK17021).

References

1. Colton, D., Kress, R.: Inverse Acoustic and Electromagnetic Scattering Theory, 3rd edn. Springer, Berlin (2013)
2. Sun, Y., Guo, Y., Ma, F.: The reciprocity gap functional method for the inverse scattering problem for cavities. Appl. Anal. **95**, 1327–1346 (2016)
3. Guo, Y., Monk, P., Colton, D.: The linear sampling method for sparse small aperture data. Appl. Anal. **95**(8), 1599–1615 (2016)
4. Guo, Y., Hömberg, D., Hu, G., Li, J., Liu, H.: A time domain sampling method for inverse acoustic scattering problems. J. Comput. Phys. **314**, 647–660 (2016)
5. Feng, L., Ma, F.: A sampling method for solving inverse scattering problems with a locally perturbed half plane. Northeast. Math. J. **19**, 1–4 (2003)
6. Dong, H., Ma, F.: Reconstruction of the shape of object with near field measurements in a half-plane. Sci. China, Ser. A Math. **51**, 1059–1070 (2008)
7. Chen, B., Ma, F., Guo, Y.: Time domain scattering and inverse scattering problems in a locally perturbed half-plane. Appl. Anal. **96**(8), 1303–1325 (2016)
8. Colton, D., Kress, R.: Integral Equation Methods in Scattering Theory. Wiley, New York (1983)
9. Banjai, L., Sauter, S.: Rapid solution of the wave equation in unbounded domains. SIAM J. Numer. Anal. **47**, 227–249 (2008)
10. Chen, Q., Haddar, H., Lechleiter, A., Monk, P.: A sampling method for inverse scattering in the time domain. Inverse Prob. **26**, 85001–85017 (2010)
11. Sayas, F.J.: Retarded Potentials and Time Domain Boundary Integral Equations: A Road-Map. Springer Series in Computational Mathematics, vol. 50. Springer, Basel (2016)
12. Hsiao, G.C., Wendland, W.L.: Boundary Integral Equations. Springer, Berlin (2008)

Adversarial Multiarmed Bandit Problems in Gradually Evolving Worlds

Chia-Jung Lee[1,2(✉)], Yalei Yang[1,2], Sheng-Hui Meng[1,2], and Tien-Wen Sung[1,2]

[1] College of Information Science and Engineering, Fujian University of Technology, Fuzhou, Fujian, China
leecj2015@foxmail.com, tienwen.sung@gmail.com

[2] Fujian Provincial Key Laboratory of Big Data Mining and Applications, Fujian University of Technology, Fuzhou, Fujian, China
{yangyalei,menghui}@fjut.edu.cn

Abstract. We study the adversarial multi-armed bandit problem, in which a player must iteratively make online decisions with linear loss vectors and hopes to achieve a small total loss. We consider a natural measure on the loss vectors, called deviation, which is the sum of the distances between every two consecutive loss functions. We propose an online algorithm and the experimental results show that the proposed algorithm can achieve a small total loss when the loss functions have a small deviation.

Keywords: Adversarial multi-armed bandit problem · Online algorithm · Deviation

1 Introduction

We study the *adversarial multi-armed bandit problem* with N possible actions. In each round, an online algorithm $\mathcal{A}$ has to choose, possibly in a probabilistic way, an action I_t to play, and then receives a corresponding loss $\ell_{I_t}^t$ according to some loss vector $\ell^t = (\ell_1^t, \cdots, \ell_N^t) \in \mathbb{R}^N$. The algorithm $\mathcal{A}$ would like to minimize its total loss, or equivalently, to minimize its regret, defined as the difference between the total loss it costs and that of the best fixed action in hindsight. Online algorithms which can achieve a regret bound of $O(\sqrt{NT})$ for T rounds have been proposed [1,2].

Many applications in daily life can be modeled as an adversarial multi-armed bandit problem, for example, the commuting problem. Assume that there are N possible routes from the home of the player to his office. Each morning, the player has to choose one of N routes to work, and then spends the corresponding commuting time. Observe that for each route, the commuting time between different days may be very similar, which implies that in this situation, the difference between these loss functions may be small.

To model this kind of loss functions, we consider the notion, called L_∞-deviation [7]. For a sequence of loss functions $\ell^1, \cdots, \ell^T$, their L_∞-deviation

J.-S. Pan et al. (eds.), *Advances in Smart Vehicular Technology, Transportation, Communication and Applications*, Smart Innovation, Systems and Technologies 86,
https://doi.org/10.1007/978-3-319-70730-3_36

is defined as $\sum_{t=2}^{T} \|\ell^{t-1} - \ell^t\|_\infty^2$ where $\|\cdot\|_\infty$ denotes the ℓ_∞-norm. Deviation can model a dynamic environment that usually changes gradually, including examples such as commuting, weather conditions and stock markets. In this paper, we would like to consider the adversarial bandit problem with the natural restriction that the loss functions have their L_∞-deviation bounded by some parameter D_∞, and propose an online algorithm which can achieve a small total loss for this problem.

1.1 Related Works

An elementary problem in machine learning is the *multi-armed bandit problem* [12], which is a special case of the *adversarial multi-armed bandit problem* where the loss value of each action is sampled independently from a fixed and unknown distribution. It is known that using the upper confidence bound (UCB) algorithm, a regret of $O(N \ln T/\Delta)$, where Δ is the minimal gap between the loss mean of the best arm and that of the other alternatives, can be achieved [3]. In fact, the multi-armed bandit problem has become a rich topic, and a series of works studies the variants of multi-armed bandit problems including [2,6,9,11] due to the fruitful results in modern applications (see [5] for a nice survey).

For the adversarial multi-armed bandit problem, when the loss functions have the quadratic variation $Q = \sum_{t=1}^{T} \|\ell^t - \mu\|_2^2$, where $\mu = \frac{1}{T}\sum_{t=1}^{T} \ell^t$ is the mean of the loss functions, Hazan and Kale [8] achieved a regret of $\tilde{O}\left(N^2\sqrt{Q}\right)$[1]. Another related measure of loss functions is called L_∞-variation [10], defined by $\sum_{t=1}^{T} \|\ell^t - \mu\|_\infty^2$, and a regret of $\tilde{O}\left(\sqrt{NV_\infty \log N}\right)$ can be achieved when the loss functions have L_∞-variation bounded by some V_∞.

2 The Adversarial Bandit Problem

For a positive integer n, we use $[n]$ to denote the set $\{1, 2, \cdots, n\}$. For a vector $x \in \mathbb{R}^n$ and an index $i \in [n]$, x_i is the i'th component of x. For $x \in \mathbb{R}^n$, let $\|x\|_\infty$ denote the ℓ_∞-norm of x. Let $\{\mathbf{e}_1, \cdots, \mathbf{e}_n\}$ be the set of standard basis for $\mathbb{R}^n$.

In this paper, we study the *adversarial bandit problem*, in which an online algorithm $\mathcal{A}$ has to play iteratively for T rounds in the following way. Assume that there are N possible actions. In each round $t \in [T]$, the algorithm $\mathcal{A}$ must choose, possibly in a probabilistic way, an action $I_t \in [N]$ to play. Simultaneously, the adversary selects a loss vector $\ell^t = (\ell_1^t, \cdots, \ell_N^t)$ without any information about the choice of the algorithm $\mathcal{A}$. For convenience, we assume that $\ell_i^t \in [0, 1]$ for any $t \in [T]$ and $i \in [N]$, and the generalization of $\ell_i^t \in [a, b]$ for any $a < b$ is straightforward. After that, the loss value of the played action $\ell_{I_t}^t$ is revealed to the algorithm $\mathcal{A}$. The online algorithm $\mathcal{A}$ would like to minimize its total loss over T rounds, which is $\sum_{t=1}^{T} \ell_{I_t}^t$.

We consider a variant of the adversarial bandit problem, with a constraint on loss vectors. For a sequence of loss functions $\ell^1, \ldots, \ell^T \in \mathbb{R}^N$, we define their L_∞-deviation by $\sum_{t=1}^{T-1} \|\ell^t - \ell^{t+1}\|_\infty^2$. We then study the adversarial bandit problem

---
[1] We use the notation $\tilde{O}(\cdot)$ to hide the dependence on poly$(\log T)$ factor.

with the restriction that the loss functions have their L_∞-deviation bounded by some parameter D_∞.

3 The Main Algorithm

In this section, we introduce the algorithm for the adversarial bandit problem with a restriction that the ℓ^2_∞-deviation of the loss vectors is small. When there is no additional restriction on the loss functions, one can achieve a small cumulated loss using the well-known EXP3 [4] algorithm. Our algorithm, described in Algorithm 1 below, is modified from the EXP3 [4] algorithm, with some modifications.

Algorithm 1. MAIN ALGORITHM

Let u be the uniform distribution over $[N]$, with each $u_i = 1/N$.
Let $\pi : [N] \to [N]$ be a random permutation.
for $t = 1$ to N **do**
    Choose $I_t = (\pi(t) \mod N) + 1$
    Compute the estimator $\bar{\ell}^t = \left(\ell^t_{I_t} - \bar{\ell}^{t-1}_{I_t}\right) \mathbf{e}_{I_t} + \bar{\ell}^{t-1}$.
    Let $\hat{p}^{t+1} = \hat{q}^{t+1} = u$.
end for
for $t = N+1$ to T **do**
    Sample an action $I_t \in [N]$ according to the distribution $\hat{p}^t$.
    After receive the feedback $\ell^t_{I_t}$, compute the estimators

$$\tilde{\ell}^t = \frac{\ell^t_{I_t}}{\hat{p}^t_{I_t}} \mathbf{e}_{I_t}, \text{ and } \bar{\ell}^t = \left(\ell^t_{I_t} - \bar{\ell}^{t-1}_{I_t}\right) \mathbf{e}_{I_t} + \bar{\ell}^{t-1}.$$

    for $i = 1$ to N **do**
        Update $q_i^{t+1} = q_i^t e^{-\eta \tilde{f}_i^t} / Z_{t+1}$, where $Z_{t+1} = \sum_j q_j^t e^{-\eta \tilde{f}_j^t}$.
        Update $\hat{q}_i^{t+1} = q_i^{t+1} e^{-\eta \bar{f}_i^t} / \hat{Z}_{t+1}$, where $\hat{Z}_{t+1} = \sum_j q_j^{t+1} e^{-\eta \bar{f}_j^t}$.
        Compute $\hat{p}_i^{t+1} = (1-\gamma)\, \hat{q}_i^{t+1} + \gamma u_i$.
    end for
end for

Note that if one can receive the entire loss function ℓ^t instead of only the loss value $\ell^t_{I_t}$ of chosen action in each round, this problem is the prediction with expert advice problem in [7]. Chiang *et al.* observed that in round t, if one can select the action I_t according to the distribution probability modified by ℓ^t, then the total loss will be small. However, ℓ^t is not available before we take the action I_t. Fortunately, when the loss functions have a small L_∞-deviation, it would be a good idea to use ℓ^{t-1} to estimate ℓ^t and choose an action according to the probability modified by ℓ^{t-1} instead of ℓ^t.

However, in the adversarial bandit problem, we only obtain the loss value $\ell^t_{I_t}$ of the played action. Now, how to construct a good estimator of the loss function ℓ^t to update the modified probability? When the L_∞-deviation of the loss functions is small, every entry of ℓ^{t-1} may be close to ℓ^t. Therefore, we use $\bar{\ell}^{t-1}$ to approximate ℓ^t at Step 10. Note that $\bar{\ell}^t$ computed there are obtained from $\bar{\ell}^{t-1}$ by replacing only its I_t'th entry with $\ell^t_{I_t}$. The parameter η is the learning rate and γ is the probability of "exploration", which will be chosen later.

4 Experiments

4.1 Setting

Recall that our algorithm would like to minimize the total loss when the loss functions have a small L_∞-deviation. Therefore, our first experiment considers the loss functions with L_∞ deviation $D_\infty < \sqrt{T}$. Set $T = 10000$ and $N = 10$. We then randomly divide T rounds into 4 intervals. Within each interval i, we pick a loss function $F_i \in [0,1]^{10}$ at random in the first round, and generate each entry of loss functions in the other rounds by adding a random value sampled independently from a Gaussian distribution $\mathcal{N}(0, 0.02)$ to the corresponding entry of F_i. The generated functions in $[a,b]^{10}$ for some $a < b$ were normalized to $[0,1]^{10}$. The L_∞-deviation of the loss functions in the first experiment is less than $D_\infty = 29$, as shown in Table 1, which is much smaller than $\sqrt{T} = 100$.

One related measure of loss functions $\ell^1, \ldots, \ell^T \in \mathbb{R}^N$, called L_∞-variation [10], is defined by $\sum_{t=1}^{T} \|\ell^t - \mu\|_\infty^2$, where $\mu = \frac{1}{T}\sum_{t=1}^{T} \ell^t$ is the average of the loss functions. Our second experiment consider the case that the loss functions have a small L_∞-variation. We sampled the loss functions independently from a multivariate Gaussian distribution on $\mathbb{R}^{10}$. The mean vector for the Gaussian is a random vector in $[0,1]^{10}$, and the covariance matrix is sampled as random but constrained to rank 2. Again, the generated functions were normalized to $[0,1]^{10}$. The bounds of the L_∞-deviation and the L_∞-variation of the loss functions in the second experiment are $D_\infty = 233$ and $V_\infty = 109$, both of which are much smaller than $T = 10000$.

Table 1. The L_∞-deviation and the L_∞-variation of the loss functions.

	Experiment 1	Experiment 2
L_∞-deviation (D_∞)	29	233
L_∞-variation (V_∞)	991	109

4.2 Experiment Results

In Algorithm 1, we need to select two parameters η and γ. Inspired by [7,10], we first fixed the learning rate $\eta = 1/\sqrt{D_\infty}$ in the first experiment, and then run Algorithm 1 with $\gamma = 0.5$, $1/\sqrt{D_\infty} \approx 0.186$, 0.1, 0.05 and $1/\sqrt{T} = 0.01$. In Fig. 1(a), we plot the total loss for the corresponding parameters as a function of the round number. Note that the Algorithm 1 uses internal randomization for choosing an action. Hence, the curve is actually the average of total loss over 100 runs for the same loss functions. Clear, when $\gamma = 0.01$, the total loss achieves the best performance. Therefore, we then fixed the parameter $\gamma = 1/\sqrt{T} = 0.01$, and run Algorithm 1 with $\eta = 0.5$, $1/\sqrt{D_\infty} \approx 0.186$, 0.1, 0.05 and $1/\sqrt{T} = 0.01$. The total losses for the corresponding parameters are shown in Fig. 1(b). Note that when $\eta = 1/\sqrt{D_\infty}$, we achieve the best performance.

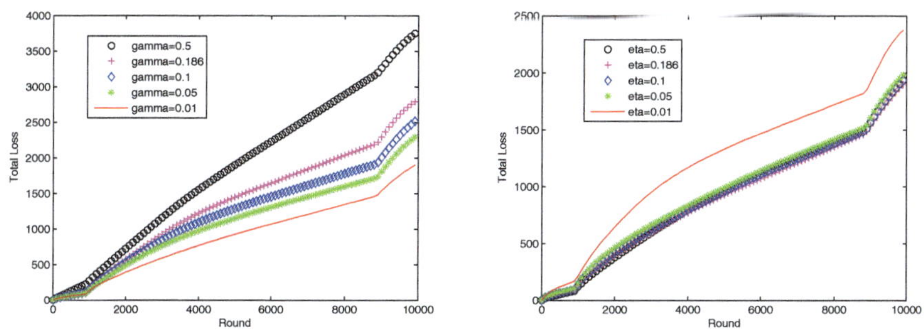

(a) $\eta = \frac{1}{\sqrt{D_\infty}} \approx 0.186$ (b) $\gamma = \frac{1}{\sqrt{T}} = 0.01$

Fig. 1. The total loss of the Algorithm 1 for different choices of the parameters γ(gamma) and η(eta) in the first experiment.

To confirm the selection of two parameters η and γ, we then proceed to the second experiment in which $D_\infty \approx 2\sqrt{T}$. As shown in Fig. 2(a), when fixed $\eta = 1/\sqrt{D_\infty}$, the curves are almost the same as $\gamma \leq 0.1$. Therefore, we remain to set $\gamma = 1/\sqrt{T} = 0.01$, and examine the total loss for different choices of η. Figure 2(b) shows that the curves are almost identical except $\eta = 0.5$.

Finally, we compare the performance of the following algorithms: the proposed Algorithm 1 with $\eta = 1/\sqrt{D_\infty}$ and $\gamma = 1/\sqrt{T}$, the EXP3 algorithm [4], and the algorithm in [10] which can achieve a small total loss when the loss functions have a small L_∞-variation. Figure 3 shows that in both of the two experiments, the Algorithm 1 can have the smallest total loss. Moreover, in the first experiment (D_∞ is much small than V_∞), the total loss of the Algorithm 1 is obviously smaller than those of the other two algorithms as shown in Fig. 3(a).

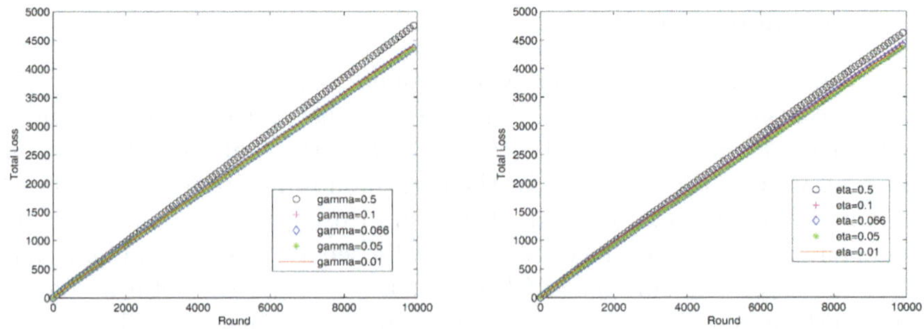

(a) $\eta = \frac{1}{\sqrt{D_\infty}} \approx 0.066$ (b) $\gamma = \frac{1}{\sqrt{T}} = 0.01$

Fig. 2. The total loss of the Algorithm 1 for different choices of the parameters γ(gamma) and η(eta) in the second experiment.

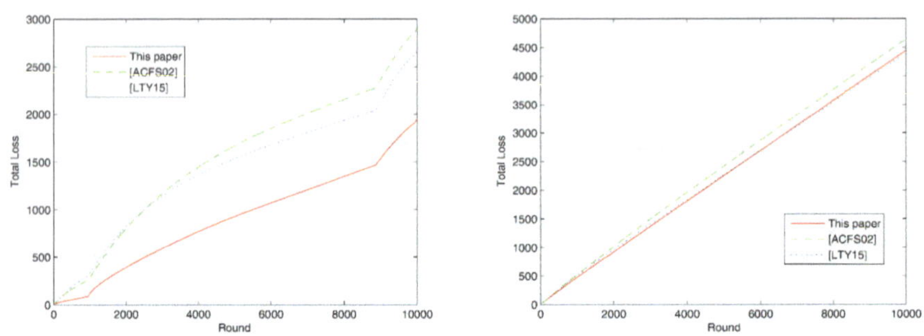

(a) The first experiment (b) The second experiemnt

Fig. 3. The left figure compares the total losses of Algorithm 1, EXP3 [4] algorithm, and the algorithm in [10] in the first experiment, while the right figure shows the total losses in the second experiment.

5 Conclusions

We propose an online algorithm for the adversarial bandit problem with the restriction that the loss functions have their L_∞-deviation bounded by some parameter D_∞. To evaluate the proposed algorithms, we consider two classes of loss functions, and the experimental results show that the proposed algorithm can achieve a small total loss. In the future, we would like to obtain some theoretical results to figure out the performance of our algorithm, and consider other variants of the adversarial bandit problem.

Acknowledgements. This work was supported by the Natural Science Foundation of Fujian Province of China under Grant No. 2014J01168, and the Education Department of Fujian Province of China under Grant No. JAT160335.

References

1. Audibert, J.-Y., Bubeck, S.: Regret bounds and minimax policies under partial monitoring. J. Mach. Learn. Rese. **11**, 2635–2686 (2010)
2. Audibert, J.-Y., Bubeck, S., Lugosi, G.: Minimax policies for combinatorial prediction games. In: COLT, pp. 107–132 (2011)
3. Auer, P., Cesa-Bianchi, N., Fischer, P.: Finite-time analysis of the multiarmed bandit problem. Mach. Learn. **47**(2–3), 235–256 (2002)
4. Auer, P., Cesa-Bianchi, N., Freund, Y., Schapire, R.E.: The nonstochastic multiarmed bandit problem. SIAM J. Comput. **32**(1), 48–77 (2002)
5. Bubeck, S., Cesa-Bianchi, N.: Regret analysis of stochastic and nonstochastic multiarmed bandit problems. Found. Trends Mach. Learn. **5**(1), 1–122 (2012)
6. Cesa-Bianchi, N., Lugosi, G.: Combinatorial bandits. J. Comput. Syst. Sci. **78**(5), 1404–1422 (2012)
7. Chiang, C.-K., Yang, T., Lee, C.-J., Mahdavi, M., Lu, C.-J., Jin, R., Zhu, S.: Online optimization with gradual variations. In: COLT, pp. 6.1–6.20 (2012)
8. Hazan, E., Kale, S.: Better algorithms for benign bandits. J. Mach. Learn. Res. **12**, 1287–1311 (2011)
9. Kleinberg, R., Niculescu-Mizil, A., Sharma, Y.: Regret bounds for sleeping experts and bandits. Mach. Learn. **80**(2–3), 245–272 (2010)
10. Lee, C.-J., Tsai, S.-C., Yang, M.-C.: Online prediction problems with variation. In: COCOON, pp. 49–60 (2014)
11. Neu, G., György, A., Szepesvári, C., Antos, A.: Online markov decision processes under bandit feedback. IEEE Trans. Automat. Contr. **59**(3), 676–691 (2014)
12. Robbins, H.: Some aspects of the sequential design of experiments. Bull. Amer. Math. Soc. **58**(5), 527–535 (1952)

A Video Coloring Method Based on CNN and Feature Point Tracking

George Guan[1(✉)], Fuquan Zhang[1,2], Gangyi Ding[1], Meng Niu[1], and Lin Xu[3]

[1] School of Software, Beijing Institute of Technology, Beijing 100081, China
guanzheng@bit.edu.cn

[2] Fujian Provincial Key Laboratory of Information Processing and Intelligent Control, Minjiang University, Fuzhou 350121, China

[3] Innovative Information Industry Research Institute, Fujian Normal University, Fuzhou 350300, China

Abstract. Black and white films were the main form of human culture records before. Colorization of those films is creative. At present, Colorization of black and white films is still handmade which is expensive and time consuming. In this paper, a framework based on CNN and particle filter tracking algorithm is proposed, which can color black and white video and try to solve the problem of dynamic frame based on context correlation. At the same time, the objective function of CNN structure and particle filter tracking are optimized. The result of colorization on videos is satisfactory.

Keywords: Colorization · Black and white films · CNN structure · Particle filter tracking · Optimized

1 Introduction

From 1826 Nicvephore's daylighting process, the human enters the photographic era, and entered the black and white film era in 1930s, color film era in 1970s. Looking back the history of the film, the famous films are numerous, but until the 1970s, the color film became the mainstream of history, of course, today there are still many directors use black and white film as a means of artistic expression. In 50 years, black and white movies such as West Without War, Roman Holiday, Kill A Robin capture 30% IMDB-TOP100 and still influence the contemporary.

Of course, the first contact with black and white movies will have a little exclusion, because they do not have beautiful gorgeous colors, which is also their directors' regret.

To this end, the Hollywood company had tried to color the film "Citizen Kane" by manually, but in fact because of the cost finally they ended in failure when they colored 172,800 frames.

J.-S. Pan et al. (eds.), *Advances in Smart Vehicular Technology, Transportation, Communication and Applications*, Smart Innovation, Systems and Technologies 86,
https://doi.org/10.1007/978-3-319-70730-3_37

With the development of computer technology, Wilson Markle[1] invented the coloring technique, which was used for the Apollo 11 photo process, and had a shocking influence. Although he created more than 3,000 software-based color samples, but the cost of coloration was still as high as $ 3,000 per frame.

Of course, with the development of neural networks, based on artificial intelligence color filling technique, Developers have achieved results on the photo coloring with a wide range of attempts. These fast coloring techniques through a variety of network methods speed up the efficiency of coloring. The results of coloring pictures directly are presented in Fig. 1, which compose the first part of our work and it seems pretty good. The first row are the gray frames, the second row are the original frames, the third row are the coloring frames by our algorithm.

Fig. 1. The first line is the gray frame, the second line is the original frame, the third line is the coloring frame by our algorithm which seems good.

Currently most of machine learning coloring means more biased in a creativity area which makes the elements look more comfortable and the whole picture looks more beautiful. Our goal is more difficult than this because we are dealing with higher dimension with time information and try to color the video. Although video can be used as a series of video frames or called image composition process, but the video and the image is essentially different from the semantic point of view. From the perspective of handwork coloring, a group of images can be colored one by one because they express different content and information. Even to improve efficiency, they can be assigned to different operators with different colors sets. On the contrary, because of the relevance in the video context, the coloring process of the next video frame always takes into account the contents of the last frame, which probably includes skin color, clothes, props, sky features (and so on) of the characters already appeared.

---

[1] Wilson Markle (born September 2, 1938) is a Canadian engineer who invented the film colorization process in 1970. His first company, Image Transform, colored pictures from the Apollo space program to make a full-color television presentation for NASA.

2 Algorithm and Structure

The video is composed of a series of video frames, and video coloring is the process of coloring each video frame. However, the adjacent video frames in a video have a strong correlation, including the correlation of the elemental color composition and the motion of the object in the video frame, which is quite different. The relevance of the image in the video should be fully used to make the results of the video coloring more realistic.

2.1 Framework

Figure 2 shows the basic process mentioned above, which is used to complete the video coloring. First build a learning network, and use the network and the image set ImageNet(30K) to learn the image features. And then the video is divided to distinguish the different sub-lens. This will be used for sequential coloring process. For a sub-lens, we will extract its first frame as a coloring starting point, and use the learning results of Convolutional Neural Network (CNN) for coloring. For the following frames, the tracking algorithm will be applied to color them based on the coloring result of the first frame. Through this process, we will complete video coloring.

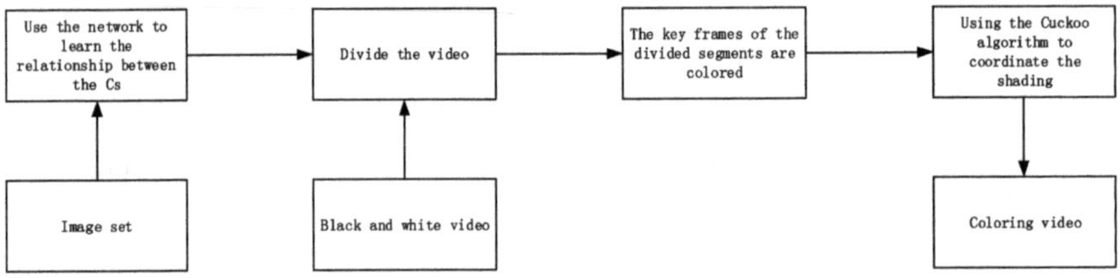

Fig. 2. Video coloring process framework

2.2 Color Space

Before describing the other parts, we need to identify the color space set, which will facilitate our description of the algorithm followed. Here we use the CIE Lab color space, where component *L* is used to represent the brightness of the pixel, of which value range is [0,100], from absolute black to absolute white; component *a* in range [−128, 127] represents the transformation from green to red component *b* in range [−128, 127] represents the transformation from blue to yellow [1, 2].

This color space can distinguish grayscale images effectively by *L* vector. If RGB or CMYK color set were used, the expression of single channel luminosity will be more complex. But in Lab space, if we have to describe the black and white video frame, we only need to pay attention to the gray scale L.

2.3 The Construction of Convolutional Neural Network

We use 8 convolutional layers in our convolutional neural network (we refer to the structure of others [3, 12]). After the specific experiment, we found it still works well without pooling layer. In addition, all the changes are carried out in the convolutional layer. We use Caffe framework to build the convolutional layers.

We have trained this CNN network, and taken the black and white video frame as input, and then learned the features of black and white video frames, through the 6 layers in the middle. And the learning results are related to the possibility of ab in the color space by function (3), and then the results are specifically mapped into color video frames.

2.4 Euclidean Distance and Objective Function

We will describe the expression of the video frames according to the color space of CIE Lab in Sect. 3.2 (When we choose, we refer to some successful advice [11]). Let R be the video frame element, G $\in R_L$ be the photometric vector, C $\in R_{a,b}$ be the color vector, and our goal is to learn the mapping relation Q between the photometric vector G and the generated color vector $\check{C}$, where $\check{C}$ means "generated", that is, $\check{C} = Q(G)$, where Q is the convolutional network. Through the convolution process, we can obtain the corresponding color vector by the photometric vector feature. Generally, objective function is defined as the calculation of the relevant distance. Our aim is to minimize the distance between the generated color vector $\check{C}$ and the corresponding color vector C. Here we use the definition of Euclidean regression to describe the distance between the luminosity $\check{C}$ in the prediction video frame and C, with T to represent the objective function.

$$\mathrm{T}(\check{C}, C) = \sum\nolimits_X \| \check{C} - \mathrm{C} \| \tag{1}$$

2.5 Information Entropy and Objective Function

The above objective function can be applied to the convex structure input image set. The convex structure input image set refers to the average of the eigenvalues calculated by the Euclidean distance when the images in the target set are abundant and as different as possible. The average is similar to the optimal solution and the eigenvalue is the optimal solution. However, for an indeterminate input image set, the overall distribution may have multiple peaks for the concave function. The above-mentioned common objective function seems not good [4, 5]. In order to further correct this, we use the cross entropy to improve the objective function [6, 7]. Entropy is actually the expected value of information and a deterministic measure of random variables. The greater the entropy, the more uncertain the value of the variables while the smaller, the more certain. For a random variable X, the expectation (E [I (x)]) of all its possible values of information is called entropy. The entropy of X is defined as:

$$H(X) = E_p log\frac{1}{p(x)} = -\sum\nolimits_{x\epsilon X} logp(x) \tag{2}$$

On the basis of this definition, we introduce the cross entropy, calculate the minimum value of the fixed distribution C and the measured distribution Č by cross entropy, and obtain the objective function T_1:

$$T_1(\check{C}, C) = CEH(\check{C}, C) = -\sum\nolimits_{x\epsilon X} p(C_x) logp(\check{C}_x) \tag{3}$$

where x ∈ X is the photometric vectors G corresponding to the color vector C. Of course, in the specific calculation, we also need to take into account the height, width, light length of the image corresponding to the photometric vector. The generation of $\check{C}$ will be finished by using the objective function and the climbing algorithm. (Some works are also using this method, but our model is simpler [8–10]). We use the ImageNet dataset of 30,000 images as input data to train the network and try to color some black and white images which has obtained some good results.

2.6 A Cuckoo Search Algorithm

In this Section, we design the algorithm to migrate the color from frame to frame migration. Because of the huge number of frames, we need a fast and effective search algorithm. we design a cuckoo colorization search algorithm.

In nature, some animals are always in a random or quasi-random way to find food. The feeding path of an animal belongs to a random walk of space, and its step is able to keep to the power distribution. The next state and action of the animals are determined by the current location of the animal and the probability of state transition which is called "Levy flight" by scientists. CS (Cuckoo Search Optimization) is a meta-heuristic optimization algorithm proposed by Yang et al. in 2009. The algorithm includes the idea of breeding strategies and introduces the random flight behavior that other populations generally have, namely the Levi flight mechanism.

The mathematical model of the Levi flight algorithm is designed as follows:

$$X_i^{t+1} = X_i^t + a \otimes Levy(\lambda) \tag{4}$$

where X_i^{t+1} represents the new nest position, X_i^t represents the position of the i-th nest nested on the tth generation, and α is the step size factor, and usually a = 1. $Levy(\lambda)$ is a search vector that represents the Levy flight distribution:

$$Levy(\lambda) \sim v = t^{-\lambda} 1 \leq \lambda \leq 3 \tag{5}$$

Therefore, the introduction of the Levi flight mechanism, fully speed up the search speed of the algorithm, but also to avoid the solution into the local optimal.

2.7 Tracking Algorithm Based on Cuckoo Algorithm

The accuracy of the estimation of the video frame elements are closely related to the number of particles. The more the number of randomly selected particles, the more accurate the estimation of the posterior probability density distribution. However, the excess numbers of particles may reduce the efficiency of algorithms. The incorporation of the cuckoo search algorithm into the particle filter algorithm can reduce the complexity of the algorithm.

The basic idea of the algorithm to achieve moving target tracking is to express a host nest by each particle and perform the nested position of the flight mechanism, and finally gather all the nest and select an optimal nest to complete the state output. So as to optimize the tracking performance to achieve the tracking of moving targets. When the particle filter arrives after the new measurement, the weight is updated by recursive.

We combine the algorithm with the video stream and track the video. We selected 50 frames of continuous video as the object, which includes the camera fixed mobile, human fixed camera movement and other 10 video to shoot, the system can complete the tracking. We follow the specific pixel according to the pixel, we track a series of video, the series of video height H is 600, width W is 800, the selected pixel for the square, the side length were 100, 80, 60, 40, 20, 10, the overall retention rate of 97.2%, able to complete the tracking process which is showed in Fig. 3.

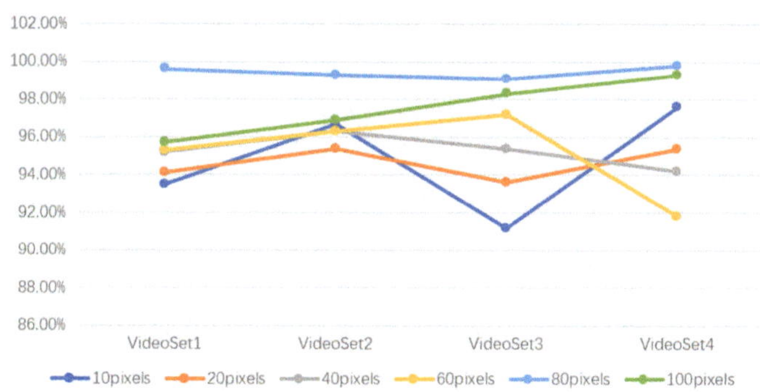

Fig. 3. The capture rate on 4 video sets which shows the rate is always higher than 90% and the algorithm can run well.

2.8 Video Segmentation Method

Section 2.7 introduces the key areas of the framework of the tracking method which is the video frame of the continuous tool. In fact, the experiment shows that coloring method is difficult to work long hours. This is in addition to tracking errors, but also because the performance elements within the video often occur non-differential changes, usually by a face immediately into another face or a landscape, in order to prevent the key point tracking method results. Together, we propose a simple but effective segmentation method.

We use the histogram representation method to segment the video. The histogram of the image is an effective tool for describing the basic composition of the image color. It is also applicable to the gray scale scene. We define the asymptotic vector after the gray scale is defined as R and set the asymptotic vector of the picture as $\check{R}$. We use R and R to describe the gray frame information of the corresponding image P. We set the threshold E and ideally Get the split formula:

$$\mathrm{Dis}\left(\mathrm{R}, \check{R}\right) = \sum\nolimits_{X} \| R - \check{R} \| \tag{6}$$

When Dis > E in the corresponding P and $\check{\mathrm{P}}$ between the video segmentation. But the experiment proved that this segmentation is not ideal, we found that segmentation algorithm is difficult to distinguish some scenes, such as multi-person dialogue can not identify the face of the switch, in a similar scene in the emergence and disappearance of elements.

We further transform the algorithm, do not pre-division of the video, but choose the appropriate time to interrupt the track and start to achieve the purpose of separation. Of course, the choice of timing is still the gray-scale frame histogram threshold changes, and still use the threshold comparison method. In contrast, we are using this for the relevant tracking target and its external expansion of the R description, and compared to the difference between the two, when the threshold is greater than E when the division, the specific implementation process as shown below (Fig. 4):

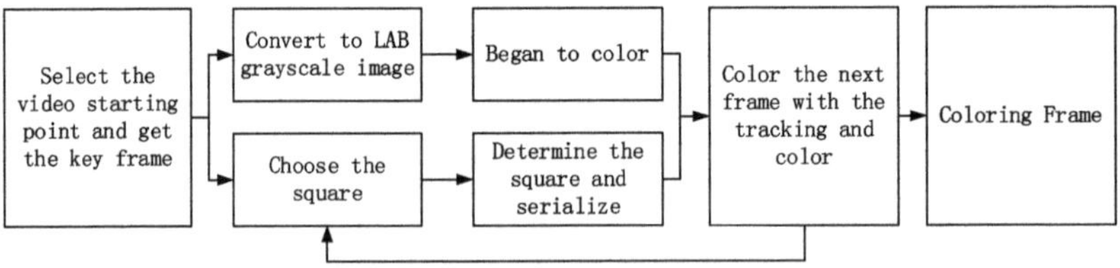

Fig. 4. The step of video segmentation method

3 Experiment and Verify

We selected three colorful videos to verify the technique. We show differences between original videos and gray videos to verify the effectiveness of our technique. We get continuous 100 frames of videos and enter them into our algorithm. The following Fig. 5 shows the results of our algorithm.

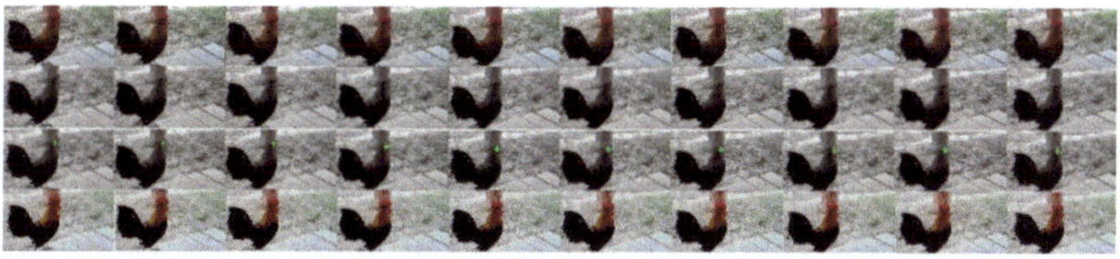

Fig. 5. The result of coloring the video. The first line of frames is original, the second line is gray frames, the third one shows tracking result, the last one is coloring video.

The main tracking elements are well performed, including a lot of features. But due to the color shift to the key, many positions are clearly expressed on the first frame of the picture, and the second pin is ambiguous on the picture. This is because the algorithm will weaken the performance of pixels around the key.

Finally, we color the classical movie Roman.Holiday.1953 which is Black and White. The Result is in Fig. 6 which proves our technique is effective.

Fig. 6. The first line is coloring movie with our technique, the second line is original movie.

4 Conclusion

From the results, CNN structure can complete the single-frame coloring process. Based on the cuckoo algorithm, the particle filter tracking algorithm can map the color to the relative feature. Intuitively, this work is gratifying. This paper presents a colorization technique which can be more effective solution to black and white video color problem.

References

1. Cheng, Z., Yang, Q., Sheng, B.: Deep colorization. In: Proceedings of the IEEE International Conference on Computer Vision, pp. 415–423 (2015)
2. Dahl, R.: Automatic colorization (2016). http://tinyclouds.org/colorize/
3. Charpiat, G., Hofmann, M., Schölkopf, B.: Automatic image colorization via multimodal predictions. In: Computer Vision ECCV 2008, 126–139. Springer (2008)
4. Ramanarayanan, G., Ferwerda, J., Walter, B., Bala, K.: Visual equivalence: towards a new standard for image fidelity. ACM Trans. Graph. (TOG) **26**(3), 76 (2007)
5. Simonyan, K., Zisserman, A.: Very deep convolutional networks for large-scale image recognition. arXiv preprint arXiv:1409.1556 (2014)

6. Bengio, Y., Courville, A., Vincent, P.: Representation learning: a review and new perspectives. IEEE Trans. Pattern Anal. Mach. Intell. **35**(8), 1798–1828 (2013)
7. Agrawal, P., Carreira, J., Malik, J.: Learning to see by moving. In: Proceedings of the IEEE International Conference on Computer Vision, pp. 37–45 (2015)
8. Jayaraman, D., Grauman, K.: Learning image representations tied to ego-motion. In: Proceedings of the IEEE International Conference on Computer Vision, pp. 1413–1421 (2015)
9. Pathak, D., Krähenbühl, P., Donahue, J., Darrell, T., Efros, A.: Context encoders: Feature learning by inpainting. In: CVPR (2016)
10. Lotter, W., Kreiman, G., Cox, D.: Deep predictive coding networks for video prediction and unsupervised learning. arXiv preprint arXiv:1605.08104 (2016)
11. Owens, A., Isola, P., McDermott, J., Torralba, A., Adelson, E.H., Freeman, W.T.: Visually indicated sounds. In: CVPR (2016)
12. Owens, A., Wu, J., McDermott, J.H., Freeman, W.T., Torralba, A.: Ambient sound provides supervision for visual learning. In: ECCV (2016)

Opinion Target Extraction for the Chinese Formal Text Based on Dependency Relations

Xiao-Yan Yang[1,2], Ge Xu[1,2], Fu-Quan Zhang[1,2,3(✉)], Xiang-Wen Liao[4], and Lin Xu[5]

[1] Department of Computer Science, Minjiang University, Fuzhou 350121, Fujian, China
8528750@qq.com, 1191@mju.edu.cn, xuge@pku.edu.cn

[2] Fujian Provincial Key Laboratory of Information Processing and Intelligent Control, Fuzhou 350121, Fujian, China

[3] School of Software, Beijing Institute of Technology, Beijing 10081, China

[4] College of Mathematics and Computer Science, Fuzhou University, Fuzhou 350108, Fujian, China
85802493@qq.com

[5] Innovative Information Industry Research Institute, Fujian Normal University, Fuzhou 350300, Fujian, China
xulin@fjnu.edu.cn

Abstract. Due to the increasing amount of opinion data on the internet, opinion mining has become a hot topic, in which extracting opinion targets is a key step. The state-of-the-art approaches only use direct dependency relation patterns to extract opinion targets and the indirect dependency relation patterns have not been used. In this paper, the dependency relations between opinion target and opinion word are defined, and direct and indirect dependency relation patterns are designed. Then, a bootstrapping approach is used to extract and evaluate both candidate patterns and opinion targets. The experimental results show that in formal text, the approach improves the performance compared with the state-of-the-art approaches for opinion target extraction.

Keywords: Bootstrapping · Opinion target · Dependency relation

1 Introduction

With the rapid development of Internet, people share their opinions. These opinion data contain valuable information, which are helpful for users in many tasks, such as, purchase decision, interest mining, personalized information service, information prediction, etc. However, manually scanning through large amounts of user-generated data, especially containing opinion, is time-consuming. In this case, opinion mining has become a hot topic.

One of the important pre-tasks in opinion mining is opinion target extraction. Its main task is to extract the entities and their attributes on which opinions have been expressed. Opinion target extraction is one of the subtasks of some large-scale evaluations, such as TREC, NTCIR and the Chinese Opinion Analysis Evaluation.

J.-S. Pan et al. (eds.), *Advances in Smart Vehicular Technology, Transportation, Communication and Applications*, Smart Innovation, Systems and Technologies 86,
https://doi.org/10.1007/978-3-319-70730-3_38

The state-of-the art approaches of extracting opinion target are based on the dependency relations [1] between known opinion words and opinion targets. Qiu [2] defined direct and indirect dependency relations between words. However, Qiu [2] only utilized direct dependency relations extract opinion targets. Zhang [3], Liu [4], Wang [5] extended the work of Qiu [2] but still utilized direct dependency relations extract opinion targets.

In formal texts, such as, news reports, official documents, articles, business letters, sentences are longer and more complicated. There are many words and phrases between the opinion word and the target that it modifies. The dependency relations can analyze long-distance relations between opinion words and opinion target. So, in this paper we define direct and indirect dependency relations and design the direct and indirect dependency relation patterns to extract opinion target for the formal text.

The results of the experiment on the COAE2011 dataset show that the proposed method can effectively utilize the dependency relations between opinion words and opinion targets, and improve the performance of opinion target extraction based on the state-of-the art method.

2 Related Work

Currently, the methods of opinion target extraction can be divided into unsupervised machine learning [6–8], supervised machine learning [9, 10], semi-supervised machine learning. Semi-supervised machine learning based bootstrapping approaches [11–14] are very successful.

Zhang [11] utilized the graph model to link and measure the relationship between the pairs of product features and opinion words. The shortcoming of this method is that only the co-occurrence rate of the candidate opinion targets and opinion words is considered, so the recall rate is low. Wei [12] used the synonyms and near-synonyms of opinion targets as candidate opinion targets. But this method brought many low frequency words, which influence the identifying performance. Song [13] utilized the bootstrapping to extract the opinion target based on the word patterns and part-of-speech (POS) patterns. However, this method only considered the word and POS information, and didn't use the syntactic information between opinion targets and opinion words. Qiu [2] proposed to utilize the dependency relations between opinion targets and opinion words to extract opinion targets based on double propagation strategy. Zhang [3] extended the work of Qiu [2], he designed some heuristic patterns to indicate opinion target candidates. Liu [4] proposed a partially-supervised alignment model based on the dependency relation patterns defined in Qiu [2]. Wang [5] proposed a combination of the pipeline framework and Qiu [2] method for opinion mining. But the direct dependency relations based on the Qiu [2] method is used in methods [3–5], and the indirect dependency relations haven't been used.

3 A Bootstrapping Approach for Opinion Target Extraction Based on Dependency Relations

3.1 The Definition of Dependency Relations Between Opinion Target and Opinion Word

In this paper, we define direct and indirect dependency relations between opinion target and opinion word as follows:

- Definition of direct dependency relations between opinion target and opinion word: opinion target depends on opinion word without any additional words in their dependency path.

Examples of direct dependency relations between opinion target and opinion word are shown in Fig. 1(a) and (b). T is the tag used to refer to opinion target. O is used to refer to opinion word. Figure 1(a) shows the opinion target is the parent node of opinion word. Figure 1(b) shows the opinion target is the child node of opinion word.

- Definition of indirect dependency relations between opinion target and opinion word: opinion target depends on opinion word through some additional words.

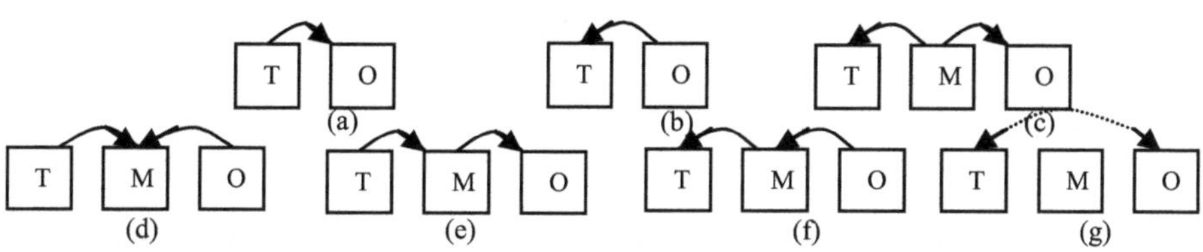

Fig. 1. The dependency relations between opinion target and opinion word

Examples of indirect dependency relations between opinion target and opinion word are shown in Fig. 1(c), (d), (e), (f) and (g). Figure 1(c) shows the opinion target and opinion word have the same parent node. Figure 1(d) shows the opinion target and opinion word have the same child node. Figure 1(e) shows the opinion target is the parent node of intermediate word, and the intermediate word is the parent node of opinion word. Figure 1(f) shows the opinion target is the child node of intermediate word, and the intermediate word is the child node of opinion word. Figure 1(g) shows the opinion target and opinion word generate dependency relations through a number of other words.

3.2 Dependency Relations Pattern Between Opinion Target and Opinion Word

According to the forms of dependency relations, this paper proposed a direct dependency relation pattern and an indirect dependency relation pattern between opinion target and opinion word.

The direct dependency relation pattern is defined in the following form (*dp*, *target*, *relate*, *opinion*). The "*dp*" refers to the annotation of dependency relation between opinion target and opinion word. The "*target*" refers to opinion target. The "*relate*" refers to the form of dependency relation between opinion target and opinion word. Here we use "−1" and "+1" to denote the two forms of direct dependency relations. "−1" is used to denote that opinion target is the parent node of opinion word, and "+1" is used to denote that opinion target is the child node of opinion word. The "*opinion*" refers to opinion word. The indirect dependency relation pattern is defined in the following form (dp_1, *target*, $relate_1$, $middle_1$) + … + (dp_{i-1}, $middle_{i-1}$, $relate_{i-1}$, $middle_i$) + (dp_i, $middle_i$, $relate_i$, *opinion*). The "dp_{i-1}" refers to the annotation of dependency relation between the $i-1$th and the ith intermediate word. The "$middle_{i-1}$" refers to the ith intermediate word. The "$relate_{i-1}$" refers to the form of dependency relation between the $i-1$th and the ith intermediate word. "−1" is used to denote that $i-1$th intermediate word is the parent node of ith intermediate word, "+1" is used to denote that $i-1$th intermediate word is the child node of ith intermediate word.

Take the sentence "这款产品最大的特点是性价比非常高。" (Fig. 2) as an example. Opinion target "性价比" is the child node of opinion word "高". The annotation of dependency relation between the opinion target and opinion word is "SBV". Using the direct dependency relation pattern, it is denoted as (SBV, 性价比, +1, 高).

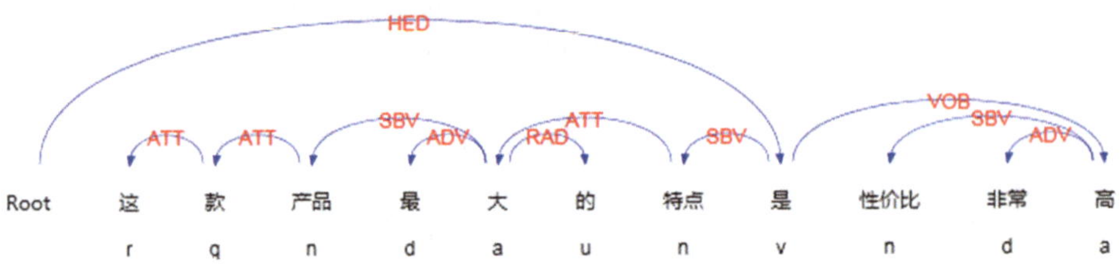

Fig. 2. Example of dependency relations between the opinion target and the opinion word

3.3 Calculate the Score of Candidate Opinion Target

We construct an a two-dimensional matrix as follows: $W = (w_{ij}), i = \{1, 2, \ldots, u\}, j = \{1, 2, \ldots, v\}$, where u refers to the number of known dependency relation patterns; v refers to the number of candidate opinion targets.

When the j-th candidate opinion target was extracted through the known opinion words and the known i-th dependency relation pattern, $w_{ij} = w_{ij} + 1$.

Equation (1) is used to calculate the score of candidate opinion target. When the score greater than the threshold value, candidate word is extracted as the opinion target.

$$sw_j = \frac{\sum_{i=1}^{u} w_{ij} * \sum_{i=1}^{u} \frac{w_{ij}}{\sum_{j=1}^{v} w_{ij}}}{u} \quad (1)$$

3.4 Calculate the Score of Candidate Dependency Relation Pattern

We construct a two-dimensional matrix as follows: $P = (p_{ij})$, $i = \{1, 2, \ldots, m\}$, $j = \{1, 2, \ldots, n\}$, where m refers to the number of opinion targets and n refers to the number of candidate dependency relation patterns.

When the j-th candidate dependency relation pattern was constructed through the known opinion word and the i-th opinion target, $p_{ij} = p_{ij} + 1$.

Equation (2) is used to calculate the score of candidate dependency relation pattern. When the score greater than the threshold value, candidate dependency relation pattern is extracted as the dependency relation pattern.

$$sp_j = \frac{\sum_{i=1}^{m} p_{ij} * \sum_{i=1}^{m} \frac{p_{ij}}{\sum_{j=1}^{n} p_{ij}}}{m} \tag{2}$$

3.5 The Bootstrapping Algorithm

The bootstrapping algorithm for opinion target extraction as follows:

Step 1: if a known opinion word matches the current word with the known dependency relation pattern, the current word was extracted as a candidate opinion target.
Step 2: calculate the score of the candidate opinion target using Eq. (1).
Step 3: if the score of candidate opinion target greater than a threshold value, the candidate opinion target is added into the opinion target set.
Step4: if a dependency relation between the known opinion word and the known opinion target exists, then the dependency relation pattern is constructed as candidate pattern.
Step 5: calculate the score of candidate dependency relation pattern using Eq. (2).
Step 6: if the score of candidate dependency relation pattern greater than a threshold value, the candidate dependency relation pattern is added into the dependency relation pattern set.
Step 7: repeat step1–step6 until no new candidate opinion targets or dependency relation patterns are generated

4 Corpus and Evaluation

The experimental data is consisted of 460 formal sentences, which come from COAE2011 financial and securities field. Each sentence in the corpus is pre-processed by word segmentation, POS tagging, dependency parsing, etc. Specifically, we employ the Language Technology Platform (LTP)[1] as the Chinese sentence parser. Then two annotators are asked to hand-annotate opinion words. This paper use Kappa value to

[1] http://www.ltp-cloud.com/intro/en/.

compute the consistence of the annotation results. The Kappa value is greater than 0.8, which shows a good agreement.

Precision, recall and F-score value are used as evaluation criteria.

$$\text{precision} = \frac{\text{true positives}}{\text{true positives} + \text{flase positives}}$$
$$\text{recall} = \frac{\text{true positives}}{\text{true positives} + \text{flase negatives}}$$
$$\text{F - score} = 2 \cdot \frac{\text{precision} \cdot \text{recall}}{\text{precision} + \text{recall}}$$

The evaluation method use strict evaluation [8]. Strict evaluation means that the extracted opinion targets match the correct opinion targets exactly.

5 Experimental Results and Analysis

The dependency relations of Qiu [2] include *ATT, SBV, COO, VOB and FOB*. The initial seed set of opinion targets in our method is {"外观", "性能", "设计", "配置", "功能", "价格", "色彩", "造型"}. The initial seed set of dependency relation patterns in our method is {("SBV", *target*, "+1", *opinion*), ("ATT", *target*, "−1", *middle*), ("ATT", *target*, "+1", *middle*) + ("VOB", *middle*, "−1", *opinion*), ("SBV", *target*, "+1", *middle*) + ("COO", *middle*, "−1", *opinion*)}.

Table 1 shows the experiment results of the comparison of our method and Qiu [2]. The precision, recall and F-score of our method are better than Qiu [2] (The strict evaluation is shown in Fig. 3). It is because that our method considers the indirect dependency relations.

Table 1. Comparison results

Evaluation	Strict evaluation (%)		
Method	Precision	Recall	F-score
Qiu [2]	36.87	27.00	31.18
Ours	45.72	37.83	41.40

For example, in the sentence "北海是最宜人居的现代化旅游城市" (Fig. 4). There exist an indirect dependency relation between the opinion word "最宜人" and the candidate opinion target "城市". The indirect dependency relation pattern is ("SBV", "城市", "−1", "居") + ("SBV", "居", "−1", "宜人") + ("ADV", "宜人", "−1", "最").

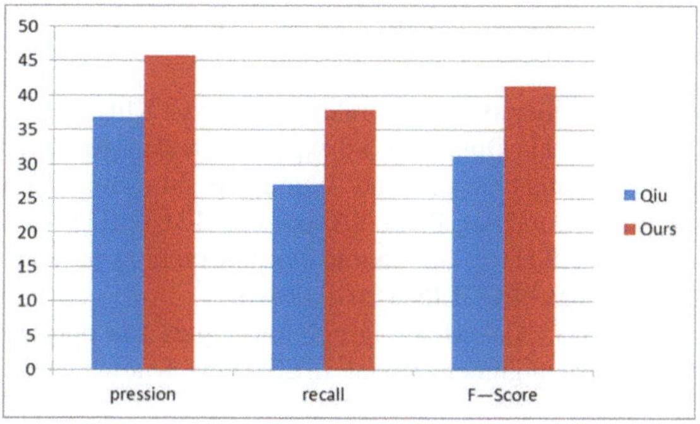

Fig. 3. Strict evaluation of Qiu [2] and ours

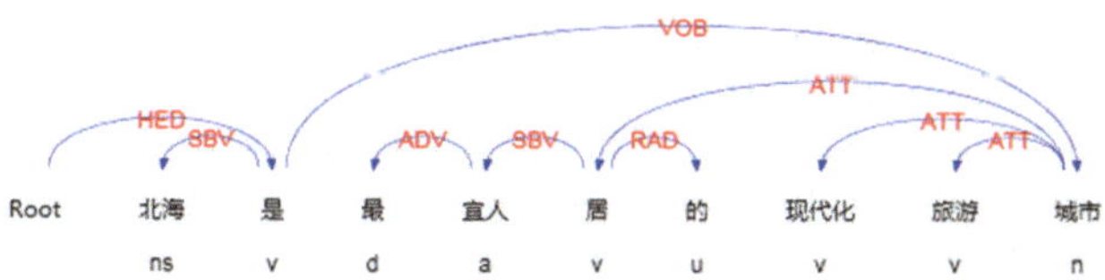

Fig. 4. Example of indirect dependency relations

In Qiu [2] method, it will not extract the opinion target "城市" if only use the direct dependency relations. In our method, the use of indirect dependency relation pattern can extract the opinion targets which have long-distance relationship with opinion words. Taking into consideration the use of the indirect dependency relation pattern is likely to be essential to the acquisition performance.

6 Conclusions and Future Works

In this paper a novel method is proposed to deal with the long-distance dependency problem of the state-of-the-art methods for extracting opinion target from the formal texts. We first define the direct and indirect dependency relations between opinion target and opinion word, and then design the direct and indirect dependency relation patterns. Finally, a bootstrapping approach is used to extract and evaluate both the opinion targets and dependency relation patterns. Experimental results based on the COAE2011 financial and securities field dataset show that the proposed method is more suitable for formal texts.

Our feature work will focus on how to improve the accuracy of the opinion target extraction.

Acknowledgements. Thanks to the Research Center for Social Computing and Information Retrieval of Harbin Institute of Technology for providing the Language Technology Platform (LTP). This work was funded by the Fujian Education Department (No. JAT160387) and also funded by the National Natural Science Foundation of China (No. 61300156) and also funded by the Research Program Foundation of Minjiang University (No. MYK17021).

References

1. Tesniere, L.: Elements de Syntaxe Structurale. Librairie C. Klincksieck, Paris (1959)
2. Qiu, G., Liu, B., Bu, J.J., et al.: Opinion word expansion and target extraction through double propagation. In: Proceedings of the 2nd Workshop on Computational Approaches to Subjectivity and Opinion Analysis, PA, USA, pp. 125–131 (2011)
3. Zhang, L., Liu, B., Lim, S.H., et al.: Extracting and ranking product features in opinion documents. In: Proceedings of COLING 2010 (2010)
4. Liu, K., Xu, L.H., Zhao, J.: Co-extracting opinion targets and opinion words from online reviews based on the word alignment model. IEEE Trans. Knowl. Data Eng. **27**(3), 636–650 (2015)
5. Wang, H., Zhang, C., Yin, H., Wang, W., Zhang, J., Xu, F.: A unified framework for fine-grained opinion mining from online reviews. In: 49th Hawaii International Conference on System Sciences, pp. 1530–1605 (2016)
6. Hu, M., Liu, B.: Mining and summarizing customer reviews. In: Proceedings of SIGKD, pp. 168–177. ACM, New York (2004)
7. Bloom, K., Garg, N., Argamon, S.: Extracting appraisal expressions. In: Proceedings of Conference on Human Language Technologies/North American Association of Computational Linguistics, pp. 308–315 (2007)
8. Zhang, L., Feng, X.: Extracting sentiment element from chinese micro-blog based on POS template library and dependency parsing. Comput. Sci. **42**(6A), 474–478 (2015)
9. Dai, M., Wang, R.Y., Li, S.S., et al.: Opinion target extraction with syntactic feature. J. Chin. Inf. Process. **28**(4), 92–97 (2014)
10. Zhang, L., Li, S., Peng, J., et al.: Feature-opinion pairs classification based on dependency. J. Univ. Electron. Sci. Technol. Chin. **43**(3), 420–425 (2014)
11. Zhang, S., Xia, Y.J., Meng, Y., Yu, H.: A bootstrapping method for finer-grained opinion mining using graph model. In: PACLIC, pp. 589–595 (2009)
12. Wei, J., Hung, H.H., Rohini, K.S.: Opinion miner: a novel machine learning system for web opinion mining and extraction. In: The 15th ACM SIGKDD International Conference on Knowledge Discovery and Data Mining, pp. 1195–1204 (2009)
13. Song, X.L., Wang, S.G., Li, H.X.: Research on comment target recognition for specific domain products. J. Chin. Inf. Process. **24**(1), 89–93 (2010)
14. Abney, S.: Bootstrapping. In: Proceedings of the Annual Meeting of the Association for Computational Linguistics, pp. 360–367 (2002)

Chaotic Sequence Generator Based on m Sequence Perturbation

Chuanfu Wang and Qun Ding(✉)

Key Laboratory of Electronic Engineering College of Heilongjiang University, Harbin, China
Qunding@aliyun.com

Abstract. Chaotic sequence after digitization is often affected by the limited word length effect. The pseudo-random sequences produced by chaotic sequence generators have short periodicity and less random characteristics. In order to solve the influence of short periodicity, m sequence theory is introduced into chaotic system, and a new scheme of chaotic sequence generator with m sequence perturbation is proposed. The hardware implementation of the new scheme effectively improves the stochastic characteristics of chaotic sequence generator without adding additional logic resources.

1 Introduction

With the advent of the information age, the analysis and processing of big data is the basis of many scientific and technological fields. With the development of network communication technology, the security of data transmission in the network has been paid more and more attention. Encryption algorithm is a mainstream way to ensure the safe transmission of information. However, as the basic module of cryptography, pseudo-random sequence generator has been applied in many fields of cryptography, especially in sequence cipher. Chaos is a new discipline [1–3] developed in recent decades. Its inherent complexity and dynamics make it have many characteristics, such as sensitivity to initial values, inherent randomness, ergodicity and determinacy. These characteristics are consistent with Shannon's "confusion" and "diffusion", which provides a theoretical basis for applying chaotic systems to pseudo-random sequence generator. Compared with traditional pseudo-random sequence generator, chaotic system is easier to implement by hardware, and has fewer logical resources, so it has a great application prospect [4–9]. The two salient features of a pseudo-random sequence generator are high speed and low resources consumption. Chaotic systems just meet these two advantages of pseudo-random sequence generators, but chaotic systems also have some disadvantages in themselves [10–12]. Because the chaotic system has limited word length effect when it is digitized, a short periodicity is generated when the chaotic system is implemented in hardware [13–15]. From the chaotic theory, the chaotic system

This work is partially supported by Natural Science Foundation of China (No. 61471158).

J.-S. Pan et al. (eds.), *Advances in Smart Vehicular Technology, Transportation, Communication and Applications*, Smart Innovation, Systems and Technologies 86,
https://doi.org/10.1007/978-3-319-70730-3_39

after digitization reduced the random characteristics of the output pseudo-random sequence. In order to overcome this shortcoming, some scholars have proposed a method of adding the perturbation in chaotic systems [16, 17]. The most common perturbation source is the m sequence. To a certain degree, the method improves the short cycle phenomenon caused by digital chaos. Although the effect of short cycle is solved, the additional m sequence generator also need many resources consumption and the inherent advantage of less occupied resource of the chaotic system is eliminated. To solve this problem, a new m sequence perturbation scheme is proposed in this paper. The new scheme does not increase the additional resources consumption of the m sequence generator, and optimizes the short period of chaotic pseudo-random sequence generator at the same time. The new scheme still retains the inherent advantages of chaotic pseudo-random sequence generator, and consumes less resources.

2 Chaotic System

There are many kinds of chaotic system, which can be divided into low-dimension chaotic system and high-dimension chaotic system according to the dimension of chaotic system equations. Although high-dimensional chaotic systems have more chaotic characteristics, they consume more resources. Compared with the high-dimensional chaotic systems, the low-dimensional chaotic system occupies less resources and has a wider application area. Therefore, we choose the Logistic map in one-dimensional chaotic system. The hardware of Logistic chaotic system is easy to implement and has good chaotic characteristics. Therefore, Logistic map is widely used in chaotic pseudo-random sequence generator.

Logistic map is a one-dimensional discrete chaotic system, which describes the insect model. The expressions for the Logistic map are very simple and easy to implement. The expressions for the Logistic map are as follows:

$$x_{n+1} = \mu x_n(x_n + 1) \tag{1}$$

μ is the parameter variable of Logistic map, and its range is $[0, 4]$. x_n is the current iteration input value of the Logistic map, and its range is $[0, 1]$. x_{n+1} is the current iteration output value of the Logistic map, and its range is $[0, 1]$.

2.1 Parameter Selection for Logistic Map

The parameters in the chaotic system have a great influence on the chaotic orbit. A chaotic system does not exhibit good chaotic characteristics in all parameter spaces. Therefore, it is very important to select the parameters of chaotic system. The influence of chaotic system parameters on the chaotic orbits can be reflected in the bifurcation diagrams of chaotic systems. The bifurcation diagram describes the relation between the current output of the chaotic system and the parameter variables of the chaotic system. The whole bifurcation diagram of the Logistic map is shown in Fig. 1, and the partial bifurcation diagram of the Logistic map is shown in Fig. 2.

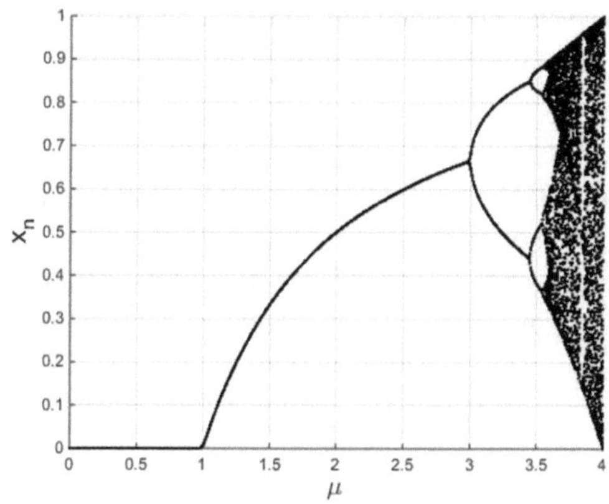

Fig. 1. The bifurcation diagram of the Logistic map, $\mu \in [0, 4]$

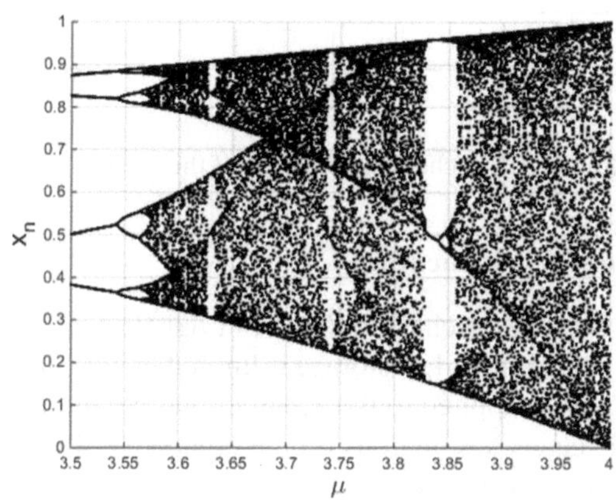

Fig. 2. The bifurcation diagram of the Logistic map, $\mu \in [3.5, 4]$

As you can see from Fig. 1, the Logistic map has a very distinct non-chaotic region. As shown in Fig. 2, the Logistic map parameter variable μ is in the range of $[3.5, 4]$, and the chaotic region is closely related to the non-chaotic region. Some scholars use the parameter μ as the initial key to increase the key space of the Logistic chaotic pseudo-random sequence generator. However, from the Figs. 1 and 2, we can see that the value space of the parameters μ in Logistic map are very small. it reduces the Logistic chaotic pseudo-random sequence generator safety. Therefore, in order to ensure that the Logistic chaotic system must be in the chaotic region and has the full ergodic property in the chaotic orbits, the μ is chosen as 4. The effect of parameters on the Logistic map chaotic orbits are shown in Table 1. The schematic diagram of the Logistic map is shown in Fig. 3.

Table 1. The effect of parameters μ on the Logistic map chaotic orbits

The value of μ	Orbits
$\mu \in (0, 1]$	Fixed point
$\mu \in (1, 3]$	Fixed point
$\mu \in (3, 3.571448]$	Period doubling bifurcation
$\mu \in (3.571448, 3.82842]$	Intermittency chaos
$\mu \in (3.82842, 3.85]$	Period trebling bifurcation
$\mu \in (3.825, 3.9]$	Explosive bifurcation
$\mu \in (3.9, 4]$	Chaos

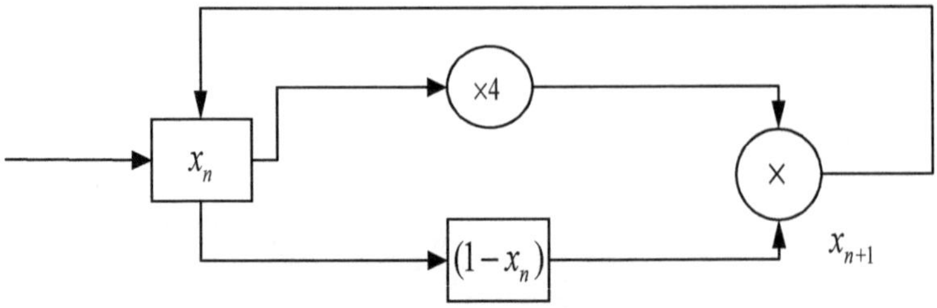

Fig. 3. The schematic diagram of the Logistic map

2.2 The Digitization of Logistic Map

All the above discussions are the characteristics of Logistic map in real domain. Because of the hardware implementation, data storage is limited in word length. Therefore, when the Logistic chaotic map is applied to the actual digital circuit, it needs to be digitized. When $\mu = 4$, Logistic chaos has obvious orbital ergodicity. 4 is a multiple of 2, which facilitates the realization of Logistic map digitization. The conversion formula between decimal and binary as follow:

$$x_{n+1} = \sum_{k=1}^{l} (y_k \cdot 2^{-k}) \tag{2}$$

The value of y_k is $\{0, 1\}$. The formula (3) is expressed in binary decimal numbers as follow:

$$x_{n+1} = 0.\ y_1 y_2 y_3 y_4 \cdots y_{l-1} y_l \tag{3}$$

The formula (3) is represented by a fixed point as follows:

$$x_{n+1} \cdot 2^l = y_1 y_2 y_3 y_4 \cdots y_{l-1} y_l \tag{4}$$

From y_i $(i = 1,\ 2,\ 3,\ 4, \ldots\ l)$, any number of bits can be used as the outputs of the Logistic chaotic pseudo-random sequence generator.

3 Logistic Map Based on m Sequence Perturbation

3.1 m Sequence

Linear feedback shift register is a very common pseudo-random sequence generator. The schematic diagram of the linear feedback shift register is shown in Fig. 4.

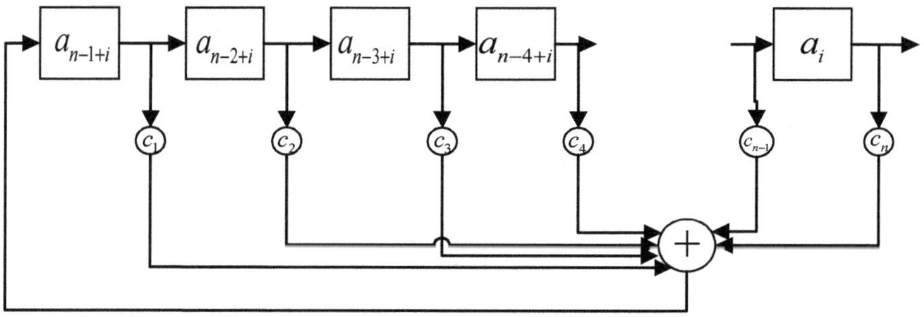

Fig. 4. Linear feedback shift register

The value of $a_i, a_{i+1}, \ldots\ldots$ and $c_i (i = 1, 2, 3 \ldots n)$ is $\{0, 1\}$. Addition in the finite field is xor operation. If the initial value of the linear feedback shift register is $a_0, a_1, \ldots a_{n-1}$, then the a_n can be obtained as follow:

$$a_n = (c_1 a_{n-1} + c_2 a_{n-2} + \cdots + c_n a_0) \tag{5}$$

The linear recurrence relation of the linear feedback shift register is as follow:

$$a_k = \sum_{i=1}^{n} c_i a_{k-i}, \quad k \geq n \tag{6}$$

The cycle length linear feedback shift register output sequence depends on the value of c_i. when the value of c_i is a primitive polynomial over finite field, linear feedback register output sequence has a maximum period, and is named m sequence. The m sequence is a pseudo-random sequence with the largest periodicity in the output sequence of the linear feedback shift register. The m sequence has very good stochastic properties, and it satisfies the three theories of Golomb randomness assumptions.

3.2 Logistic Map Based on m Sequence Perturbation

Adding perturbation is a common method to enhance chaotic characteristics of chaotic systems after digitization. There are many ways to increase the disturbance, and there are many alternatives for the source of disturbance. The m sequence is the most widely used for saving the required resources consumption of pseudo-random sequence generators. The chaotic pseudo-random sequence generator based on m sequence perturbation is shown in Fig. 5.

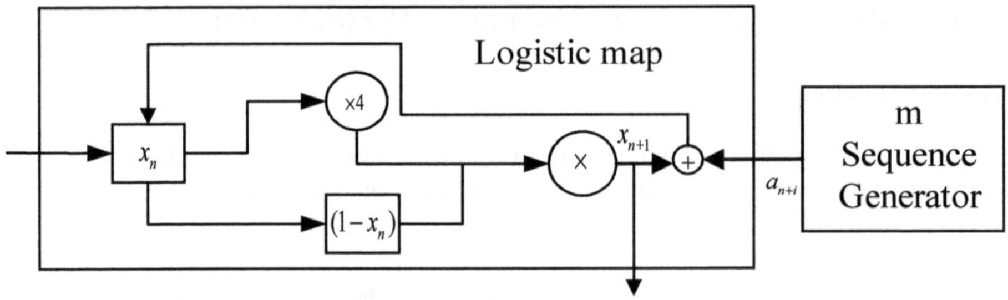

Fig. 5. The Logistic map pseudo-random sequence generator based on m sequence perturbation

As shown in Fig. 5, a_{n+i} as the output the m sequence generator, is the input of the Logistic map. The scheme adds a m sequence generator based on the Logistic chaotic pseudo-random sequence generator. Although the initial value of the m sequence can provide additional key space, the m sequence itself is very vulnerable. Berlekamp-Massey algorithm can easily find the initial value of m sequence. So, the scheme does not actually increase the key space of sequence generator. Since the scheme does not actually increase the key space of the pseudo-random sequence generator, and adds the additional resources of the m sequence generator. So, a new chaotic sequence generator based on m sequence perturbation is proposed in this paper. The principle block diagram of a new Logistic map sequence generator based on m sequence perturbation is shown in Fig. 6.

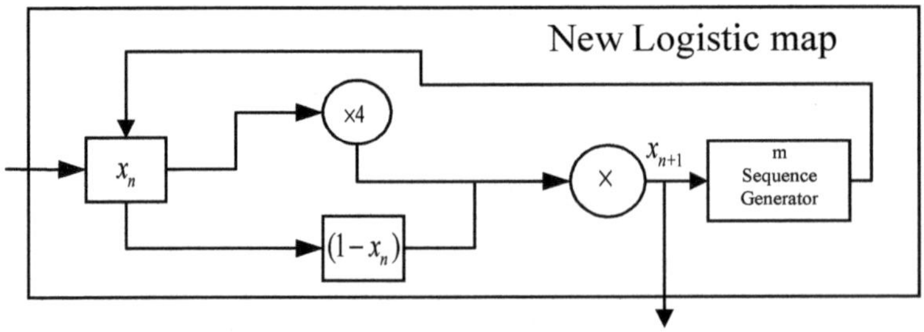

Fig. 6. New Logistic map sequence generator based on m sequence perturbation

By fixed-point methods, x_{n+1} can be represented by $y_1y_2y_3y_4 \cdots y_{l-1}y_l$. Then make $y_1y_2y_3y_4 \cdots y_{l-1}y_l$ equal to $a_{l-1}a_{l-2}a_{l-3}a_{l-4} \cdots a_1a_0$, and get a_l by the m sequence generator. Finally, assign $a_la_{l-1}a_{l-2}a_{l-3} \cdots a_2a_1$ to $y_1y_2y_3y_4 \cdots y_{l-1}y_l$.

$$y_1y_2y_3y_4 \cdots y_{l-1}y_l = (c_1y_1 + c_2y_2 + \cdots + c_ly_l)\, y_1y_2y_3 \cdots y_{l-1}y_{l-1} \tag{7}$$

Using the m sequence generator as part of the Logistic map iteration reduces the attack which can be generated by the m sequence generator alone. While reducing the resource consumption of the m sequence generator, the m sequence generator and the logistic map pseudo-random sequence generator are fused closely. This new scheme of logistic map pseudo-random sequence generator has higher security and more compact hardware implementation. The hardware implementation of new logistic map pseudo-random sequence generator based on m sequence perturbation in FPGA is shown in Fig. 7, and the output waveform simulation is shown in Fig. 8.

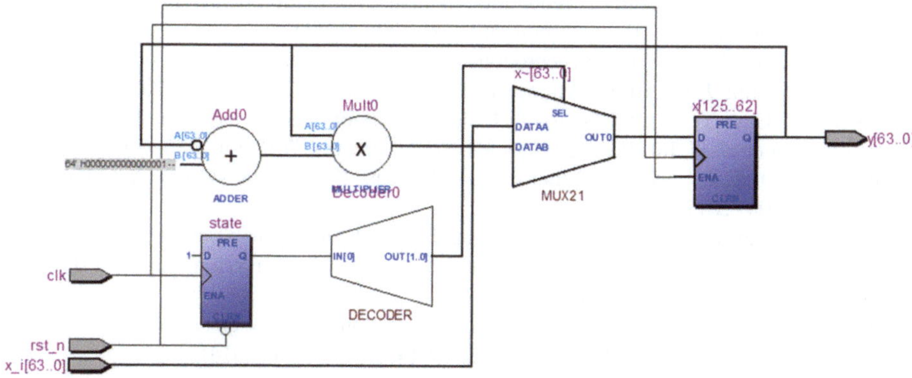

Fig. 7. The hardware implementation of new logistic map pseudo-random sequence generator

Fig. 8. The simulation of output waveform

4 Conclusion

With the development of information security, reducing the hardware consumption of pseudo-random sequence generator is a very important factor in the design of pseudo-random sequence generator. The new m sequence perturbation scheme in this paper eliminates the additional hardware resources consumption caused by the m sequence generator itself, and fuse the m sequence generator with the Logistic map perfectly. Pseudo-random sequence generator is the most widely used basic security module in the field of information security. A pseudo-random sequence generator with smaller resources consumption can integrate into many fields of information security, and break the limitations of the previous pseudo-random sequence generator which requires more hardware resources. So, the new pseudo-random sequence generator has a very large application prospect.

References

1. Lorenz, E.N.: Deterministic nonperiodic flow. J. Atmos. Sci. **20**, 130–141 (1963)
2. Lorenz, E.N.: The Essence of Chaos. Meteorology Press (1997). pp. 7–11
3. Li, T.Y., Yorke, J.A.: Period three implies chaos. Am. Math. Mon. **82**(10), 985–992 (1975)

4. Zhang, Z., Ding, Q.: A new super resolution algorithm in ship-borne radar. J. Internet Technol. **14**(1), 153–159 (2013)
5. Pan, J., Ding, Q., Du, B.: A new improved scheme of chaotic masking secure communication based on Lorenz system. Int. J. Bifurcat. Chaos **22**(5), 56–64 (2012)
6. Ding, Q., Wang, J.: Design of frequency-modulated correlation delay shift keying chaotic communication system. IET Commun. **5**(7), 901–905 (2011)
7. Li, W., Yang, Z., Ding, Q.: Synchronous digital signal communication system based on chaotic masking. J. Inf. Hiding Multimed. Signal Process. **7**(2), 304–316 (2016)
8. Zhang, Q., Guo, Y., Li, W., Ding, Q.: Image encryption method based on discrete Lorenz chaotic sequences. J. Inf. Hiding Multimed. Signal Process. **7**(3), 576–586 (2016)
9. Wang, C., Ding, Q.: SM4 key scheme algorithm based on chaotic system. Acta Phys. Sin. **66**(2), 020504 (2017)
10. Ding, Q., Pang, J., Fang, J., Peng, X.: Designing of chaotic system output sequence circuit based on FPGA and its possible applications in network encryption card. Int. J. Innov. Comput. Inf. Control **2**(3), 449–456 (2007)
11. Pan, J., Ding, Q., Ning, L., Zheng, Y.: Different random distributions research on Logistic-based sample assumption. Math. Probl. Eng. **2014**(8), 144–151 (2014)
12. Zheng, Y., Pan, J., Song, Y., Cheng, H., Ding, Q.: Research on the quantifications of chaotic random number generator. Int. J. Sens. Netw. **15**(1), 139–143 (2014)
13. Wang, H., Song, B., Pan, J., Ding, Q.: FPGA design and applicable analysis of discrete chaotic maps. Int. J. Bifurcat. Chaos **24**(4), 1–15 (2014). 1450054
14. Pan, J., Ding, Q., Zheng, Y., Wang, H., Ning, L.: Random process representations and some researches of chaos. Int. J. Sens. Netw. **14**(4), 241–250 (2013)
15. Du, B., Geng, X., Chen, F., Pan, J., Ding, Q.: Generation and realization of digital chaotic key sequence based on double K-L transform. Chin. J. Electron. **22**(1), 131–134 (2013)
16. Tao-Hu, L.I., Gao, B.S.: A long-period chaotic sequence generator based on logistic-map **44**(20), 77–80 (2014)
17. Li, F.G.: Effects of noise on periodic orbits of the logistic map. Open Phys. **6**(3), 539–545 (2008)

Research on the Complexity of Binary Chaotic Sequences

Liu Chunyuan[1,2], Ding Qun[1(✉)], and Xu Wei[3]

[1] School of Electronic Engineering, Heilongjiang University, Harbin 150080, China
Qunding@aliyun.com
[2] School of Computer and Information Engineering, Heilongjiang University of Science and Technology, Harbin 150022, China
[3] School of Computer Science and Technology, Heilongjiang University, Harbin 150080, China

Abstract. Binary sequence of chaos has been widely applied in research and studies on secret communication. Binary chaotic pseudo-random sequence is obtained through assessing and quantifying chaotic sequence and the complexity of binary sequence is got hardly. In this paper, the chaotic sequence linear complexity is discussed, and a new criterion is proposed on binary sequence, based on the approximate entropy. Simulations indicate that the method is effective, and the corresponding theories are proved right.

Keywords: Complexity · Binary sequence · Approximate entropy

1 Introduction

Chaos is classic and complex phenomenon in the nonlinear dynamics system. In recent decades, it has been widely applied in research and studies on secret communication [1–3]. Because chaotic sequence has broadband and high sensibility to original state, it can take the place of traditional pseudorandom sequence, and be applied in military and commercial spread spectrum communication system, which pose high requirements on secrecy. Different areas involve research on complexity. The complexity of finite-length sequence refers to level of similarity with random sequence. It is the measurement of difficulty in recovering the entity with partial sequence. Therefore, it is an important indicator of anti-interference and anti-interception ability of spread spectrum sequence in secret communication system. Hence, in-depth research and studies on complexity of chaos contain great importance. The potential value of chaos in spread spectrum communication has attracted widespread attention [4–6].

Currently, as for spread spectrum sequence design, the Berlekamp-Massey algorithm is adopted to assess the sequence complexity. However, research shows

This work is partially supported by Natural Science Foundation of China (No. 61471158).

J.-S. Pan et al. (eds.), *Advances in Smart Vehicular Technology, Transportation, Communication and Applications*, Smart Innovation, Systems and Technologies 86,
https://doi.org/10.1007/978-3-319-70730-3_40

that this algorithm can't effectively assess the complexity of binary sequences quantified by chaotic random sequence. As a consequence, a type of complexity criterion is required. Based on the research concerning chaotic sequence complexity, the paper also utilizes the information quantity induced by chaotic motion to measure the complexity of chaotic pseudo-random sequence. Approximate entropy (ApEn) is used as the standard to measure complexity. Theoretical and experimental results indicate that this method can used observed data to assess the complexity of binary sequence [7,8].

2 ApEn Method for Complexity Analysis

Literature [9] discusses an effective ApEn method to calculate and measure ApEn. Theories and experiments demonstrate: ApEn method assesses the possibility in signal generation by measuring sequence complexity. In theory, the main feature of approximate entropy is the positive proportional relationship between time sequence complexity and approximate entropy. In order to differentiate complexity levels of chaotic pseudo-random sequences, inherent features of chaotic system should involved in the process. Neighboring tracks of chaotic system refer to the number of tracks separated at the exponential speed within a period of time. If the number M is large, the complexity also increases. For chaotic motion, M indicates exponential increase over the course of time, i.e.

$$M \propto e^{Kt} \tag{1}$$

The constant K is metric entropy, which reveals the generation of chaotic motion. For regular motions, $k = 0$. For random motions, $K = +\infty$. For chaotic motion $0 < K < +\infty$, when K increases, randomness raises, complexity enhances, and it is less likely to be recovered. K-S entropyis an effective method of measurement. But this method requires large observed sample sequences with large calculation amount. The method of ApEn measurement solves the problem. Specific steps are stated in the following:

(1) For sequence sample space $\{x(1), x(2), x(3), \ldots, x(N)\}$ with the length of N, sequence $\{x_i\}$ forms the m dimension vector, i.e.:

$$X(i) = [x(i), x(i+1), \ldots, x(i+m-1)] \tag{2}$$

$i = 1, 2, \ldots, N - m + 1$.

(2) $d[X(i), X(j)]$ is defined as the maximum difference value in the corresponding elements between $X(i)$ and $X(j)$, i.e.:

$$d[X(i), X(j)] = \max_{k=1,2,\ldots,m} |x(i+k-1) - x(j+k-1)| \tag{3}$$

For each i value that calculates $X(i)$ and remaining vector $X(j)$, the distance $d[X(i), X(j)]$ of $(j = 1, 2, \ldots, N - m + 1)$.

(3) According to the given threshold value $r(r > 0)$, the ratio of $d[X(i), X(j)] \leq r$ for each i value to the total number of vectors $N - m + 1$ is recorded as $C_i^m(r)$

$$C_i^m(r) = \frac{1}{N-m+1} num\{d[X(i), X(j)] \leq r\} \quad i = 1, 2, \ldots, N-m+1 \quad (4)$$

(4) Obtain the logarithm of $C_i^m(r)$, and then calculate the meaning value of all i, which is recorded as $\Phi^m(r)$, i.e.:

$$\Phi^m(r) = \frac{1}{N-m+1} \sum_{i=1}^{N-m+1} \log C_i^m(r) \quad (5)$$

(5) The dimension is added by 1, which becomes $m+1$. Steps (1)–(4) are repeated to obtain $\Phi^{m+1}(r)$ and $C_i^{m+1}(r)$

$$C_i^{m+1} = \frac{1}{N-m} num\{d[X(i), X(j) \leq r]\} \quad i = 1, 2, \ldots, N-m \quad (6)$$

$$\Phi^{m+1}(r) = \frac{1}{N-m} \sum_{i=1}^{N-m} \log C_i^{m+1}(r) \quad (7)$$

(6) Approximate entropy of the sequence is:

$$\text{ApEn}(m, r, N) = \Phi^m(r) - \Phi^{m+1}(r) \quad m \geq 2 \quad (8)$$

Maximum value of vector dimension m is determined by the length N of observed space. When m becomes greater, ApEn approaches the metric entropy. The distance parameter r determines resolution ratio of the algorithm. Less r signifies greater higher resolution of ApEn.

3 Complexity of Binary Chaotic Random Sequence

Chaotic pseudo-random sequence is obtained through assessing and quantifying chaotic iteration sequence. Hence, two types of complexity performance ratio need to be measured: complexity of chaotic system and chaotic pseudo-random sequence. For complexity of chaotic system, chaotic sequence, generated by iterative mapping, can be used in Eq. (8) to calculate ApEn. It reveals the complexity of chaotic movement. Binary sequences can also be calculated by Eq. (8). However, as pseudo-random sequence has discrete value, which is a special scenario, calculation method is different.

Literature [10] conducts detailed discussion on the selection of binary sequence m. Ornstein and Weiss' research demonstrates: For K-scale random sequence with length N, in order to assess the probability of sequence block with m, length increase of observed sequence is decided by exponential growth of space K. Hence, the minimum length of observed section $K^{2^m} < N$, i.e.

$$m_{\max} = \max(m : K^{2^m} < N) \quad (9)$$

When calculating complexity of chaotic movement system, parameter r is the distance of measurement, which determines the precision. When r increases, ApEn will decrease. Hence, when $r \to 0$, ApEn approaches the real situation. Meanwhile, parameter m is not only decided by N, but also r. For random variables independently distributed in (0, 1), the length of observed sequence $(1/r)^{2^m} < N$.

Because calculating ApEn requires less r, the least distance of discrete sequence as the value of r. If $r = 0$ is selected for binary pseudo-random sequence, then Eq. (4) can be converted to:

$$C_i^m(r) = \frac{1}{N-m+1} num\{d[X(i) = X(j)]\} \quad i = 1, 2, \ldots, N-m+1 \tag{10}$$

The calculation equation of binary random sequence ApEn is identical to Eq. (8).

Take the binary sequence induced by logistic quantification. Figure 1 is the simulation results of $N = 300$ ($m_{\max} = 1$) and $N = 5000$ ($m_{\max} = 2$). The results indicate that when $m \geq 2$, $N = 300$, ApEn and $N = 5000$ show significant differences. It verifies the validity of Eq. (1). The value of m is dependent on N.

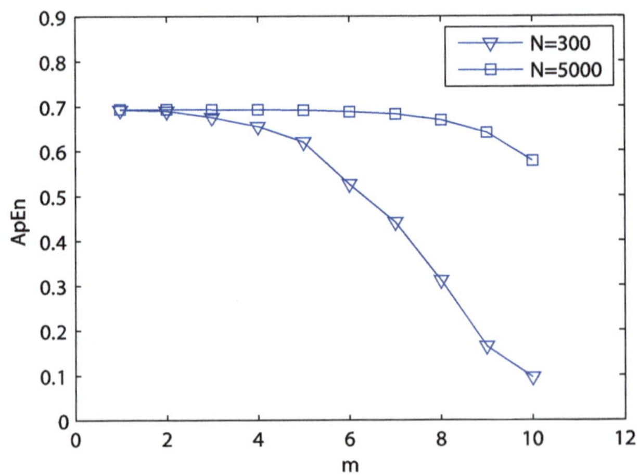

Fig. 1. ApEn of binary random sequence.

4 Simulation and Analysis

According to the above analysis, approximate entropy needs to be used to conduct test, comparison and analysis on chaotic sequence and binary random sequence of chaotic system induced and quantified by Logistic.

The chaotic mapping of Logistic defined as the following:

$$x_{n+1} = \mu x_n(1 - x_n) \quad 0 \leq x \leq 1, 0 < \mu \leq 4 \tag{11}$$

x_n is the value of logistic chaos. μ is the system parameter. When $\mu \in (3.57, 4]$, the system is in chaotic system. The chaotic signal indicates random phenomenon the time domain. Figure 2 is the bifurcation diagram of logistic chaos system.

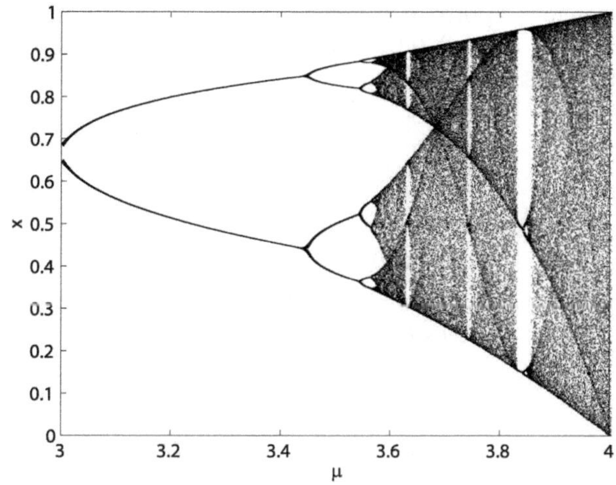

Fig. 2. Bifurcation diagram of logistic chaos time domain.

For chaotic quantification, the irreversible transfer function $T[x(n)]$ is used to convert intrinsic value sequence in chaos equation to 0-1 sequence. The irreversible transfer function $T[x(n)]$ is defined as the following:

$$T[x(n)] = \begin{cases} 0 & x(n) \in \bigcup_{k=0}^{2^m-1} I_{2k}^m \\ 1 & x(n) \in \bigcup_{k=0}^{2^m-1} I_{2k+1}^m \end{cases} \tag{12}$$

According to different original value $x(n)$ in continuous and equal sub-intervals of the irreversible transfer function $T[x(n)]$, 0 or 1 is selected as the value. $m > 0$, which is also arbitrary integer. $I_0^m, I_1^m, I_2^m, \ldots$ are the 2^m continuous and equal sub-interval in [0, 1]. In order to achieve function, Eq. (12) is converted as the following:

$$T[x(n)] = \begin{cases} 0 & 2^m x(n) \in [2k,(2k+1)) \\ 1 & 2^m x(n) \in [(2k+1),(2k+2)] \end{cases} \tag{13}$$

Approximate entropy $\mathrm{ApEn}(m, r, N)$ is used to conduct complexity analysis on binary sequence of quantified logistic chaos sequence. Results are shown in Table 1. For parameter selection, m is the dimension of distance vector. According to the above analysis, the value of N includes 300, 5000 and 20000 for binary random sequence. $r = 0$, and m respectively includes 1, 2, 3, 4 and 5. When approximate entropy is used to assess the complexity of binary sequence, if the distance vector m and precision vector r are accurate, result of complexity test is not highly dependent on parameter N. However, if the sample space N is large enough, only precision vector r impacts approximate entropy. For quantified binary sequence, approximate entropy algorithm can effectively assess the complexity, which is around 0.69. Besides, the maximum Lyapunov index of logistic mapping is also 0.68.

Table 1. ApEn of binary sequence

m	ApEn $= (m, 0, N)$		
	$N = 300$	$N = 5000$	$N = 20000$
1	0.6920	0.6931	0.6930
2	0.6897	0.6930	0.6929
3	0.6748	0.6929	0.6927
4	0.6544	0.6923	0.6923
5	0.6196	0.6912	0.6914

For ApEn algorithm adopted tested for random sequence and mapping induced by logistic, ApEn parameters should involve $m = 2$, $r = 0.1$ SDx, $N \geq 1000$. In order for logistic mapping to generate complex chaotic sequence, control parameter $\mu = 4$ can be selected. Figure 3 shows the different μ values of approximate entropy when original values are $x_0 = 0.4$, $m = 2$, $r = 0.1$ SDx,

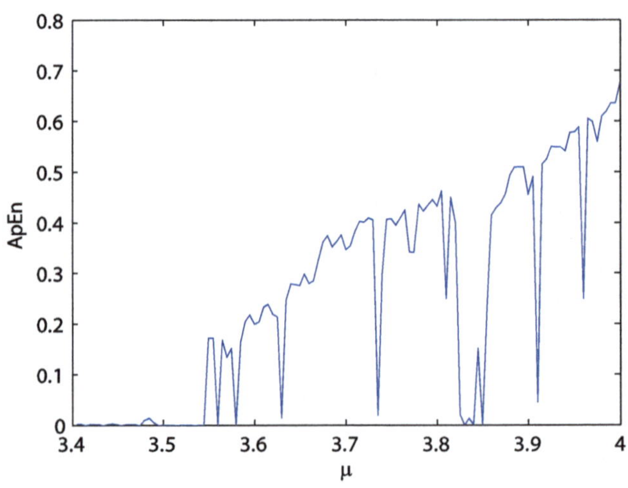

Fig. 3. ApEn of logistic map on

SDx = 0.348. When $\mu = 4$, value of approximate entropy should also approach the maximum value of Lyapunov index.

5 Conclusion

The paper analyzes the limitation of K-S entropy algorithm assessing complexity of chaotic pseudo-random sequence. Besides, approximate entropy is used to determine the random metric property of chaotic movement, and analyze the parameters for using ApEn to assess complexity of binary sequence. Besides, the paper also adopts transfer function to convert random sequence to binary sequence. Theoretical and experimental results demonstrate that this method can effectively assess the complexity of chaotic system and binary sequence. Research results provides theoretical and experimental basis for the application of chaotic sequence in information security. The complexity of digitized chaotic sequence contains great significance in the research on chaotic sequence complexity.

References

1. Schneier, B.: Secrets Digital Security in Networked World. Wiley, New York (2000)
2. Cai, J.-P., Li, Z., Song, W.-T.: Analysis on the chaotic pseudo-random sequence complexity. Acta Phys. Sin. **52**, 1871–1876 (2003)
3. Sviridova, N., Sakai, K.: Human photoplethysmogram: new insight into chaotic characteristics. Chaos, Solitons Fractals **8**(77), 53–63 (2015)
4. Li, L.: Bifurcation and chaos in a discrete physiological control system. Appl. Math. Comput. **12**(252), 397–404 (2015)
5. Dachselt, F., Schwarz, W.: Chaos and cryptography. IEEE Trans. Circuits Syst. I: Fundam. Theor. Appl. **48**(12), 1498–1509 (2001)
6. Naderi, B., Kheiri, H.: Exponential synchronization of chaotic system and application in secure communication. Optik - Int. J. Light Electron Opt. **5**(127), 2407–2412 (2016)
7. Kohda, T., Tsuneda, A.: Statistics of chaotic binary sequences. IEEE Trans. Inf. Theory **43**(1), 104–112 (1997)
8. Grassberger, P., Procaccia, I.: Estimation of the Kolmogorov entropy from a chaotic signal. Phys. Rev. A **28**(4), 2591–2593 (1983)
9. Pincus, S.M.: Approximate entropy as a measure of system complexity. Proc. Natl. Acad. Sci. **88**, 2297–2301 (1991)
10. Ornstein, D.S., Weiss, B.: How sampling reveals a process. Ann. Probab. **18**(3), 905–930 (1990)

Designing a Lightweight Cipher Algorithm Based on Chaos Theory and LoRaWAN

Chunlei Fan and Qun Ding(✉)

Electronic Engineering College of Heilongjiang University, Harbin, China
Qunding@aliyun.com

Abstract. Aiming at a large number of Sensor Nodes is restricted by hardware resources about the application of Internet of Things so that the AES and CMAC algorithm of LoRaWAN protocol cannot run normally. A novel lightweight block cipher algorithm is proposed with the purpose of overcoming the disadvantage, which combines Logistic chaotic mapping with Arnold transformation. The cipher algorithm considers that two important factors, namely hardware resources consumption and security. It is compatible with LoRaWAN protocol and suitable for limited hardware resources of sensor nodes.

Keywords: Lightweight algorithm · Arnold · Internet of things · Logistic mapping

1 Introduction

With the development of the wireless sensor network and Internet of Things, long range (LoRa) is widely concerned. LoRa combines with digital spread spectrum, digital signal processing and forward error-correcting coding technology, which possesses a large number of characteristics, including low speed, wide coverage, low-cost and low-power consumption, etc. LoRa technology is widely applied in Low Power Wide Area Network. LoRa alliance designs LoRaWAN protocol with the purpose of providing a wireless networking mode for Low Power Wide Area Network. Security mechanism of LoRaWAN protocol adopts Advanced Encryption Standard and CMAC message authentication algorithm to ensure security and integrity of wireless data frame. However, the limited hardware resources of sensor nodes are unsuitable for heavyweight cipher algorithm. Besides, chaotic theory has good pseudo randomness, initial value sensitivity and resembles noise features so that chaotic systems were widely used in secret communication and cryptography field.

At present, researchers have carried out a number of investigations and have already gained some achievements in this respect. References [1] proposes a hybrid stream cipher based on chaos. Reference [2, 3] analyzes a secure sink node architecture for wireless sensor network. Reference [4] researches chaos block cipher for wireless sensor

This work is partially supported by Natural Science Foundation of China (No. 61471158).

J.-S. Pan et al. (eds.), *Advances in Smart Vehicular Technology, Transportation, Communication and Applications*, Smart Innovation, Systems and Technologies 86,
https://doi.org/10.1007/978-3-319-70730-3_41

network. However, these research achievements are unsuitable for LoRaWAN protocol or difficult to achieve by programming.

Considering the above situation and problems, this paper puts forward a novel symmetric encryption algorithm to overcome the shortcomings. Through the security analysis and performance simulation test, the algorithm obtained good encryption speed and security. In addition, the block cipher algorithm is fully compatible with LoRaWAN protocol, which has a certain reference value in the Internet of Things.

2 LoRaWAN Topology Structure

LoRaWAN is a Low Power Wide Area Network protocol intended for wireless battery operated Things in a regional, national or global network. LoRaWAN targets key requirements of Internet of Things such as secure bi-directional communication and localization services. The LoRaWAN protocol provides seamless interoperability among smart Things without the need of complex local installations and gives back the freedom to the user, developer, businesses enabling the roll out of Internet of Things [5].

LoRaWAN topology architecture is typically laid out in a star-of-stars topology in which gateways is a transparent bridge relaying messages between end-devices and a central network server in the backend. Gateways are connected to the network server via standard IP connections while end-devices use single-hop wireless communication to one or many gateways. All end-point communication is generally bi-directional, but also supports operation such as multicast enabling software upgrade over the air or other mass distribution messages to reduce the on air communication time. LoRaWAN topology structure is shown in Fig. 1.

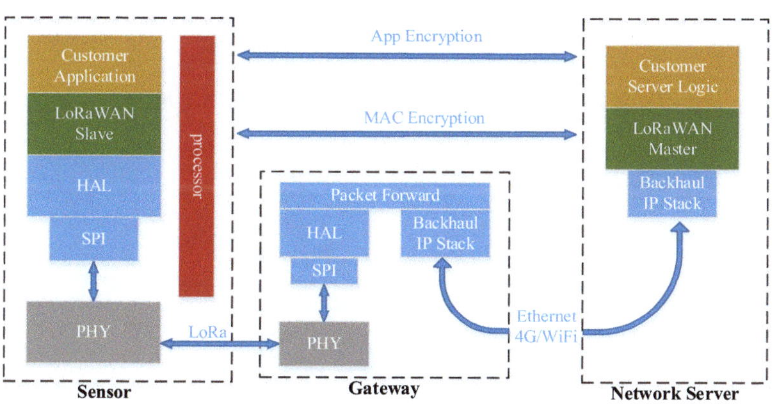

Fig. 1. LoRaWAN topology structure

National wide networks pay attention to internet of things such as critical infrastructure, confidential personal data or critical functions for the society has a special need for secure communication. LoRaWAN solves this problem by AES and CMAC algorithm with a large consumption of hardware resources. Therefore, the design and implementation of lightweight cipher algorithm is of great significance.

3 Design of a Novel Block Cipher Algorithm

3.1 Block Cipher Algorithm Implementation Scheme

Aiming at LoRaWAN protocol with AES and CMAC algorithm would consume a large amount of hardware resources for the problems. In this paper, a novel cipher algorithm is proposed, which combines Logistic chaos mapping with Arnold transformation. This cipher algorithm has many characteristics, such as large key space, small consumption of hardware resources and high security. The schematic diagram of a novel block cipher algorithm is shown in Fig. 2.

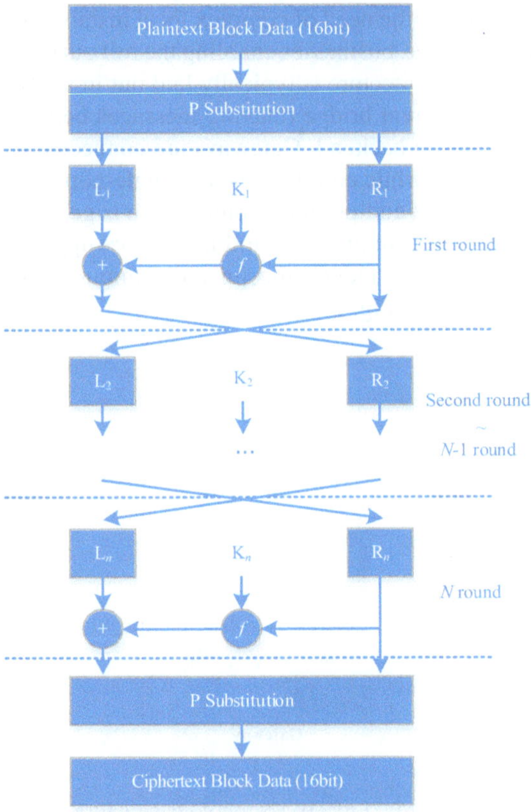

Fig. 2. Schematic diagram of a novel block cipher algorithm

The basic design idea of a novel block cipher algorithm is described as below. To begin with, plaintext is divided into 16 bits data block, each data block is spread through

P replacement operation. Block data for 16 bits is divided into high and low portion with 8 bits. High and low portion are represented by L_i and R_i. Round keys K_i and R_i serve as the input parameters of *f* function, R_{i+1} value is obtained by output result of *f* function for first round xor L_i. As the same time, R_i turns into L_{i+1}, the value of *i* range from one to *N*. This process is called one round Feistel operation, which is shown below by mathematical formula description:

$$\begin{cases} L_{i+1} = R_i \\ R_{i+1} = L_i \oplus f(R_i, K_i) \end{cases} \tag{1}$$

The rest operation can be done in the same Feistel manner [6, 7] and to generate the final ciphertext. P substitution of a novel block cipher algorithm adopts Arnold transform with the purpose of achieving favorable diffusion effectiveness. Classical Arnold transform is a two-dimensional reversible mapping [8, 9]. The definition can be expressed as:

$$\begin{bmatrix} x' \\ y' \end{bmatrix} = A\begin{bmatrix} x \\ y \end{bmatrix} (\text{mod } 1), A = \begin{bmatrix} 1 & 1 \\ 1 & 2 \end{bmatrix} \tag{2}$$

$$\begin{bmatrix} x' \\ y' \end{bmatrix} = A\begin{bmatrix} x \\ y \end{bmatrix} (\text{mod } N), A = \begin{bmatrix} 1 & 1 \\ 1 & 2 \end{bmatrix} \tag{3}$$

Among $0 \le x, y \le 1, 0 \le x', y' \le 1$ and $x \in [0, N-1], y \in [0, N-1]$. In this paper, the formula (2) can be converted into formula (3) with the purpose of convenient operation. Schematic diagram of Arnold transform for 16 bits diffusion is shown in Fig. 3. P replacement is achieved by one round Arnold transform with the purpose of diffusing bits data.

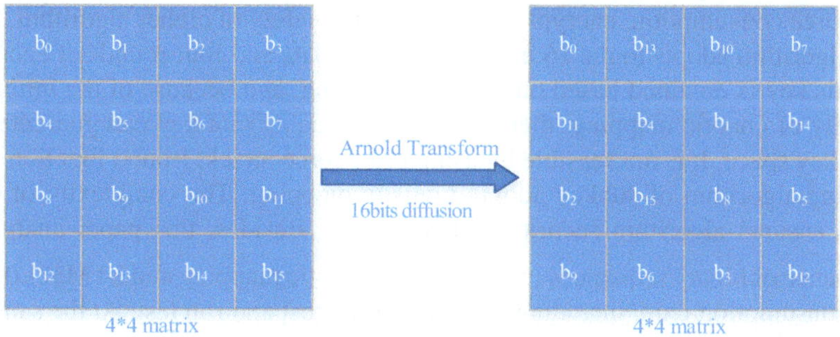

Fig. 3. Schematic diagram of Arnold transform for 16bits diffusion

The chaos mapping has features such as the sensitivity of initial value and pseudo randomness. Therefore, Logistic chaotic mapping serves as *f* function to enhancing the complexity and security of block cipher algorithm. Logistic mapping is a classical model of chaotic system, and its simple equations as well as good performance are widely as chaotic cipher algorithm. The mapping is defined as:

$$x_{n+1} = \mu x_n (1 - x_n), \mu \in (0, 4], x_n \in (0, 1) \quad (4)$$

Among μ is called as branch parameter, when the value area of μ is $[3.5699456, 4]$, Logistic mapping work on chaotic state and show a complex dynamic characteristics. Substituting initial value x_0 into the chaotic equation, the chaotic binary sequences were generated based on repeated iterative operation. In this algorithm, Root key K_1 serves as x_0 for Logistic iterative operation. The original chaotic sequences $\{x_n\}$ are quantified into binary sequences $\{s_n\}$ as output sequences of *f* function. R_{i+1} is generated through xor each 8 bits of sequences $\{s_n\}$ with L_i. The each interval 8 bits of original chaotic sequences $\{x_n\}$ serves as subkey K_i for round encryption operation. Chaos mapping has a variety of ways to quantize chaotic sequences, this paper chooses relatively simple quantitative method to avoid complex computations. It is defined as follows:

$$x_n = \begin{cases} 0, s_n < 0.5 \\ 1, s_n \geq 0.5 \end{cases} \quad (5)$$

When Logistic mapping show chaotic state, iterative computation value x_n will traverse on (0,1) interval. It can get good chaotic binary sequences by the quantitative method.

3.2 Security and Performance Analysis

The influencing factors of block cipher algorithm has five important aspects, including the size of block data, the length of key, the number of round encryption, generating algorithm of subkey and the complexity of round function. In this paper, considering the resource consumption, encryption speed and the safety of cipher algorithm, and the cipher algorithm adopts to the 16 bits size of block data and four rounds of encryption. Due to its key is obtained based on Logistic mapping, and because of the initial value sensitivity of chaotic mapping, therefore initial value as root key can guarantee large enough key space. It is not feasible by using brute force attack. Round function and the generating algorithm of subkey adopts Logistic mapping. The complexity of chaotic systems ensures algorithm security and reliability. Besides, this paper analyzes the performance of binary sequences $\{s_n\}$ based on Logistic mapping and STM32L051 low-power consumption chip, including autocorrelation test and run test. At the same time, initial value is defined as a float type data.

Autocorrelation Test. Autocorrelation is an important property of pseudo random binary sequences, and the perfect pseudo random sequences are δ function. Suppose $x(n)$ is a chaotic binary sequences, $R_x(m)$ is the autocorrelation function of the sequences, the definition of autocorrelation function as following:

$$R_x(m) = \sum_{n=-\infty}^{\infty} x(n)x(n+m) \tag{6}$$

In this experiment, the sequences length are 10^6, the matlab simulation diagram is shown in Fig. 4. It can be seen from the diagram that the sequences in the case of the calculation accuracy of float type is close to the δ function [10], which shows good pseudo random property.

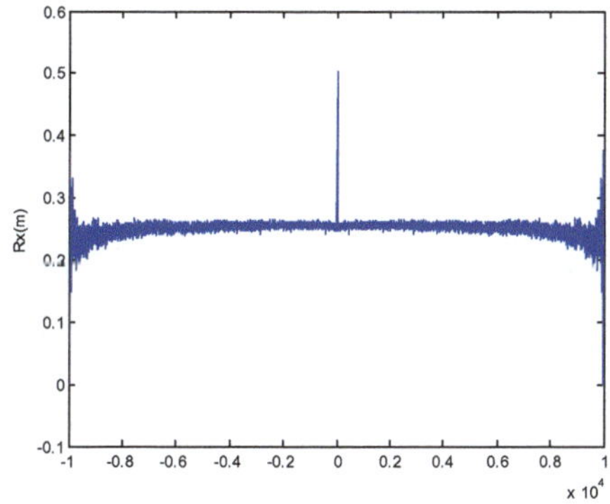

Fig. 4. Simulation diagram of autocorrelation test

Run Test. The focus of this run test is the total number of runs in the sequences, where a run is an uninterrupted sequence of identical bits [11]. The purpose of the run test is to determine whether the number of runs of ones and zeros of various lengths is as expected for a random sequence. The run test results of binary sequences are shown in Table 1.

Table 1. Run test results

Sequence length	Length of runs			
	1	2	3	4
$1 * 10^4$	0.4950	0.2535	0.1194	0.0640
$5 * 10^6$	0.5004	0.2490	0.1255	0.0625

The 1-length run nearly satisfies the theoretical value 1/2 from the results of Table 1, and 4-length run nearly satisfies the theoretical value $1/2^4$ by this analogy. Therefore, the binary sequences can better satisfy the run test requirement.

4 Encryption Process Based on LoRaWAN and a Novel Cipher Algorithm

4.1 Over-the-Air Activation of Sensor Node

For over-the-air activation based on LoRaWAN, from a sensor node's point of view, the join procedure consists of two MAC messages exchanged with the server, namely a join request and a join accept. Message structure of two MAC messages is shown in Fig. 5.

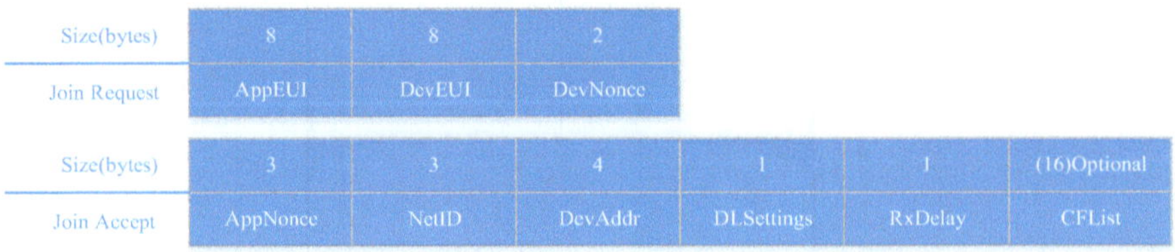

Fig. 5. Message structure of over-the-air activation

The join procedure is always initiated from the sensor node by sending a join-request message. The AppEUI is a global application ID in IEEE EUI64 address space that uniquely identifies the application provider of the sensor node. DevNonce is a random value. The join-request message is not encrypted. The network server will respond to the join-request with a join-accept if the senor node is permitted to join a network. The join-accept message contains an application AppNonce, a network identifier (NetID), a sensor node address (DevAdddr), a delay between TX and RX (RxDelay) and an optional list of channel frequencies (CFList) for the network the sensor node is joining. The AppNonce is a random value or some form of unique ID provided by the network server and used by the senor node to derive the two session keys NwkSKey and AppSKey as follow:

$$\begin{aligned}&\text{NwkSKey} = \text{logistic\_iteration(Rootkey,Count-Nwk)}\\&\text{AppSKey} = \text{logistic\_iteration(Rootkey,Count-App)}\\&\text{Count - Nwk} = \text{AppNonce} \oplus \text{NetID} \oplus (\text{0x01|DevNonce})\\&\text{Count - App} = \text{AppNonce} \oplus \text{NetID} \oplus (\text{0x02|DevNonce})\end{aligned} \tag{7}$$

Count-Nwk and Count-App are the iteration number of Logistic mapping. The join-accept message itself is encrypted with the Rootkey as follows:

$$\text{chaos\_encrypt(Rootkey,join-accept)} \tag{8}$$

4.2 Frame Payload Encryption of Sensor Node

If a data frame carries a payload, FRMPayload must be encrypted the cyclic redundancy check is calculated. As a default, the encryption is done by the LoRaWAN layer for all FPort. The key *K* used depends on the FPort of the data message (Table 2):

Table 2. FPort List

FPort	K
0	NwkSKey
1…255	AppSKey

The fields encrypted are FRMPayload. For each data message, the algorithm defines a sequence of block A_i for i = 1…k with k = ceil(len(FRMPayload)/8) (Fig. 6):

Fig. 6. The sequence of block A_i

The direction field (**Dir**) is 0 for uplink frames and 1 for downlink frames. The blocks A_i are encrypted to get a sequence S of blocks S_i. Encryption of the payload is done with the purpose of getting ciphertext.

$$\begin{aligned} S_i &= \text{chaos\_encrypt}(K, A_i) \text{ for } i = 1 \ldots k \\ S &= S_1|S_2| \ldots |S_k \\ \text{ciphertext} &= \text{truncating}(S) \oplus \text{FRMPayload} \end{aligned} \quad (9)$$

5 Conclusion

In recent years, due to the phenomenal growth of Internet of Things application, hardware resources and security of sensor nodes are widely concerned. This paper designs a novel lightweight block cipher algorithm based on the premise. This algorithm adopts chaotic dynamic theory and brief cipher structure to improve algorithm security and encryption speed. In addition, the proposed scheme is compatible with LoRaWAN protocol and suitable for limited hardware resources of sensor nodes. There are also significant research and reference values for the security of Internet of Things.

References

1. Chen, T.M., Jiang, R.R.: New hybrid stream cipher based on chaos and neural networks. J. Acta. Phys. Sin. **4**(62), 040301 (2013)
2. Muhammad, S., Furqan, Z., Guha, R.: Wireless sensor network security: a secure sink node architecture. In: Performance Computing and Communications Conference, Phoenix, USA, pp. 371–376 (2005)
3. Shahare, P., Chavhan, N.: An approach to secure sink node's location privacy in wireless sensor networks. In: International Conference on Communication Systems and Network Technologies, Washington, USA, pp. 748–751 (2014)

4. Chen, S., Zhong, X.X., Wu, Z.Z.: Chaos block cipher for wireless sensor network. J. Sci. China. Ser. F-Inf. Sci **51**(8), 1055–1063 (2008)
5. The LoRaWAN Topology Structure [EB/OL]. https://www.lora-alliance.org
6. Choy, J.: Cryptographic properties and application of a generalized unbalanced Feistel network structure. J. Crypt. Commun. **3**(3), 141–164 (2011)
7. Coron, J., Holenstein, T., Kunzler, R., et al.: How to build an ideal cipher: the indifferentiability of the Feistel construction. J. Cryptology **29**(1), 61–114 (2016)
8. Li, Q., Liao, X., Qu, G.Q., et al.: Adaptive steganography algorithm in digital image based on Arnold transform. J. Commun. **37**(6), 192–198 (2016)
9. Li, Y.J., Zhang, R.Z., Ge, J.H., et al.: Periods of the 3-arnold transformation and its application in image encryption. J. Univ. Electr. Sci. Technol. China **44**(2), 289–294 (2015)
10. Du, B.X., Ding, Q., Geng, X.L.: Generation and realization of digital chaotic key sequence based on K-L transform. In: Chaos-Fractals Theories and Application, Hangzhou, China, pp. 385–389 (2011)
11. Rogelio, H.B., Roxana, R.R.: Cycle detection for secure chaos-based encryption. J. Commun. Nonlinear Sci. Numer. Simul. **16**(8), 3203–3211 (2011)

Research on Multiple Classifiers Combination Method for Remote Sensing Images

AiPing Jiang(✉), Shengjie Xiao, LongYun Wei, and YanLin Zhu

Key Lab of Electronic and Communication Engineering,
Heilongjiang University, Harbin, People's Republic of China
13613655791@163.com

Abstract. Remote sensing technology is widely used in land surveying and earth science research, such as the study of the earth's oceans, glaciers, hydrology, ecology, geology and so on, is still in the military, intelligence, commercial, economic and other aspects can also be used to select the appropriate training samples, supervised classification with five kinds of single classifier, the classification accuracy comparison of different methods. It was found that the maximum classification accuracy of the two plots was the highest with the maximum likelihood method, but the classification accuracy was not the highest for 1 of the cultivated areas. Therefore, two decision fusion based multiple classifiers combination algorithms are proposed, and the results of single classifier are processed by using ENVI and IDL software. Compared with the classification results of a single classifier, the overall classification accuracy of multiple classifiers is improved by 2.5%, and the accuracy of 1 of the cultivated land in the study area is increased by 15.5%. Methods the combination of multiple classifiers can use different characteristics of single classifier, so as to compensate for the lack of a single classifier.

Keywords: Remote sensing classification · Multiple classifiers combination · Decision fusion image

1 Foreword

With the continuous development of remote sensing images technology, remote sensing technology is to obtain the information of objects or phenomena [1]. In contrast with the field observation, this technique does not require physical contact with objects when acquiring information [2]. The term "remote sensing" usually refers to using the sensor technology based on satellite or spacecraft to detect and

This work was supported in part by the National Natural Science Foundation China under Grant 61302074, in part by the Natural Science Foundation of Heilongjiang Province under Grant QC2013C061, in part by the Modern Sensor Technology Research and Innovation Team Foundation of Heilongjiang Province under Grant 2012TD007, in part by the Postdoctoral Research Foundation of Heilongjiang Province under Grant LBH-Q15121.

J.-S. Pan et al. (eds.), *Advances in Smart Vehicular Technology, Transportation, Communication and Applications*, Smart Innovation, Systems and Technologies 86,
https://doi.org/10.1007/978-3-319-70730-3_42

analyze things on the earth, including the earth's surface [3], atmosphere and ocean and so on [4]. Information is acquired in different ways by sensors, if distinguish them in accordance with the classified object, there are several classifications: Pixel-based classification, Object-based classification, mixed pixel-based classification. However, all these have shortcomings, that is, each region has more errors in extracting information, and the extraction is not accurate, so we put forward a new method of multiple classifiers combination algorithm, the modified algorithm includes voting decision method and consensus decision algorithm. The advantage of these two algorithms is that the information of each area can be accurately determined so as to achieve the purpose of classification.

2 An Overview of Multiple Classifiers Combination

2.1 The Necessity of Multiple Classifiers Combination

In the process of target identification, the general method is to use different classifier to identify the target, then to choose an optimal result from the several results obtained as the final recognition results. However, the best classifier is not optimal for target of all characteristic, different classifiers can complement each other. Therefore, the pattern recognition method is applied to the classifiers combination, the purpose is to allow different classifiers complement each other, so that the recognition accuracy can be improved [5].

2.2 Combination Methods of Multiple Classifiers Combination

According to the form of classifiers combination, they can be broadly divided into 3 types:

(1) Serial mode

Connect the classifiers together in a serial way, we must first consider the sequential order before the classifiers, and then put the classifiers together, then enter the training set for classification, finally, the classification results are obtained through combination decision [6]. The serial mode of the multiple classifiers combination is shown in Fig. 1.

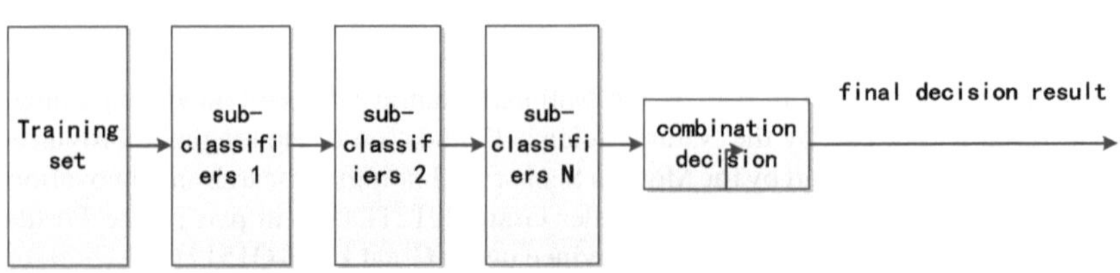

Fig. 1. Serial combination of multiple classifiers

(2) Parallel mode

Combine the categories in a parallel way, classifiers do not interfere with each other, each classifier is processed separately, the final classification results are obtained by combination decision [7]. The advantage of concurrent mode is that when a classifier is out of order, it won't make a big difference to the final result, the advantage of using different information layers between classifiers. Parallel combination mode of multiple classifiers combination is shown in Fig. 2.

(3) Embedded mode

In the classification process, a classifier is embedded into another classifier.

(4) Bagging algorithm

This method uses a series of classifiers to mix the learning results of each classifier by some rules (voting method, weighted voting, etc.), and achieves better recognition effect than the single learner. Bagging is the abbreviation of Bootstra Aggregating, Bagging algorithm consists of two steps, namely Bootstrap and aggregating. The Bootstrap method is a sampling with reset, this means that n sample sets are extracted from the original sample training set to form a new sample set, the number of samples extracted can be arbitrarily defined [8]. The Aggregating method is integration, voting method is generally used.

Take out t samples from the database of given training data. Using the sample obtained from the training sample database, machine learning algorithms can be independently trained on each data. After the training is finished, the classifier C_t is obtained, these classifiers are similar to the cabinet lineup mentioned in the example. Each classifier C_t is then used to classify the unknown samples, choose the optimal classification results as the final classification results through each classifier C_t. Bagging algorithm tree is shown in Fig. 2.

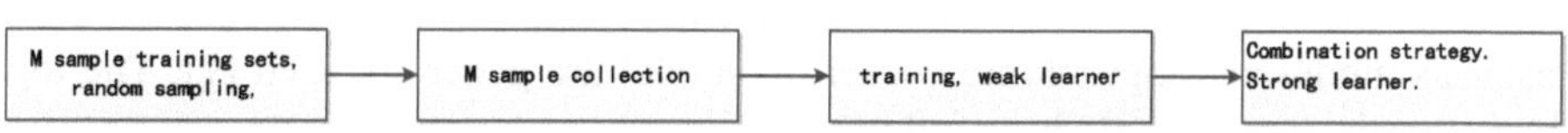

Fig. 2. Flow chart of Bagging algorithm

2.3 Multiple Classifiers Combination Algorithm

Classifiers combination algorithm includes voting method, Bayes rule, fuzzy integration, decision tree, theory of evidences, dynamic classifier selection method and so on.

(1) Voting method

In the voting method, there are voters and candidates, so when the voting method is applied to the multiple classifiers combination, each voter is the base classifier, the candidate is the result of each base classifier [9]. The premise of voting method is majority judgments are better than few judgments, which can solve the difference between the results of base classifiers, and achieve the purpose of improving classification accuracy.

Assuming the remote sensing image as a particular pattern space P, there are M disjoint categories in the image, so this remote sensing image can be represented as:

$$P = C_1 \cup C_2 \cup \cdots C_i \cdots \cup C_M, C_i \cap C_j = \Phi(i \neq j; i,j = 1,2,\cdots,M) \tag{1}$$

First, using R classifiers e_R to classify this pattern space P, then each classifier can obtain a category number j_K of classification result, that is $e_K(X) = j_K$, among this $j_K \in \wedge \cup \{M+1\}$. Next, the single classifiers are made up of a classifier E, so that $E(X) = j, j \in \wedge \cup \{M+1\}$, among this j = M+1 represents that the classifier cannot determine the category of J, cannot endow unclassified categories.

The general voting decision rules are expressed as follows:

$$e(X) = \begin{cases} j & (\sum_{k=1}^{R} T_k(X \in C_j) = \max_{i \in \Lambda} \sum_{k=1}^{R} T_k(X \in C_i) > \alpha \cdot R) \\ M+1 & others \end{cases} \tag{2}$$

In (2), $(X \in C_j)$ is input, T_k is classified output votes. Among them, $0 < \alpha \leq 1$. If $\alpha = 1$, the Eq. (2) is the most conservative unanimous voting rules, that is when the results of all base classifiers are completely the same, the final result can be obtain. If $\alpha = 0.5$, the above equation is majority voting rule, which is once the results of more than half of the base classifier is consistent, majority consistent results can be the final result. If α is an infinitesimal positive number, then the formula (2) is the relative majority voting, that is when a category is voted many times, this category will be classified as definite categories.

(2) Voting method

As a theory method of a multiple classifiers combination, the consensus principle is suitable for the single classifiers combination whose output is class probability. Linear consensus model and logarithmic consensus model are generally used [7]. The logarithmic consistence formula is as follows:

$$T_j(X) = \prod_{i=1}^{N} (p_i(C_j|X))^{\lambda_{ij}} \tag{3}$$

The linear consistence formula is as follows:

$$T_j(X) = \sum_{i=1}^{N} p_i(C_j|X) * \lambda_{ij} \tag{4}$$

The output pixel X belongs to the membership degree, reliability or probability of each category. $T_j(X)$ is identified by multi-classifiers, the pixel X belongs to membership degree of class j. $p_i(C_j|X)$ is identified by classifier i, the pixel X belongs to the reliability or probability of class j. As the weight represents degrees of importance, λ_{ij} is the recognition accuracy of the classifier i to class j.

3 Multiple Classifiers Combination Classification Based on Decision Fusion

There are many kinds of remote sensing image classification methods, and each method has its own characteristics and advantages. The advantage of the neural network method is that there is no need to know the prior information about remote sensing images. However, if the multiple variable statistical model of the remote sensing images can be determined, statistical method has better classification accuracy than neural network.

3.1 Decision Fusion

5 supervised classification algorithms for remote sensing images are presented in this paper: Minimum Distance (MD), Maximum Likelihood (ML), Mahalanobis Distance (MaD), Nerve Net (ANN), Support Vector Machine (SVM) are data sources of decision data fusion. The design is as the followed 2 sets of decision fusion strategies:

Strategy 1: Select 3 results from 5 single classifiers results through permutation and combination as the input of the decision data fusion, so there are 10 permutations and combinations. Respectively (1) MD, ML, MaD;(2) MD, ML, ANN; (3) MD, ML, SVM; (4) MD, MaD, ANN;(5) MD, MaD, SVM;(6) MD, ANN, SVM;(7) ML, MaD, ANN; (8) ML, MaD, SVM;(9) ML, ANN, SVM;(10) MaD, ANN, SVM. Then consistently classify classification results of 3 single classifiers through voting method on the value of the same pixel, if 2 or 3 classifiers have the same classification results, then it will be the final result. If the results of the 3 classifications are not the same, the final classification result is the highest overall classification accuracy (OA).

Assuming that C_i is the pixel value of the remote sensing image of the i classification method, C is the final combined classification results, among them, $f = OA$. Therefore, the rule of strategy 1 can be expressed as:

(1) If $C_1 = C_2$ or $C_1 = C_3$ or $C_1 = C_2 = C_3$, then $C = C_1$. If $C_2 = C_3$, then $C = C_2$.
(2) If $f(C_k) = \max(f(C_i)), i = 1, 2, 3$, then $C = C_k$

The type of the single classifier classification result file is general raster data format used by ENVI. ENVI raster file contains two files, one is a simple binary file which contains the single classifier classification results data, its suffix is.dat. The other is a relevant header file of the ASCII, ENVI header text files provide image size, embedded header files, data formats, and other related information. As long as you save the image as a ENVI raster format, ENVI will create a new header file.

Processing with decision fusion: For example, when making decision fusion for combinations of MD, ML, and MaD, firstly using voting rule identify pixel meeting the voting rules. The overall classification accuracy of ML is the highest, ML classification results are applied to the remaining pixels.

Save classification results: Compared classification accuracy of 5 single classifiers to the same features type, select the single classifier with good classification accuracy to different features, improve its classification disadvantage of other features on the basis of ensuring the advantages of some features classification. For example, in the study area 1, among 5 single categories, overall classification accuracy of the ML is the highest. For the 5 single classifiers, its producer accuracy of Forest land, water body, artificial mulch, cultivated land 2 are all out of the front. While compared with the MaD, its producer accuracy of cultivated land 1decreases by 15.86%. Therefore, it can be considered that on the base of ensuring ML for water body, forest, artificial mulch, improve ML producer accuracy to cultivated land 1 using MaD classification results to cultivated land 1.

4 Simulation Result

After the classification result of the single classifiers in study area 1 were processed by decision fusion, using the same sample in the third chapter, we evaluate the accuracy of the results of multiple classifiers combination. Overall accuracy degree and Kappa coefficient of the final result are shown in Chart 1, the classification results are shown in Fig. 3 (Table 1).

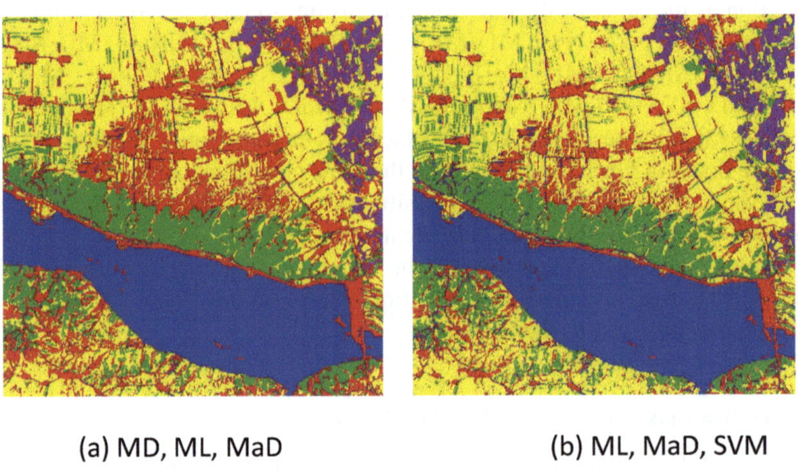

Fig. 3. Classification results of Research Area 1 based on Strategy 1

After the classification results of 2 single classifiers in the study area were processed by decision fusion, using the verification ROI file used in the third chapter, we evaluate the accuracy of the results of multiple classifiers combination. Overall accuracy degree and Kappa coefficient of the final result are shown in chart 3.

Using the method of strategy 2, study area 1 was processed by multiple classifiers combination, ML and MaD classification results were selected to obtain the combined results. Using the verification ROI file used in the third chapter, we evaluate the accuracy of the results of multiple classifiers combination. Overall accuracy degree, Kappa coefficient, confusion matrix and other accuracy of various categories of the final result are shown in chart 3, the combined classification results are shown in Fig. 4 (Table 2).

Table 1. Classification accuracy of Research Area 1 based on Strategy 1

Number	Combination	Overall classification accuracy (%)	Kappa coefficient
1	MD, ML, MaD	94.71	0.93
2	MD, ML, ANN	93.86	0.92
3	MD, ML, SVM	94.35	0.93
4	MD, MaD, ANN	94.25	0.93
5	MD, MaD, SVM	93.52	0.92
6	MD, ANN, SVM	93.80	0.92
7	ML, MaD, ANN	94.25	0.93
8	ML, MaD, SVM	95.00	0.94
9	ML, ANN, SVM	93.82	0.92
10	MaD, ANN, SVM	94.51	0.93

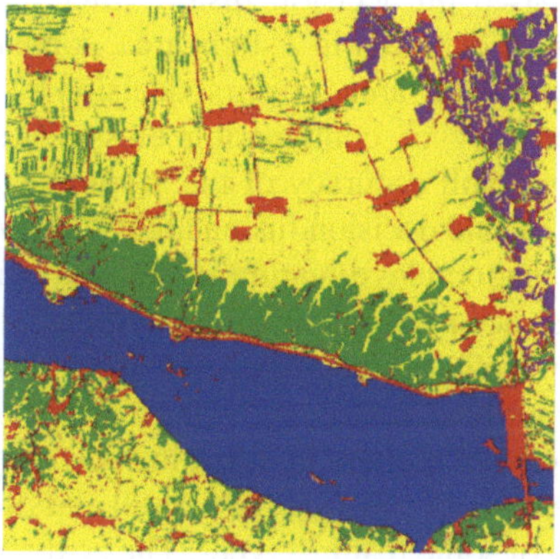

Fig. 4. Classification results of Research Area 1 based on Strategy 2

Table 2. Classification accuracy of Research Area 1 based on Strategy 2

Classification	Forest	Water	Land 1	Land 2	Artificial	Accuracy (%)
Forest	1197	0	4	0	0	99.67
Water body	0	1125	0	0	14	98.77
Cultivated land 1	9	0	1148	1	56	94.56
Cultivated land 2	2	0	0	404	2	99.02
Artificial mulch	3	7	21	29	1041	94.55
Producer accuracy (%)	98.84	99.38	97.87	93.09	93.53	
Overall classification accuracy = 97.08%, Kappa coefficient = 0.96						

5 Conclusions

Remote sensing image classification technology is widely used in various fields, with the development of image processing technology, a lot of image classification methods are proposed, each method has some classification characteristics and different classification abilities for different types of features. How to use the characteristics of different classifiers to achieve complementary advantages has become the focus of people's research. Two regions of Harbin, Hulan and A Cheng were selected as the study area, and one of the research areas was mainly agricultural land, while the other one was mainly urban land. Five supervised classification methods are used: Minimum Distance (MD), Maximum Likelihood (ML), Mahalanobis Distance (MaD), Nerve Net (ANN), Support Vector Machine (SVM). Based on the five supervised classification methods, supervised classification of two research areas was conducted, and the accuracy of classification results was compared. After classifying the two study areas, the overall classification accuracy of the maximum likelihood was the highest, but the classification accuracy of cultivated land 1 was not the highest [12]. Thus it can be seen, different classifiers have different classification characteristics, so we can use their respective advantages to improve the classification accuracy. Propose two multiple classifiers combination algorithms based on decision fusion, after respectively combining the classification results of single classifier in 2 study area, compare it with single classifier classification results for each study area. The first decision fusion algorithm has slightly improved the overall accuracy of combination classification in the study area 1, improved by 0.05%. The overall accuracy of the 2 combinations classification in the study area 2 reduces. The second decision fusion algorithm has obviously improved the overall accuracy of combination classification in the study area 1, improved by 2.5%. In particular, the producer accuracy of the cultivated land 1 in the study area 1 was increased by 15.5%, the overall accuracy of combination classification results in study area 2 has been slightly improved, improved by 0.15%. The result shows, the classifier based on multiple classifiers has higher classification accuracy than the classifier based on single classifier, on the basis of ensuring the overall classification accuracy, improve the classification accuracy of some features types. The next step is to do further research on the classification algorithm, correct the deficiencies, optimize algorithm, and improve the efficiency of classification.

Acknowledgment. The Paper is supported by the National Natural Science Foundation of China (No. 61601174), Science and Technology Innovative Research Team in Higher Educational Institutions of Heilongjiang Province (No. 2012TD007)

References

1. Chang, F.C., Huang, H.C.: A survey on intelligent sensor network and its applications. J. Netw. Intell. **1**, 1–5 (2016)
2. Gao, Q., Zuo, Y., Zhang, J., Peng, X.H.: Improving energy efficiency in a wireless sensor network by combining cooperative MIMO with data aggregation. IEEE Trans. Veh. Technol. **59**, 3956–3965 (2010)

3. Islam, M.R., Kim, J.: On the cooperative MIMO communication for energy-efficient cluster-to-cluster transmission at wireless sensor network. Ann. Telecommun. **65**, 325–340 (2010)
4. Karlof, C., Wagner, D.: Secure routing in wireless sensor networks: attacks and countermeasures. Ad Hoc Netw. **1**, 293–315 (2003)
5. Choudhury, P.P., Bagchi, P., Sengupta, S., Ghosh, A.: On effect of compromised nodes on security of wireless sensor network. Ad Hoc Sens. Wirel. Netw. **9**, 255–273 (2010)
6. Chen, X., Makki, K., Kang, Y., Pissinou, N.: Sensor network security: a survey. J. IEEE Commun. Surv. Tutorials **11**, 52–73 (2009)

Smart Wireless Access Technology

An Implementation of Relay Function Based on LTE Technology for Multi-UAV Communication

Wen-Chung Tsai[1], Nien-Ting Huang[1], Shih-Hsuan Lin[2], and Mao-Lun Chiang[1(✉)]

[1] Department of Information and Communication Engineering, Chaoyang University of Technology, Taichung City, Taiwan R.O.C.
{azongtsai,s1053060,mlchiang}@cyut.edu.tw

[2] Aeronautical Systems Research Division, National Chung-Shan Institute of Science and Technology, Taichung City, Taiwan R.O.C.
peter0401.lin@gmail.com

Abstract. This paper presents a design of relay function for multi-UAV communication and evaluates its performance in an implemented simulation environment. The proposed relay function design is based on the popular and commonly used LTE technology. Accordingly, the implementation can be low-cost and performance-efficient. Through the use of relay function, UAVs can forward data messages among UAVs when direct communication is available between a ground console and a remote UAV. Consequently, UAVs can be dispatched to areas that cannot reach before to monitor environment and collect information, then transmit data backward to ground consoles by the proposed relay function in real time.

Keywords: Unmanned aircraft vehicle · Relay communication · Long term evolution

1 Introduction

It has been for a period of time to apply wireless communication to flight control of UAV [1, 2]. In these years, with advances of communication technology, wireless applications change rapidly. Therefore, for an adoption selection of different communication technologies, the pros and cons should have a further discussion in respects of flight distance, communication ability, implementation cost, and so on. There are many alternative solutions can be selected, such as applying an existing public commercial telecommunication service, building a private network environment using existing products, or even creating a proprietary communication infrastructure by self-reliance. Among these solutions, in order to choose an optimal one with considering the progresses of technologies and requirements of applications, ahead analyses and further discussions are necessary.

Currently, there are many 4G-LTE (Long Term Evolution) [3] products made in Taiwan, such as mobile phone, network adapter, base station, etc. In addition to the production capability, Taiwan has sufficient design abilities and essential intellectual

J.-S. Pan et al. (eds.), *Advances in Smart Vehicular Technology, Transportation, Communication and Applications*, Smart Innovation, Systems and Technologies 86,
https://doi.org/10.1007/978-3-319-70730-3_43

properties including software and hardware to develop such advanced products. Accordingly, it is possible to enable an independent capability for development of military weapons in respect of UAV communication technologies by leveraging existing advantages of LTE productions and technologies of Taiwan. Therefore, this paper implemented a design of relay function for multi-UAV communication based on the LTE architecture. By use of the proposed relay function, UAVs can forward data messages one by one when direct communication is not available between a ground console and a remote UAV. As such, UAVs can be dispatched to areas of interest to do more worthy explorations such as environment observation and data collection.

2 Background

This chapter first evaluates advantages and disadvantages among using Wi-Fi, public and proprietary LTE networks in the flight control of UAV. Next, we will discuss the necessary of enabling a relay function for the UAV communication.

2.1 Using LTE for UAV Communication

Recently, regarding the usage of LTE for multiple UAV communications, there are two implementations [4]. In which, one utilized a public LTE telecommunication service and the other built an own LTE network infrastructure. Comparing different criteria in respects of function and performance, we listed pros and cons among Wi-Fi (802.11ac), Public LTE, and Own LTE in Table 1. In which, we observe that, criteria of Own LTE overcome those of Wi-Fi and Public LTE, except for that the Own LTE requires to operate at a legal (i.e., licensed) radio band. That is indeed a critical issue for applying to industry or academic. However, for the military purpose, it is feasible.

Table 1. Comparisons among Wi-Fi (802.11ac), public LTE and own LTE

Criteria	Wi-Fi	Public LTE	Own LTE
Roaming	Not support	Seamless	Seamless
Mobility	Not support	<350 km	<350 km
Safety	WPA2-PSK	SNOW 3G	SNOW 3G
Data Rate (Still)	<866.7 Mbps	<1 Gbps	<1 Gbps
Data Rate (Move)	Not support	<100 Mbps	<100 Mbps
Range	~300 m	~3 km	~3 km
Legality	Free ISM band	Shared bandwidth	License required
QoS	Poor	Good	Good
Spectrum	2.4/5 GHz	0.7–2.6 GHz	0.7–2.6 GHz

2.2 Enabling Transmission Relay Function to UAV

Commonly, UAVs are controlled by terrestrial base stations. However, a base station may be difficult or even impossible to set up at certain places, such as mountain and ocean. Accordingly, as Fig. 1 shows, if an UAV cannot be directly controlled by the ground based station (we call it as console later), a feasible solution is that multiple UAVs can be employed as relays to transmit signals from a console to the remote UAV. As such, UAVs can be dispatched to areas that cannot reach before to execute tasks such as environmental observation and military attack. Furthermore, collected information can immediately be transmitted backward to the console to be able to take appropriate reactions [5]. As well, the author of paper [6] proposed that UAV can be used as a relay to transmit regulative control messages to an anti-ship missile to improve its hit rate. In addition to these military purposes mentioned ahead, UAV with relay transmission capability can be applied to some emerging applications such as public safety [7] and emergency rescue [8].

Fig. 1. An application scenario of enabling long-distance flights over ocean by relay function of multi-UAV communication.

3 Design Methodology

In this section, on the basis of applying LTE technology [9], we proposes a feasible solution to enable relay transmission capability for UAVs. Accordingly, proposed design architecture and simulation environment will be introduced.

3.1 Proposed Architecture for UAV Relay Transmission

Conforming to LTE architecture, an UAV acts as an UE (User Equipment) and a console refers to an eNB (Base Station). An UAV needs to establish a connection with a console before starting a regular data transmission. Based on this LTE communication architecture, an UAV (as an UE) cannot establish a connection with another UAV (as an UE). Referring to Fig. 2, in order to achieve the function of relay transmission, a reduced-function eNB (e.g., Femtocell) is proposed to be embedded into an UAV. Consequently, for such an UAV with certain ability as a base station, it can be able to accept a connection request from another UAV.

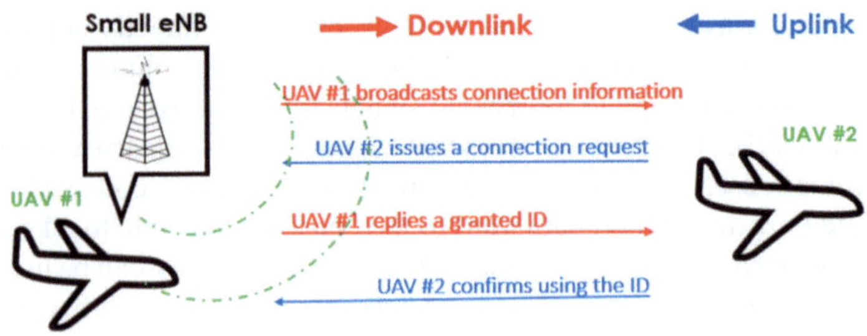

Fig. 2. Process of establishing a connection between UAVs with relay transmission capability.

3.2 Enabling Relay Transmission Capability for UAV Communication

For an UAV can operate both functions of UE and eNB, such an UAV can be enabled a basic relay transmission capacity by a proper configuration as shown in Fig. 3. Which displays both transmission directions of DL (Downlink) and UL (Uplink) among console and UAVs. To take the DL transmission in Fig. 3 as an example, the Data1 (with a transmission destination of UAV #1) and Data2 (with a transmission destination of UAV #2) represent two data inputs to the console. Within the console, based on the available radio bandwidth assigned by a given scheduling policy, for each time unit, eNB #0 encapsulates Data1 and Data2 into a packet conforming to LTE standards of PDCP (Packet Data Convergence Protocol), RLC (Radio Link Control) and MAC (Medium Access Control). Next, the packet is transmitted to the UAV #1, in which, Data1 is picked up immediately and Data2 is forwarded to its internal eNB program to be relayed to UAV #2.

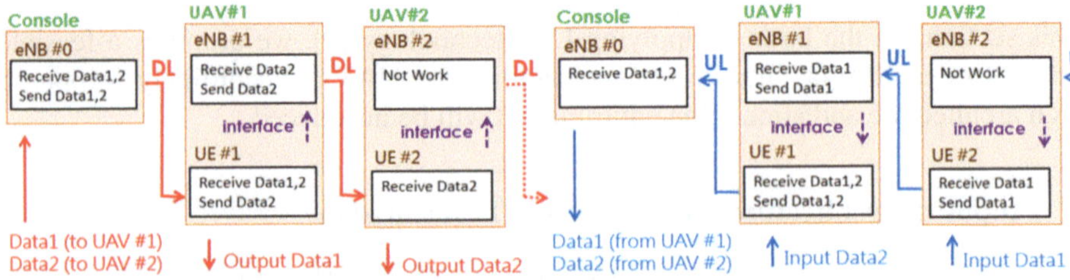

Fig. 3. Data flow of relay transmissions for DL and UL.

To put it clearly, in UAV #1, for the packet received from the console, the UE program decapsulates the contents of packet conforming to the LTE standards of MAC, RLC, and PDCP, respectively. Finally, Data1 and Data2 are extracted from the payload of the received packet. Since the transit destination for Data2 is UAV #2. Accordingly, Data2 is forwarded to the eNB program through an internal interface in UAV #1. Where, again Data2 is encapsulated as a packet conforming to the LTE standards of PDCP, RLC, and MAC, respectively. Finally, the packet with Data2 will be transmitted to UAV #2 and so forth.

4 Experimental Result

This section first introduces the settings of simulation environment including channel bandwidth and resource allocation for both DL and UL transmissions. Next, experimental results will be demonstrated and analyzed.

4.1 Bandwidth Setting and Resource Allocation

In order to simulate the real scenario of application that radio spectrum is limited, a resource scheduler in the MAC design of console can be configured to set an upper bound of bandwidth resource in respect of the actual bandwidth allocated to the application. According to the LTE protocol standard (c.f., Table 3.6-1 [10] of 3GPP TS36.101), we selected to set the value of ITBS (Index of Transfer Block Size) to 26, and set NPRB (Number of Physical Resource Block) to 100. Additionally, TDD (Time Division Duplex) multiplexing mechanism was applied, and DL and UL are set to have equal transmission opportunity for each time unit of LTE (1 ms of a subframe).

4.2 Data Throughput Analyses in DL Relay Transmission

To valid the function correctness of relay transmission in the DL direction as shown in Fig. 3, in simulations, we used VLC Media Player [11] to transmit and receive video streams among console and UAVs. In the first experiment, four identical video streams were issued from a console to different destinations of UAV #1, UAV #2, UAV #3, and UAV #4, respectively. First, all of the four video streams were received by UAV #1. In which, the video streams with destination of UAV #1 was picked up and the other three video streams were relayed to the next UAV (i.e., UAV #2). Similarly, in UAV #2, the video stream with destination of UAV #2 was picked up and the other two video streams were relayed to the next UAV (i.e., UAV #3) and so forth until there is no video stream required to be relayed. Average throughputs in the console and each UAV is listed in Table 2.

Table 2. DL data throughputs in the console and each UAV.

Average data throughput (Unit: KByte/s)				
Console	UAV #1	UAV #2	UAV #3	UAV #4
775.08	775.00	581.35	387.53	193.87

Figure 4 shows the throughput distributions for the console and four UAVs. The throughput of the console is similar to that of UAV #1. The reason is that both the console and UAV #1 shall carry out the transmissions of the four video streams. Besides, we observed that there are some variations in the distributions of each throughput. Due to the use of buffers in each UAV to regular the data flows, the variations becomes slighter and lead to smoother throughput distributions from UAV #1 to UAV #4. In addition, for a more detail data analysis, we separated the throughput into amounts of pickup data (i.e., destination is the UAV itself) and relay data (i.e., destination is other UAV) as shown in Fig. 5. In which, we observed that the amount of data throughput of pick data for each UAV is even similar and around 193 Kbytes. The reason is that the four video streams are extracted from the identical video file and our implemented relay function operated without loss of any video frame.

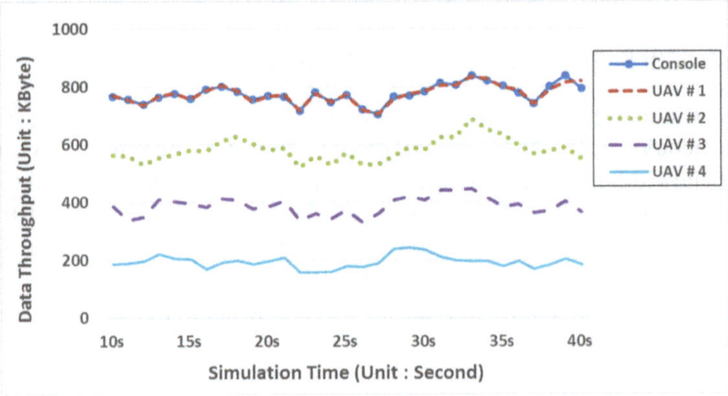

Fig. 4. DL throughput distributions in the console and each UAV.

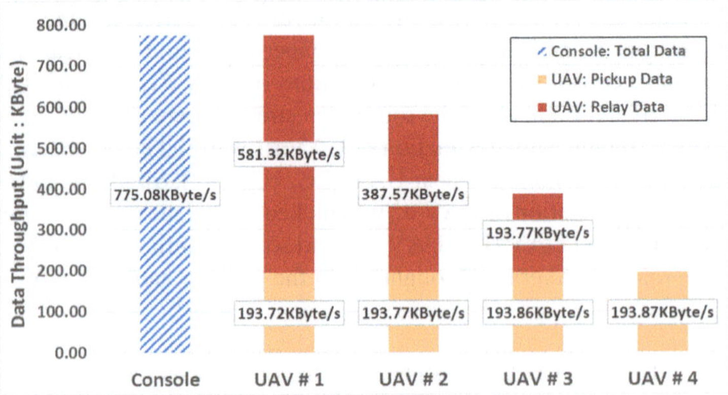

Fig. 5. Comparisons of pickup data and relay data in DL.

4.3 Data Throughput Analyses in UL Relay Transmission

In real applications, data could be collected by each UAV and then transmitted backward to the console, thus, data transmissions can be in an uplink direction. As such, each UAV not only must encapsulate its input data as LTE packets, but also to be required to relay packets received from other UAVs. To valid that our implemented design can functionally and correctly operates in the UL transmission direction, related experiment results are demonstrated as follows. There four identical video streams were individually issued from each of the four UAVs. Especially, the destinations of the four video streams were identical to the console. In progress, UAV #4 transmits a video stream with a destination of console to UAV #3. Then, UAV #3 must encapsulate the received video stream from UAV #4 and its input video stream as datagrams of different logic channels into a LTE packet and then transmit the packet to UAV #2 and so forth. Finally, the console receive the packet (i.e., MAC PDU) consisting with parts of the four video streams using as four logic channels (MAC SDUs). Average throughputs in the console and each UAV is listed in Table 3.

Table 3. UL data throughputs in the console and each UAV.

Average data throughput (Unit: KByte/s)				
Console	UAV #1	UAV #2	UAV #3	UAV #4
776.14	772.42	579.15	387.52	193.71

It is similar to the experimental result of DL transmission as shown in Table 2 and Fig. 4. In the UL transmission direction, as shown in Fig. 6, the throughput of the console is similar to that of UAV #1. The reason is that both the console and UAV #1 shall carry out the transmissions of the four video streams. Besides, due to the data flows regulated by UAV #4 and UAV #3, we observed that the throughput variation of UAV #2 in Fig. 6 is slighter than that of Fig. 4. In addition, for a more detail data analysis, we separated the throughput into amounts of pickup data (i.e., destination is the UAV itself) and relay data (i.e., destination is other UAV) as shown in Fig. 7.

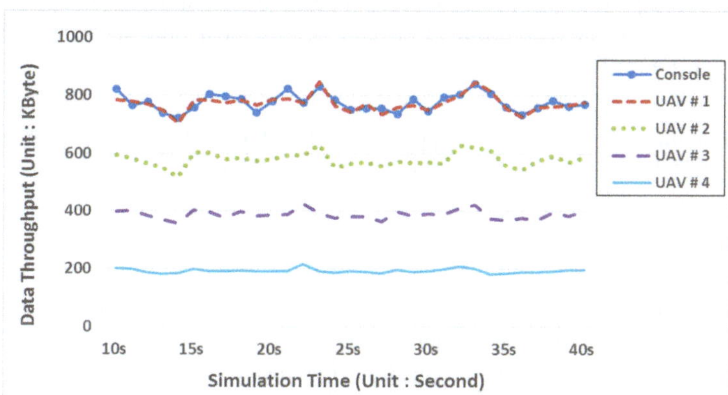

Fig. 6. UL throughput distributions in the console and each UAV.

In Fig. 7, we observed that the amount of data throughput of pick data for each UAV is even similar and around 193 Kbytes. The reason is that the four video streams are extracted from the identical video file and our implemented relay function operated without loss of any video frame.

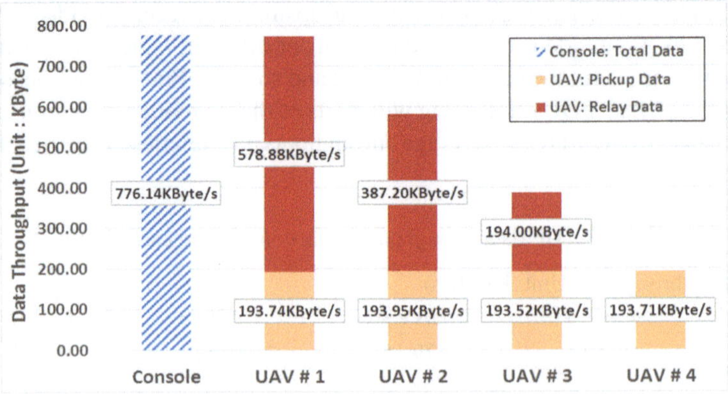

Fig. 7. Comparisons of pickup data and relay data in UL.

5 Conclusions

In this paper, we implemented a design of relay function for multi-UAV communication conforming to the LTE architecture. Function and performance were valid by simulations of both DL and UL transmissions among a console and four UAVs. By leveraging existing advantages of LTE productions and technologies, applying the proposed relay function for multi-UAV communication can be low-cost and performance-efficient, especially for the military purpose.

Acknowledgments. This work was supported by the MOST, ROC, under grant numbers of MOST 104-2221-E-324-001-MY2 and 106-2221-E-324-007-MY2.

References

1. Taylor, J.W.R., Munson, K.: Jane's Pocket Book of Remotely Piloted Vehicles: Robot Aircraft Today. Collier Books, London (1977)
2. Lan, Z.S.: The Design and Realization of Long Endurance Unmanned Aerial Vehicle. Master thesis, National Cheng Kung University (2014) http://ir.lib.ncku.edu.tw/handle/987654321/150255
3. IMT-Advanced (4G) Mobile wireless broadband on the anvil - New ITU radio interface standards to revolutionize mobile communication. ITU Press Release 21 (2009). http://www.itu.int/newsroom/press_releases/2009/48.html
4. Maiwald, F., Schulte, A.: Using LTE networks for UAS-communication. In: 36th European Telemetry and Test Conference, pp. 3958–3963 (2016)
5. Zhou, Y., Cheng, N., Lu, N., Shen, X.: Multi-UAV-aided networks aerial–ground cooperative vehicular networking architecture. In: IEEE Vehicular Technology Magazine, pp. 36–44 (2015)
6. Su, C.S.: Discussion on the techniques of UAV remote control system for the remote relay guided anti-ship missile. In: Hsin Hsin Quarterly, vol. 41, no. 4 (2013)

7. Chandrasekharan, S., Gomez, K., Hourani, A.A., Kandeepan, S., Rasheed, T., Goratti, L., Reynaud, L., Grace, D., Bucaille, I., Wirth, T., Allsopp, S.: Designing and implementing future aerial communication networks. IEEE Commun. Mag. **54**(5), 26–34 (2016)
8. Tuna, G., Nefzi, B., Conte, G.: Unmanned aerial vehicle-aided communications system for disaster recovery. J. Netw. Comput. Appl. **41**(5), 27–36 (2014)
9. Radio Access Network (E-UTRAN); Overall description, 3GPP Technical Specification 36.000 version 13.4.0, Release 13 (2016). http://www.etsi.org/deliver/etsi_ts/136300_136399/136300/13.04.00_60/ts_136300v130400p.pdf
10. Physical Layer Procedures, 3GPP Technical Specification 36.213 version 13.2.0, Release 13 (2016). http://www.etsi.org/deliver/etsi_ts/136200_136299/136213/13.02.00_60/ts_136213v130200p.pdf
11. VLC Media Player. http://www.videolan.org/vlc/

Tendency of Connected VIP and ITS Evolution

Tang-Hsien Chang(✉), Ying-Chih Lu, and Xiao-Ting Dai

School of Transportation, Fujian University of Technology, Fuzhou, China
thchang@fjut.edu.cn, Ethan424@gmail.com, Ting_fz@163.com

Abstract. The fifth generation of mobile communication (5G) will emerge on the horizon estimated before 2020. It will affect our daily activities to become more convenient and diversified. The development of Internet of Things (IoT) is also accessing to our circumstances as a network of objects and/or things. Then, huge data are generated, uploaded onto the cloud to be analyzed, organized, synthesized and outputting valuable messages as well as information for specified applications. Nevertheless, including Internet of Vehicles (IOV), Connected Vehicles (CV), particularly rely on short range communication technologies for data transmission, such as WiFi direct, Bluetooth (iBeacon), RFID, WAVE (IEEE802.11p), LTE V, NB IoT and LoRa. The fundamental demand of public is expected to connect directly among Vehicles, Infrastructure and Persons (cVIP), without bypassing any ISP. This in fact leads to complete the need to a seamless circumstance. This paper discusses which technology for connecting VIPs in 5G era will win the market. Meanwhile, the change will obviously and strictly affect traffic and transportation engineering. Advanced ITS (Intelligent Transportation Systems) to enter 5th generation of transportation era (Tx 5.0) will be presented.

Keywords: 5G · IoT · ITS · cVIP · LTE-V · WiFi · DSRC

1 Introduction

The development of mobile communication technology has been advanced rapidly since 1990s. It began from the 2G on communicating of voice and messages in 1990s, 3G Cloud compilation in 2000s, to the present 4G cloud services. It is well known that at every verge of mobile communication generation, it brought with a huge business opportunity. While we are enjoying the responsiveness and convenience of 4G, the global mobile phone communication industry has started to heavily invest in upgrading the 4G technology into 5G. It is expected the 5G era will emerge on horizon by 2020. Up to that point, the 5G mobile communication will make our living more convenient and diversified (Fig. 1).

On the other hand, along the rapid development of Internet of Things (IoT), IoT is a network of objects and/or things. Through the networks, huge big data are collected, analyzed, synthesized and uploaded onto the clouds as valuable applications at anytime and anyplace. The network connectivity is also opening up a new era of technology and economic growth. Moreover, it enables to achieve greater value and service.

J.-S. Pan et al. (eds.), *Advances in Smart Vehicular Technology, Transportation, Communication and Applications*, Smart Innovation, Systems and Technologies 86,
https://doi.org/10.1007/978-3-319-70730-3_44

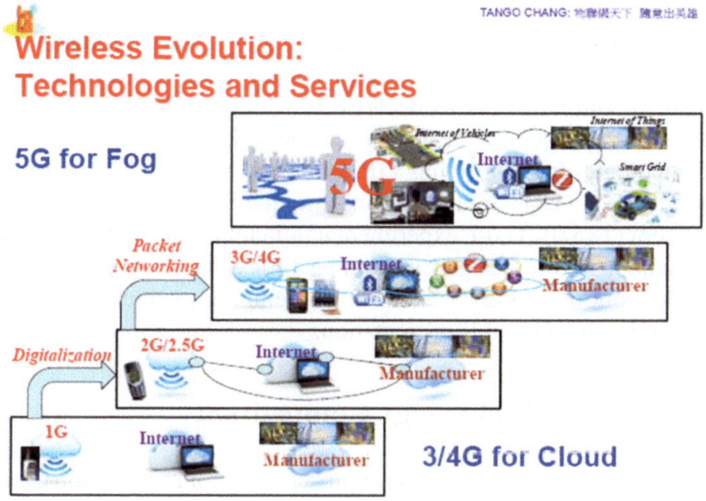

Fig. 1. The change 4G to 5G

With the mobile communication technology and IoT network, many objects and things are networking. Regardless it is Internet of Things (IoT), Internet of Vehicles (IoV), or Connected Vehicles (CV), short range communication technologies are indeed playing an important role of channels to transmit information through developing the small self-networking as talk/whisper to each other, such as with Wi-Fi direct, Bluetooth (iBeacon), RFID, WAVE/ DSRC (IEEE802.11p), LTE-V and LoRa.

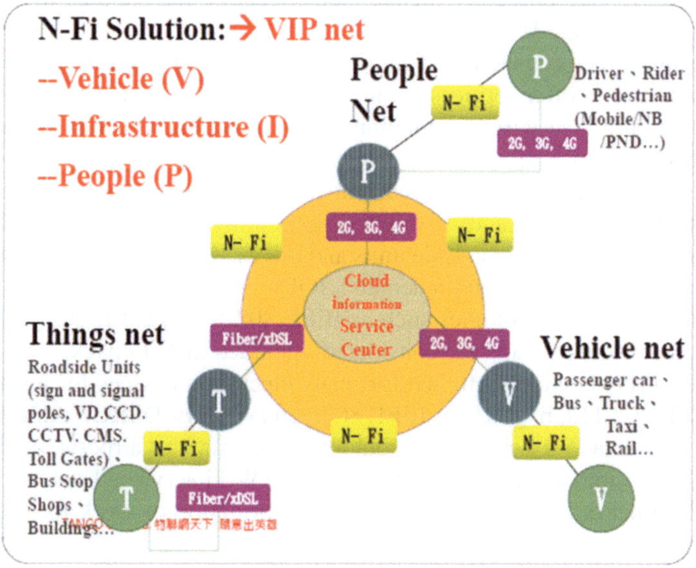

Fig. 2. Connected VIP frame for N-Fi system [4]

Herein, the general need of the public is directly to **c**onnect among **V**ehicles, **I**nfrastructure and **P**ersons (named as cVIP), not entirely or always through long range cell communication systems. Those direct linkages in connecting VIP will also be getting overcome and smooth when 5G technology is matured. The N-Fi (Near Field Informatics) portion of Fig. 2 depicts the need to seamless. This presentation will address the important technologies for connecting VIPs in 5G era. Meanwhile, the presentation will focus on the advanced ITS (Intelligent Transportation Systems) to the 5th generation of transportation era (Tx5.0). Those experimental difficulties encountered in the past, may potential/tremendously be changed in coming future, for examples, accident reconstruction, traffic control on the basis of human being, not vehicles will be come true. The foresee will be shared in this presentation.

2 Communication Technology Competition

To successfully building a great city in current, a proper communication technology that is cheap and commonly shared is important. Although 3G and 4G mobile systems are in service, we find that they are insufficient for the demands, particularly related to IoT and ITS like V2I/V2V. We pay much concerning that 3G/4G still cannot directly link between/among smart phones, vehicle onboard units, roadside devices like signal controllers, sign poles, marking restricting message, and indoor facilities. This leads to the need for an ad hoc network to fulfill the gap, particularly also for social networking. A Near Field Informatics (N-Fi) system is presented herein to compensate the loose. To maintain the system, a connected VIP net (cVIP) is proposed. The character V, I, and P, respectively denotes Vehicles, Infrastructure, and Person. "Infrastructure" could be roadside units (such as traffic signal controllers (TSC), changeable message signs (CMS), toll gates, bus stops, convenience shops, schools, and building facilities, everything, (illustrated in Fig. 2) [1–4].

Considering direct connection with ad hoc networking from user's demands, the thumb of barrier is to offer services for moving particles (cars and/or people). Indeed, the product should have wireless communication function with fast access and easy connect characteristics, and/or be wearable or in-hand, and with common protocols and inexpensive. The technology to concurrently support static equipment and moving particles, MONET conversion to VANET is studied, where VANET is set for communication among vehicles and roadside units and drives message, information that match travelers' demands (information on demand, IOD).

Some short-range technologies such as IEEE 802.11(x), IEEE 802.15(series), and RFID are presently in use, and are considered/ included in N-Fi pack.

IEEE 802.11(x) is widely utilized for many electronic devices for instances, WiFi devices (IEEE 802.11a, b, g, n) but IEEE 802.11p (WAVE/ DSRC). Almost all smartphones are plug-in a WiFi chip. Many outdoor zones such as airports, public stations, squares in smart cities around the world offer free WiFi service. Then, one can easily access the Internet or other network to obtain information or work for their business online.

However, it is hard to access WiFi when one is moving 20–30 kph and over. IEEE 802.11p in USA is planned and allocated for onboard V2V + V2I to overwrite WiFi's weakness [5, 6]. Conversely, IEEE 802.11p is not built-in smartphones. Therefore, it has obstacle to be popular wide. Nevertheless, Bluetooth (IEEE 802.15.1) has built in smartphones, and currently many auxiliary positioning products/devices are supported by Bluetooth, for example beacon-like devices. Another IEEE 802.15 (series) is Zigbee (IEEE 802.15.4) [7]. Bluetooth has a 1–2 Mbps rate, but Zigbee only 250 Kbps. Whatever, Zigbee enables to broadcast packages available for high speed moving particles since Zigbee puts authentication redundant. Sub1Giga (Sub1G) is also provided for low rate transmission, like LoRa (433, 866, 915 MHz), for some IoT applications, like water meter, electric power meter. Passive RFID is also very popular for relative short-range conditions, approximately 10 cm or less, such as for contactless tickets or door-keys. Extended passive RFID's reader allows a detection for tags mounted on high speed vehicles, which the detection range can reach 10–15 m. Active two-way RFID is utilized for the range of ten to hundred meters accomplished by using Zigbee compatible techniques. The comparison with their attributes is revealed in Table 1.

Table 1. Short range communication technologies and their attributes

Communication technology	Bluetooth	ZigBee/ Sub1G/LoRa	Wi-Fi	RFID	Wave/ DSRC	LTE-V
Moving constraint	<5 kph	50–100 kph	<30 kph	30–80 kph	<100	>40 kph
Service range, 0 dB	10–100 m	10–1000 m	10–100 m	<10 m	100	>100 m
Connection speed	5–10 s	<1 s	1–10 s	<1 s	<1	<1 s
Bandwidth	~1 Mbps	250 Kbps	>1 Mbps	–	27 M	1 Gbps
Transmission	Two ways	Two ways	Two ways	One way	Two ways	Two ways
Smartphone	Build-in	Extra	Build-in	Extra	x	Build-in

Synthetically, due to smartphone build-in, WiFi and Bluetooth are the most potential technologies for ad hoc social networking and large message packets delivery. Therefore, a tightly (and perhaps even loosely) coupled network using Zigbee, WiFi, and/or Bluetooth has its potential currently selected in N-Fi.

Regardless of that LTE in long range communication working in 4G mobile phones has its achievement, LTE-V designed for another short range communication will get into the market in near future. It is designed for high speed moving carriers. Supposed that LTE-V is operated with subcarrier modes from LTE channels, it obviously will be integrated into smart phones. Therefore, a smart phone will simultaneously work/ access for long range base stations and short range access points, and satisfying relative-highly moving conditions. Then, due to LTE's wide bandwidth, under the consideration of cost down and thinness in the smartphone module, it has potential to replace all the function of current WiFi, Bluetooth and other short range technologies. Surely, the communication devices in market will be altered tremendously. Till that moment, LTE + (LTE-V) will become the major stream among all devices (cVIP), including IoT, IoV (V2X).

3 Transportation 5.0

Intelligent Transportation System (ITS) is currently the worldwide focusing work in public engineering construction. It includes Advanced Traveler Information System (ATIS), Advanced Traffic Management System (ATMS), Advanced Public Transportation System (APTS), Electronic Toll Collection (ETC), Emergency Management System (EMS), Advanced Commercial Vehicle Operation (ACVO), and Advance Vehicle Control and Safety System(AVCSS) etc. In fact, a successful ITS system rather relies on communication systems. On the point of view of LTE + (LTE-V) (5G) development and revolution, we can foresee the change of ITS due to that all roadway, roadside equipment, such as sign, signal, marking, CMS, bus stop, vehicle detector, light, … etc., can equipped with LTE-V, and conjunct directly with onboard vehicle devices and smartphones. This picture just illustrates in Fig. 1. This leads to overcome many past difficult tasks. The coming change of ITS development is denoted as Transportation 5.0 era (Tx5.0). An example of parking auditing is presented from manual auditing (1st generation), mechanic auditing (2nd generation), electronic auditing (3rd generation), free flow automatic auditing (4th generation) and then Ubiquitous auditing (5th generation), which with intelligence, networking with all possible nodes, ways and things (see Fig. 3). During the Tx5.0 era, smart phones and LTE-V elsewhere, the message/information will be ubiquitous. Therefore, many difficult problems such as accident reconstruction, calamity message, information delivery, payment, control, management… etc. will become more efficient, real-time, precise and correct. The author selects some important projects being discussed in the followings.

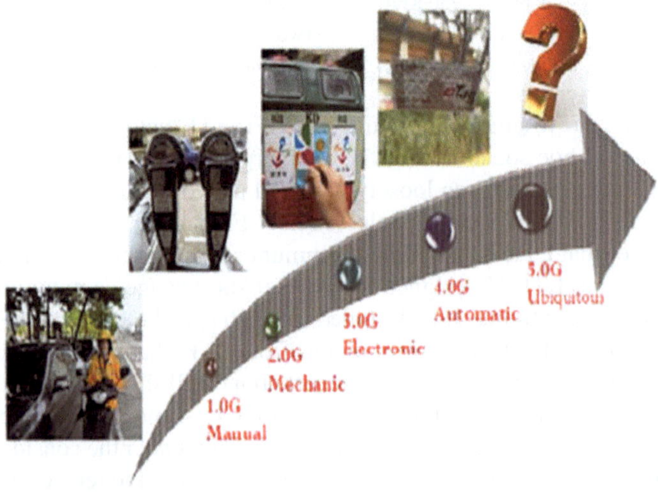

Fig. 3. The change of transportation era-an example of parking auditing

Accident Reconstruction Practice

Currently, there is a hard task to identify the fact of an accident. The factor to the record becomes history while policeman arriving at the site. The situation is unable to rewind for inspection. Accidents are most in process justified with experiences by experts' knowledge. A light bias always exists in some cases since it lacks of evidence. In coming future, implementing LTE + (LTE-V) may give the opportunity to rewind the precise trajectory in terms of time and location of respecting vehicles and persons. Based on surrounding fixed bases (like LTE-V beacons) and moving objects (OBUs, smartphones), reconstructing the theme by precisely positioning algorithm is possible. Using precise positioning to rewind the trajectory is the most important in justifying the accident fact. And, this tool will reduce bias of judgements.

Vehicle Based to Human Based

In past, traffic officers always collect traffic data by equipped loop detector, radar, supersonic/IR detector, CCTV (closed circuit television camera) with image processing. However, all are on vehicle base for traffic control or management, not on human base. From near future, LTE-V may provide the opportunity that leads traffic control and management getting into human base, due to LTE + (LTE-V) smartphone in traveler hand. The controller of traffic signals generates the timing plan in terms of not only vehicle variables, but human attributes. A very interesting issue is pcu (passenger car unit) used in traditional planning and control formulations, how it to make a transition to passengers in a car unit.

iCaution

If LTE + (LTE-V) plugged in smartphone, it will easily build up emergent messages broadcasting platform, especially for the purpose of local disasters (slope sliding, water submerge, ...), accidents, car flat, etc. These can be accomplished via a triangular pyramid with a coded module of short-range communication (see Fig. 4).

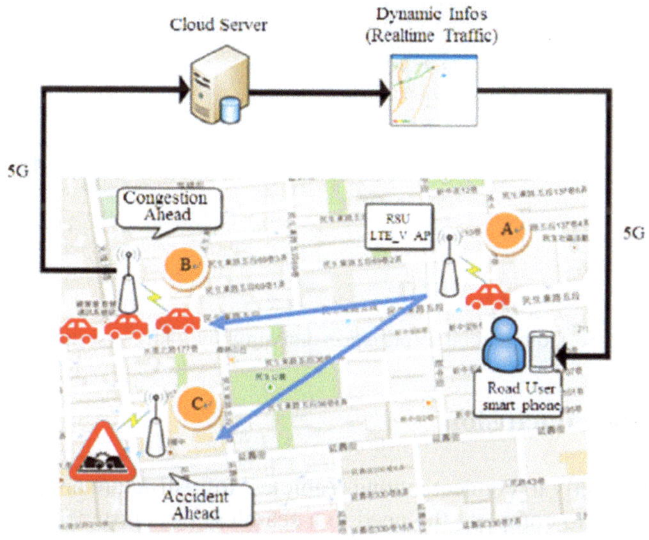

Fig. 4. Emerging traffic messages broadcasting by LTE-V AP [4] (simulated)

ETC

ETC plays an important role in ITS. Figure 3 has depicted the revolution. Herein, we will discuss the cashing issue. ETC can be the tool for traffic control if knowing the vehicle and person's position in terms of time what others. The pricing can be functioned with roadway efficiency (time cost), energy exhaustion or emission (destroy cost and pollution cost) and safety level of service (traffic impact and social cost). This implies pricing differential and comes to more equality in toll collection. On another way, traffic management office can issue eco-money (for example discount from base fare) if a vehicle obeys control message. The eco-money can be cumulated to afford to pay for other trip expenditure. Or, in business, eco-money is like coupon, can be exchanged, purchase commodities. The purpose is to promote green transportation to people, love our earth [8].

By the way, LTE + (LTE-V) will make obscure between transportation and other business, for examples, commercial activity into transportation/ traffic space, online game into traffic fields (see Fig. 5).

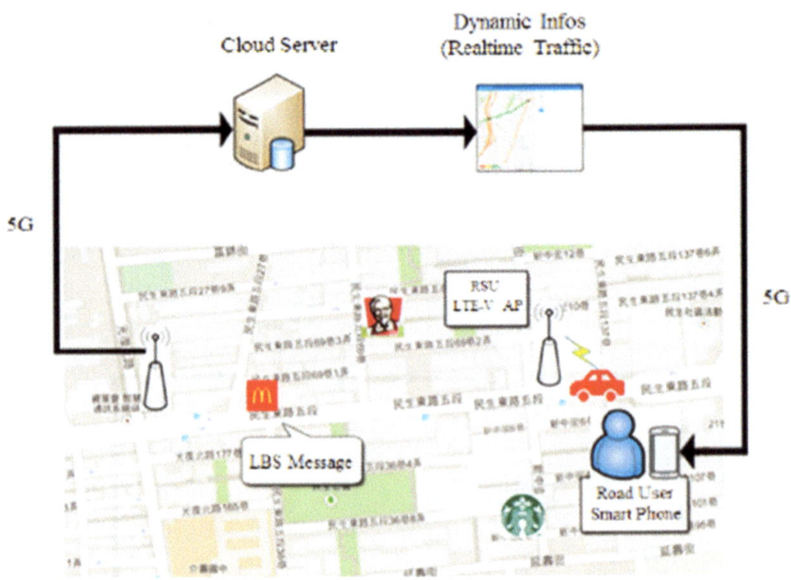

Fig. 5. Commercial information delivered by LTE-V (simulated)

4 Summary/Conclusion

People expect directly to connect among Vehicles, Infrastructure and Personal devices (cVIP). This leads to complete the need to a seamless circumstance. The presentation discusses and analyzes the communication technology evolution inducing market and application field change in near future. The author foresees LTE + (LTE-V) will win the market due to which bundles smartphone core. LTE-V will substitute IEEE 802.11 series as well as 802.15 series etc. Moreover, The author foresees advanced ITS will achieve to Transportation 5.0 era. This overcomes many current difficult problems and easily be solved, getting into more efficient, safer, greener and balanced society.

Acknowledgement. This paper is digested from the scientific project sponsored by Fujian University of Technology.

References

1. Chang, T.-H.,: The revolution of ITS—introduction to PVT-net with N-Fi technology, an IOT/ Telematics system. In: International Symposium on ITS Research, National Taiwan University, Taipei, Taiwan, 11 June 2011
2. Chang, T.-H., Yang, L.-K., and Tsai, J.-J.: A Telematic builder for V2V and V2I via N-Fi system. In: 19th World Congress on ITS, Vienna, Austria, 22–26 October 2012
3. Chang, T.-H.: Near field informatics with a PVT-net building up seamless life. In: 14th International Conference on Computing in Civil and Building Engineering, Moscow, 27–29 June 2012
4. Chang, T.-H., Yang, L.-K. Hsieh, B.-C.: Building iCaution and traffic game in smart cities. In: The 3rd International Conference on Internet of Vehicles (IOV), Nadi, Fiji, 7–10 December 2016. Internet of Vehicles – Technologies and Services, Springer (2017)
5. VII Architecture and Functional Requirements, ITS Joint Program Office, US Department of Transportation, 20 July 2005
6. From VII to IntelliDrive. http://www.its.dot.gov/press/2010/vii2intellidrive.htm/
7. ZigBee Alliance. http://www.zigbee.org/
8. Chang, T.-H., Tseng, J.-S., Ye, Y.-H.: Green safety index representing traffic level of services for online APP. IET Intell. Transp. Syst. **10**(6), 421–427 (2016)

Simulation and Modeling of a Solidly Mounted Resonator

Sheng-Hui Meng[1(✉)], An-Chi Huang[2], and Ying-Chung Chen[2(✉)]

[1] Fujian Provincial Key Lab of Big Data Mining and Applications, Fujian University of Technology, Fuzhou 350118, China
menghui@fjut.edu.cn

[2] Department of Electrical Engineering, National Sun Yat-Sen University, Kaohsiung, Taiwan
abc94299@gmail.com, ycc@mail.ee.nsysu.edu.tw

Abstract. Solidly mounted resonators (SMRs) have a stable structure, high frequency range, and good process yield, and thus are increasingly being used in various applications, including radiofrequency (RF) communication devices. In this study, we simulated SMR structures and thin films of the piezoelectric materials AlN and ZnO using the commonly used RF design program Advanced Design System.

Keywords: Solidly mounted resonators (SMR) · Advanced design system (ADS) · Aluminium nitride (AlN) · Zinc oxide (ZnO)

1 Introduction

Given the developments in wireless communications as well as the increasing demands of consumers for mobile phones with improved features, device quality and stability have become important parameters.

This paper presents the results of Advanced Design System (ADS) simulations of solidly mounted resonators (SMRs) with different piezoelectric layers.

The thin-film bulk acoustic resonator consists of a piezoelectric layer sandwiched between upper and lower electrodes. To prevent the dissipation of acoustic energy to the substrate, three configurations are typically used to minimise acoustic losses. As the process yield of the cavity etching is a crucial factor in the case of air-gap-isolated and VIA-isolated (back-trench) resonators, stacked acoustic components were simulated in this study, owing to their high working frequency, large power capacity, low loss, and small volume. Alternately stacked Mo/SiO_2 reflector layers with high and low acoustic impedances with respect to the 1/4 λ mode were studied, with aluminium nitride (AlN) [1] and zinc oxide [2] thin films used as the piezoelectric layers.

2 Simulation Design

The software ADS (Agilent) is the industry standard for designing RF and communication circuits. It is a powerful program capable of yield analysis and design optimisation and can thus help greatly improve the design efficiency of complex circuits.

J.-S. Pan et al. (eds.), *Advances in Smart Vehicular Technology, Transportation, Communication and Applications*, Smart Innovation, Systems and Technologies 86,
https://doi.org/10.1007/978-3-319-70730-3_45

[3] the design function of ADS, which allows for a high degree of freedom and convenience, was used to establish an SMR component library based on the Mason [4] model. The commonly used SMR acoustic materials were incorporated into the library, and their performances when used for the components were simulated. The acoustic characteristics of these commonly used materials are listed in Table 1.

Table 1. Properties of AlN and ZnO for piezoelectric materials

Parameters	AlN	ZnO
Acoustic impedance (10^6 kg/m^{2}s)	34.0	36.0
Bulk acoustic wave velocity (m/s)	10400	6630
Mass density (kg/m^3)	3270	5680
Electromechanical coupling coefficient (Kt^2)	6.5	7.5
Inherent material loss	Very low	Low
Dielectric constant (ε_r)	9.5	9.2
CMOS compatibility	Compatible	Incompatible
Sputtering rate	High	Middle

3 Experimental Section

Figure 1 is a schematic illustration of a Bragg reflector. Commonly used reflector materials include Mo, tantalum pentoxide (Ta_2O_5), and W; these are high-acoustic-impedance materials. Silicon dioxide (SiO_2), on the other hand, has a low acoustic impedance. The reflector layer was assembled in the 1/4 λ mode by staggering together high-acoustic-impedance Mo (Z_{Mo} = 64.2 × 10^6 kg/(m^{2}s)) and low-acoustic-impedance SiO_2 (Z_{SiO_2} = 64.2 × 10^6 kg/(m^{2}s)) to achieve an acoustic impedance ratio $\left(Z_{Mo}/Z_{SiO_2}\right)$ = 4.5 [5]. This resulted in a Bragg reflector layer with excellent acoustic resonance. Three Mo/SiO_2 layers were deposited.

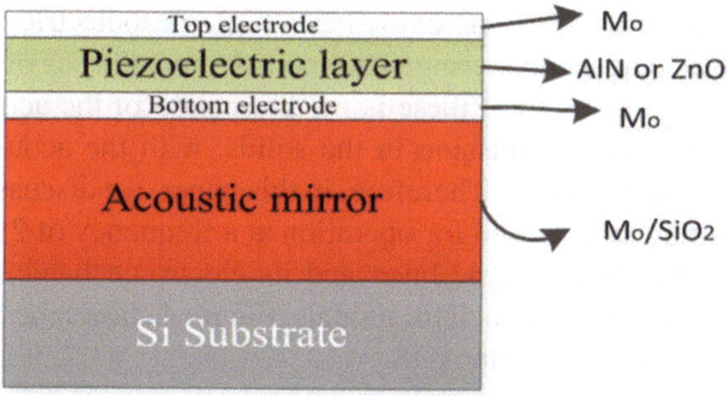

Fig. 1. Schematic illustration of a Bragg reflector

Table 2 lists the characteristic acoustic parameters of the thin film materials. Common electrode materials such as Mo, Al, Ti, and Pt provide good electrical contact by sandwiching the piezoelectric acoustic layer between the upper and lower electrodes, and function as mechanical loads for the device. The electrode materials should have good acoustic coupling and low ohmic loss; thus, Mo was chosen for the bottom electrode and Pt for the top electrode. Figure 2 is the ADS model of the Mo and Pt electrodes.

Table 2. Acoustic property parameters of thin film materials

	Acoustic impedance (10^6 kg/m^{2}s)	Mass density (kg/m^3)	Bulk acoustic wave velocity (m/s)
Mo	64.2	5680	6290
Ti	27.3	4500	6071
Al	13.7	2695	5100
SiO_2	13.8	2650	5200
Si	5.13	2330	2200

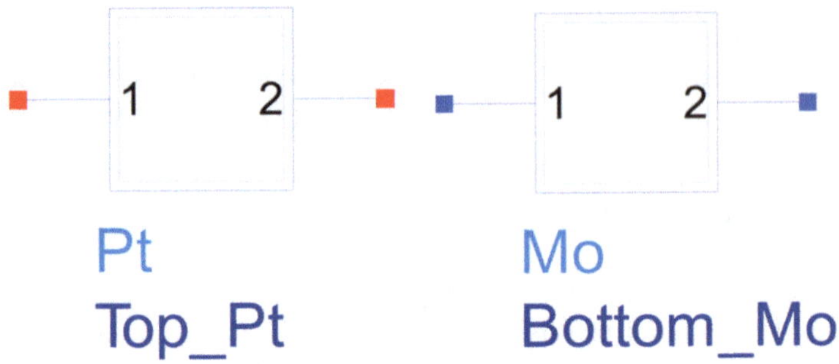

Fig. 2. ADS model of the Mo and Pt electrodes.

The ADS simple simulation of piezoelectric layer is shown in Fig. 3. Aluminium nitride and zinc oxide, which were assembled in the 1/2 λ mode, are excellent piezoelectric materials and exhibit unique advantages. Table 2 shows the characteristic parameters for the piezoelectric materials AlN and ZnO. AlN has a high acoustic velocity and CMOS compatibility, while zinc oxide has a high piezoelectric coefficient, which results in low acoustic energy scattering. The acoustic impedances of air and solids are almost zero in the ideal state; thus, these is no dissipation of the acoustic wave to the substrate or air during its transmission in the solids, with the acoustic energy being confined to the piezoelectric layer. Therefore, in this paper, we discuss the piezoelectric and reflector layers that can be used for operation at a frequency of 2.5 GHz. The resonance frequency of the piezoelectric layer and its electromechanical behaviour were simulated using the Mason and Leach [6] models. Figure 4 shows the S parameter value as well as the target settings used in ADS.

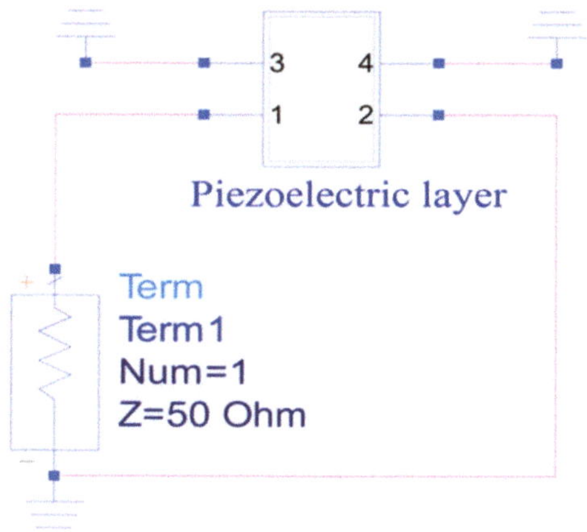

Fig. 3. ADS simple simulation of piezoelectric layer

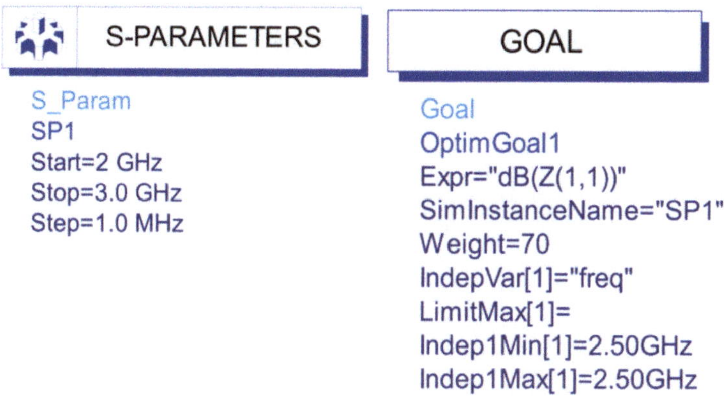

Fig. 4. S parameter value as well as the target settings used in ADS.

4 Results and Discussion

According to the data of thin film material, piezoelectric material and the design theory of SMR. In this experiment, we use the traditional piezoelectric Mason model to do the simulation of the half-wavelength mode piezoelectric layer of AlN, ZnO and quarter-wavelength mode 3 pairs of reflective layers Mo/SiO_2 in 2.5 GHz [7].

Set AlN piezoelectric layer is 2.08 um, top electrode and bottom electrode are 100 nm individually. The Fig. 5 shows the S_{11} frequency response diagram of the AlN for the ADS simulation layer of 3 for the piezoelectric layer, and the center point of the SMR resonant frequency is 2.51 GHz. The frequency of S_{11} ADS simulation for ZnO piezoelectric layer 3 of reflective layer response, set the ZnO piezoelectric layer is 1.27 um, the top electrode and the bottom electrode of individual 100 nm, the center point of SMR resonant frequency is 2.50 GHz in Fig. 6.

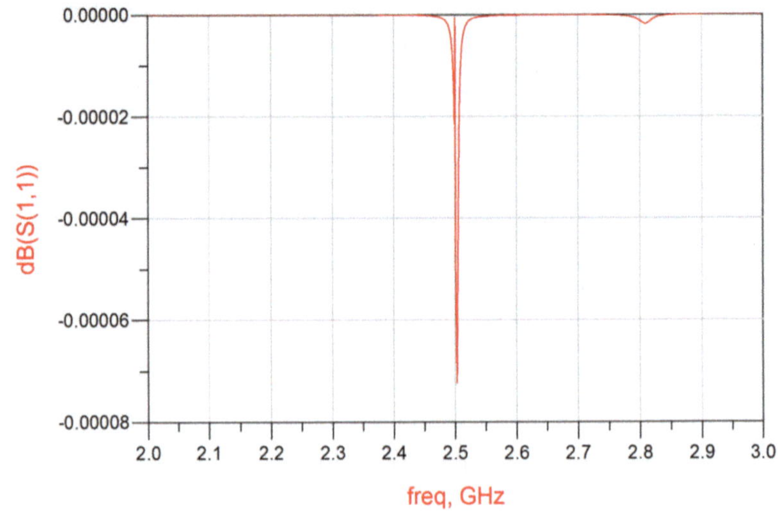

Fig. 5. Frequency response of the SMR with AlN

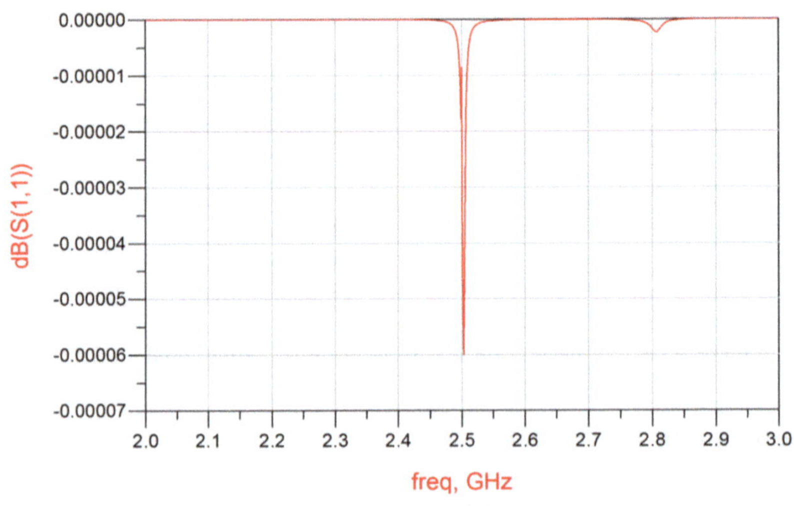

Fig. 6. Frequency response of the SMR with ZnO

5 Conclusions

Summary of this experiment with ADS simulation of micro resonator solid study. The commonly used piezoelectric materials of high and low acoustic impedance of AlN, ZnO and the reflective layer are discussed. It is possible to predict the performance of the design structure and the feasibility, also we can adjust the special parameters of materials from the simulation software.

References

1. Chen, D., et al.: The AlN based solidly mounted resonators consisted of the all-metal conductive acoustic Bragg reflectors. Vacuum **85**(2), 302–306 (2010)
2. Rughoobur, G., et al.: Room temperature sputtering of inclined c-axis ZnO for shear mode solidly mounted resonators. Appl. Phys. Lett. **108**(3), 034103 (2016)
3. Roy, A., Barber, B.P., Prasad, K.: Modeling of RF-MEMS BAW resonator. In: 23rd IEEE International Conference on VLSI Design, 2010, VLSID 2010, pp. 170–175 (2010)
4. Mason, W.P.: Electromechanical Transducers and Wave Filters. D. Van Nostrand Co., New York (1948)
5. Chung, C.-J., et al.: Fabrication and frequency response of solidly mounted resonators with 1/4 λ mode configuration. Thin Solid Films **516**(16), 5277–5281 (2008)
6. Bauer, B.B.: Equivalent circuit analysis of mechano-acoustic structures. Trans. IRE **2**(4), 112–120 (1954)
7. Shih, W.-C., et al.: Simulation of solidly mounted resonator using mason model and its implementation. Sens. Mater. **29**(4), 405–410 (2017)

Joint I/Q Mismatch Calibration, Compensation and Channel Equalization Approach for STBC 2×2 MIMO OFDM Transceivers

Chih-Feng Wu[1(✉)], Yi-Hung Lin[2], Muh-Tian Shiue[3], and Jeng-Shyang Pan[1]

[1] School of Information Science and Engineering, Fujian University of Technology, Fuzhou 350118, China
tfeng.wu@gmail.com

[2] Graduate Institute Engineering and Department of Electrical Engineering, National Taiwan University, Taipei 10617, Taiwan

[3] Department of Electrical Engineering, National Central University, Taoyuan 32001, Taiwan

Abstract. The joint I/Q mismatch calibration, compensation and channel equalization approach are proposed for space-time-block-code (STBC) 2×2 multiple-input-multiple-output (MIMO) orthogonal frequency division multiplexing (OFDM) transceivers. The I/Q mismatches of transmitter and receiver are extracted by using the *look-back* calibration during the system power-on period. The results of *look-back* calibration is used to construct a reparation block and to further compensate the I/Q mismatch. In addition, based on MMSE criterion, the joint approach is employed to simultaneously combat the residual remote/end-user transmitter I/Q mismatch and the channel impairment at end-user/remote side.

Keywords: I/Q mismatch · Equalization · STBC · SISO · MIMO · OFDM

1 Introduction

In order to increase the throughput and the robustness of wireless transmission over the multipath frequency-selective fading channel, the multiple-input-multiple-output (MIMO) and OFDM technology have been combined and adopted in IEEE 802.11n wireless local area network (WLAN) [1]. Considering high-integration issue of wireless transceiver, the zero intermediate-frequency (IF) design can be employed to instead of the radio frequency (RF) transceiver for MIMO-OFDM system.

However, a drawback of zero-IF design is I/Q mismatch. For zero-IF transceiver as illustrated in Fig. 1, the I/Q mismatch will be concurrently manifest at the remote and the end-user transceiver. The former studies [2,3] presented the special designed preamble or pilot arrangement to estimate and compensate

J.-S. Pan et al. (eds.), *Advances in Smart Vehicular Technology, Transportation, Communication and Applications*, Smart Innovation, Systems and Technologies 86,
https://doi.org/10.1007/978-3-319-70730-3_46

the I/Q mismatch. Consequently, both the estimation and the compensation approaches can not be directly used to the realistic system such as MIMO-OFDM WLAN transceiver [1]. In addition, the work [4] fulfilled the estimation and the compensation of I/Q mismatch without taking the effect of remote Tx I/Q mismatch in account. In contrast to single-input-single-output (SISO) OFDM transceiver, the computational complexity of I/Q mismatch for MIMO-OFDM transceiver is extend by the multiple antennas of received signal.

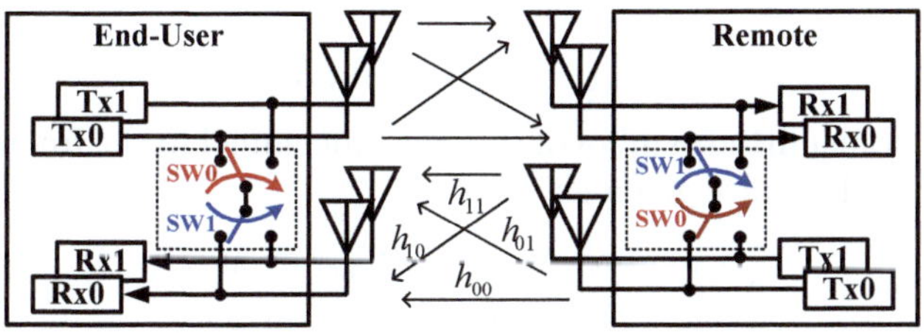

Fig. 1. Block diagram of 2×2 MIMO-OFDM transceivers.

In this paper, the joint I/Q mismatch calibration, compensation and channel equalization approach is proposed to deal with I/Q mismatch and channel impairment for STBC 2×2 MIMO-OFDM transceivers. The look-back calibration is used to concurrently estimate the related parameters of I/Q mismatches for transmitter (Tx) and receiver (Rx) either end-user or remote transceiver in the system power-on period. According to the calibrated results estimated from look-back calibration, the reparation blocks are presented to simultaneously compensate I/Q mismatch of STBC MIMO-OFDM transceiver for end-use and remote sides. Furthermore, the residual I/Q mismatches of Tx and Rx can be viewed as a portion of channel impairment. The remote/end-user Tx I/Q mismatch and the channel impairment are jointly combated by the equalization process at the end-user/remote receiver.

The paper is organized as follows. The signal model of I/Q mismatch is described in Sect. 2. The loop-back calibration of I/Q mismatch is depicted in Sect. 3. Then, the residual I/Q mismatch and channel impairment are jointly derived in Sect. 4. The simulation results of system evaluation are illustrated in Sect. 5. Finally, the concluding remarks are described in Sect. 6.

2 Signal Model

Considering a SISO OFDM transceiver [5], the received signal suffered from channel impairment and I/Q mismatch in the frequency domain is described as

$$Y_{iq,k} = \mathrm{DFT}\{y_{iq}\} = \alpha_{rx} H_k R_{iq,k} + \beta_{rx} H^*_{-k} R^*_{iq,-k} \tag{1}$$

where k is a subcarrier index. H_k is the kth subchannel frequency response. y_{iq} is the received time-domain signal. In addition, α_{rx} and β_{rx} are given as

$$\alpha_{rx} = \cos \Delta\theta_{rx} - j\Delta G_{rx} \sin \Delta\theta_{rx} \quad (2)$$

$$\beta_{rx} = \Delta G_{rx} \cos \Delta\theta_{rx} + j \sin \Delta\theta_{rx} \quad (3)$$

where $\Delta\theta_{rx}$ and ΔG_{rx} are the phase and the amplitude (gain) mismatch between I- and Q-path of Rx. The $R_{iq,k}$ is defined as the kth subchannel of received symbol suffered from only Tx I/Q mismatch in the frequency domain and further given as

$$R_{iq,k} = \alpha_{tx} R_k + \beta_{tx} R^*_{-k} \quad (4)$$

where R_k is a transmitted symbol of the kth subchannel in the frequency domain. Likewise, α_{tx} and β_{tx} are given as

$$\alpha_{tx} = \cos \Delta\theta_{tx} + j\Delta G_{tx} \sin \Delta\theta_{tx} \quad (5)$$

$$\beta_{tx} = \Delta G_{tx} \cos \Delta\theta_{tx} - j \sin \Delta\theta_{tx} \quad (6)$$

where ΔG_{tx} and $\Delta\theta_{tx}$ express the amplitude (gain) and the phase mismatches of Tx. According to the ***superposition theorem***, the signal model suffered from I/Q mismatch can be used to derive the calibration and compensation approach for STBC 2 × 2 MIMO-OFDM transceiver.

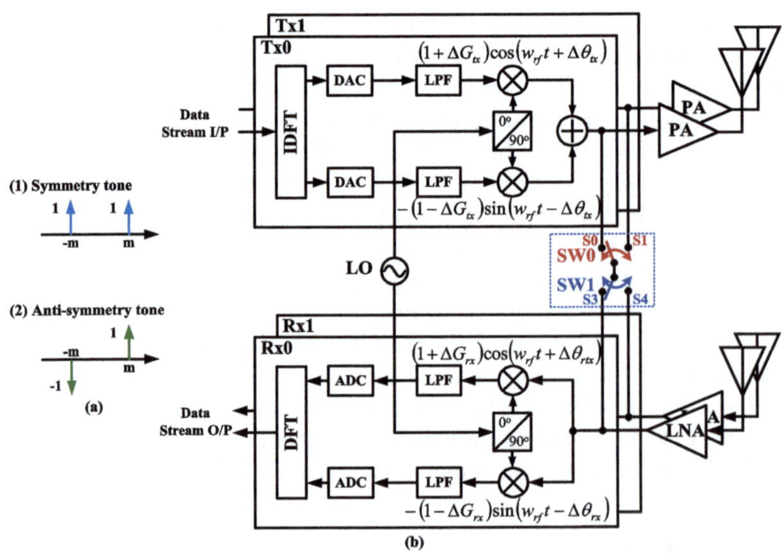

Fig. 2. Architecture of look-back calibration for zero-IF 2 × 2 MIMO-OFDM transceiver: (a) calibration signal and (b) zero-IF 2 × 2 MIMO-OFDM transceiver.

Considering a zero-IF 2 × 2 MIMO-OFDM transceiver, the architecture of look-back calibration as shown in Fig. 2 can be used in end-user- and remote-side. Without loss of generality, the following derivation and statement does not emphasize for end-user- or remote-side. In order to maintain a cost-efficient design of transceiver, a single local-oscillator (LO) is employed to be the carrier of up- and down-converter. Therefore, the related parameters of I/Q mismatch, such as amplitude (gain) and phase mismatches of Tx and Rx mixers, are regarded as equivalent and then given as

$$\begin{cases} \Delta G_{tx} = \Delta G_{tx0} = \Delta G_{tx1} = \Delta G_{rx} = \Delta G_{rx0} = \Delta G_{rx1} \\ \Delta \theta_{tx} = \Delta \theta_{tx0} = \Delta \theta_{tx1} = \Delta \theta_{rx} = \Delta \theta_{rx0} = \Delta \theta_{rx1} \end{cases} \tag{7}$$

Therefore, the look-back calibration of I/Q mismatches of Tx and Rx can be easily realized with cooperating two switches as shown in Fig. 2. In the system power-on period, the ***look-back*** calibration is mainly employed to estimate the I/Q mismatch. In addition, based on the estimated results of I/Q mismatch, the reparation blocks are used to compensate I/Q mismatch in the real data transmission. After *look-back* calibration, the residual I/Q mismatches of Tx and Rx can be viewed as a portion of channel impairment. The residual Tx I/Q mismatch and the channel impairment are jointly combated by the equalization process at the receiver in the joint tracking process. The detail operation of the proposed algorithm is described as the following sections.

3 Look-Back Calibration

The related parameters of I/Q mismatch as shown in Eq. (7) can be obtained in the ***look-back*** calibration in the system power-on period. The ***look-back*** calibration has to be worked with **SW0** and **SW1**, as illustrated in Fig. 2, performing the I/Q mismatch calibration for both end-user- and remote-side. For instance, **SW0** and **SW1** are changed to "**S0**" and "**S3**", respectively, to do the loop-back calibration from **Tx0** to **Rx0** (**Tx0–Rx0**). Furthermore, **SW0** is switched to "**S1**" to do the loop-back calibration **Tx1–Rx0**. Similarly, we can use the same approach to do the loop-back calibrations including **Tx0–Rx1** and **Tx1–Rx1**. Without loss of generality, we use a single path to show the **look-back** calibration in the following derivation. The **look-back** calibration is constructed with 2 processes such as **symmetry-** and **anti-symmetry-process** stage.

3.1 Symmetry-Tone Process

Two signals with equivalent amplitude are assigned in two symmetry tones at m and $-m$, as shown in Fig. 2(a), where m is an index of symmetry subcarrier. The signal $x_{n,S}$ at IDFT output can be described as

$$x_{n,S} = \sum_{k=m,-m} X_k e^{j\frac{2\pi kn}{N}} = 2\cos\frac{2\pi mn}{N} = 2\cos\omega_{bb}n \tag{8}$$

where ω_{bb} represent the equivalent baseband frequency. Clearly, Eq. (8) shows that $x_{n,S}$ only has real-part signal. The output amplitudes $V_{ii,P}$ and $V_{iq,P}$ at the I- and Q-path, respectively, of ADCs are given as

$$V_{ii,P} = K(1 + \Delta G_{tx})(1 + \Delta G_{rx}) \cos(\Delta\theta_{rx} - \Delta\theta_{tx}) \tag{9}$$

$$V_{iq,P} = K(1 + \Delta G_{tx})(1 - \Delta G_{rx}) \sin(\Delta\theta_{rx} + \Delta\theta_{tx}) \tag{10}$$

where K denotes a path gain. The subscripts "ii" and "iq" represent the path from Tx I-path to Rx I- and Q-path, respectively.

3.2 Anti-symmetry-Tone Process

Two signals with equivalent anti-amplitude are arranged in two symmetry tones at m and $-m$ as illustrated in Fig. 2(a). Therefore, the signal $x_{n,AS}$ at IDFT output can be derived as

$$x_{n,AS} = \sum_{k=m,-m} X_k e^{j\frac{2\pi kn}{N}} = j2 \sin \frac{2\pi mn}{N} = j2 \sin \omega_{bb} n \tag{11}$$

Apparently, Eq. (11) reveals that the $x_{n,AS}$ just has imaginary-part signal. Similarly, the output amplitudes V_{qi} and V_{qq} at I- and Q-path, respectively, of ADCs are derived as

$$V_{qi,P} = K(1 - \Delta G_{tx})(1 + \Delta G_{rx}) \sin(\Delta\theta_{rx} + \Delta\theta_{tx}) \tag{12}$$

$$V_{qq,P} = K(1 - \Delta G_{tx})(1 - \Delta G_{rx}) \cos(\Delta\theta_{rx} - \Delta\theta_{tx}) \tag{13}$$

where the subscripts "qi" and "qq" denote the path from Tx Q-path to Rx I- and Q-path, individually.

3.3 I/Q Mismatch Estimation

As mentioned above, we can employe Eqs. (9), (10), (12) and (13) to estimate the related parameters of I/Q mismatch in the look-back calibration.

A. Amplitude Mismatch Estimation: Based on the derivations of Eqs. (7), (9) and (13), the maximum ratio γ can be derived as

$$\gamma = \frac{V_{ii,P}}{V_{qq,P}} \approx \frac{(1 + \Delta G_{tx})(1 + \Delta G_{rx})}{(1 - \Delta G_{tx})(1 - \Delta G_{rx})} \approx 1 + 4\Delta G_{tx} \tag{14}$$

where $\cos(\Delta\theta_{rx} - \Delta\theta_{tx})$, as shown in Eqs. (9) and (13), is close to 1 since $\Delta\theta_{tx}$ and $\Delta\theta_{rx}$ are tiny. Therefore, $\Delta\hat{G}_{tx}$ and $\Delta\hat{G}_{rx}$ are derived as

$$\Delta\hat{G}_{tx} = \Delta\hat{G}_{rx} \approx \frac{\gamma - 1}{4} \tag{15}$$

B. Phase Mismatch Estimation: Similarly, based on Eqs. (7), (9), (10), (12) and (13), we can use the following two terms, η_+ and η_-, to find the phase mismatch. The derivations are described as

$$\eta_+ = \frac{V_{iq,P}}{V_{ii,P}} \approx \frac{(1-\Delta G_{rx})\sin(\Delta\theta_{rx}+\Delta\theta_{tx})}{(1+\Delta G_{rx})\cos(\Delta\theta_{rx}-\Delta\theta_{tx})} \approx \Delta\theta_{rx}+\Delta\theta_{tx} \tag{16}$$

$$\eta_- = \frac{V_{qi,P}}{V_{qq,P}} \approx \frac{(1+\Delta G_{rx})\sin(\Delta\theta_{rx}+\Delta\theta_{tx})}{(1-\Delta G_{rx})\cos(\Delta\theta_{rx}-\Delta\theta_{tx})} \approx \Delta\theta_{rx}+\Delta\theta_{tx} \tag{17}$$

Obviously, $(1+\Delta G_{rx})$ and $(1-\Delta G_{rx})$ are cancelled each other in Eqs. (16) and (17) since ΔG_{rx} is far less than 1 and approximates to zero. Besides, $\sin(\Delta\theta_{rx}+\Delta\theta_{tx})$ is equivalent to $(\Delta\theta_{rx}+\Delta\theta_{tx})$ iff both $\Delta\theta_{rx}$ and $\Delta\theta_{tx}$ are tiny and close to zero. Consequently, $\Delta\hat{\theta}_{tx}$ and $\Delta\hat{\theta}_{rx}$ can be acquired as

$$\Delta\hat{\theta}_{tx} = \Delta\hat{\theta}_{rx} \approx \frac{\eta_+ + \eta_-}{4} \tag{18}$$

C. Reparation Structure: Based on the results of look-back calibration derived in Eqs. (15) and (18), the amplitude (gain) and phase mismatches estimated from **Tx0**–**Rx0**, **Tx1**–**Rx0**, **Tx0**–**Rx1** and **Tx1**–**Rx1** are obtained with cooperating **SW0** and **SW1**, individually. Therefore, by averaging of four results from loop-back calibration, the estimated amplitude (gain) and phase mismatches can be used to construct the reparation structure for compensating the I/Q mismatch. Considering the I/Q mismatch of Tx mixer, the reparation block for **Tx0** and **Tx1**, as shown in Fig. 3, can be employed to pre-compensate the I/Q mismatch. Similarly, in view of the I/Q mismatch of Rx mixer for **Rx0** and **Rx1**, the reparation block, as illustrated in Fig. 4, can be used to recompense the I/Q mismatch.

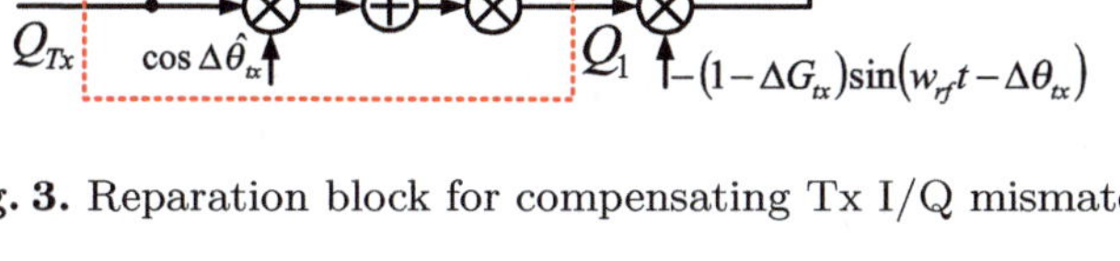

Fig. 3. Reparation block for compensating Tx I/Q mismatch.

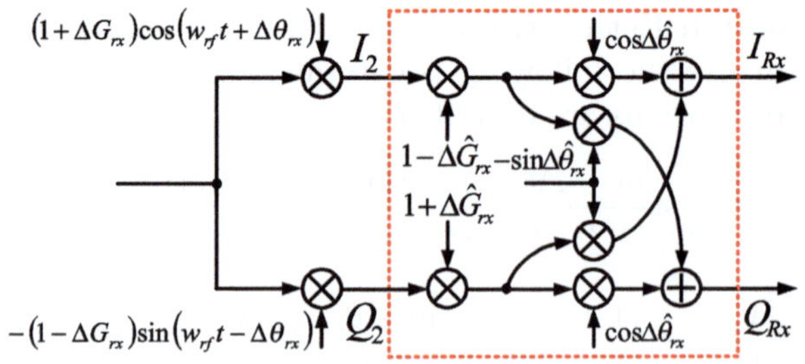

Fig. 4. Reparation block for compensating Rx I/Q mismatch.

4 Joint Tracking Process

After the loop-back calibration, the residual I/Q mismatch still exits in the Tx- and Rx-path. Fortunately, the residual I/Q mismatch can be viewed as a portion of channel impairment and, then, combated with the tracking process of channel equalization on each subchannel after channel estimation. The kth subchannel response suffered from I/Q mismatch, as derived in Eq. (1), in the frequency domain for SISO OFDM transceivers is given as

$$H_{\alpha,k} = \alpha_{tx} H_k \ \text{ and } \ H_{\beta,k} = \beta_{tx} H^*_{-k} \tag{19}$$

After completing ***look-back*** calibration and building reparation block for compensating I/Q mismatch, the both residual I/Q mismatches of Tx and Rx are very small and, then, can be further neglected in the following derivation. The received signal vector from other transmitter as illustrated in Fig. 1 is described as

$$\mathbf{Y} = \mathbf{HX} = [Y_{00,k}\ Y^*_{00,-k}\ Y^*_{01,k}\ Y_{01,-k}\ Y_{10,k}\ Y^*_{10,-k}\ Y^*_{11,k}\ Y_{11,-k}]^{\mathrm{T}} \tag{20}$$

where $\mathbf{X} = [X_{0,k}\ X^*_{0,-k}\ X^*_{1,k}\ X_{1,-k}]^{\mathrm{T}}$ denotes the vector of transmitted signal. Based on the MMSE criterion, a joint tracking process is proposed to minimize the MSE on each subchannel and to further get the right data decision on each subchannel. The cost function of joint tracking process is defined as

$$\arg\min_{\mathbf{W}} J(\mathbf{W}) = \mathcal{E}_k = E[\mathbf{E}^H \cdot \mathbf{E}] \ \text{ and } \ \mathbf{E} = \hat{\mathbf{X}} - \tilde{\mathbf{X}} \tag{21}$$

where $\mathcal{E}_k$ is MSE of the kth subchannel. $\mathbf{W}$ is coefficient matrix. $E[\cdot]$ denotes an expectation operator. Besides, $\hat{\mathbf{X}} = [\hat{X}_{0,k}\ \hat{X}_{1,k}]^{\mathrm{T}}$ and $\tilde{\mathbf{X}} = [\tilde{X}_{0,k}\ \tilde{X}_{1,k}]^{\mathrm{T}}$ are the decision signal and the equalized signal vectors, respectively. $\mathbf{E} = [E_{0,k}\ E^*_{1,k}]^{\mathrm{T}}$ is the decision error vector. Furthermore, $\hat{X}_{p,k} = Q\{\tilde{X}_{p,k}\}\ \forall\, p = \{0,1\}$, where $Q\{\cdot\}$ is a slicer. Theoretically, the subchannel SNR is defined as $SNR_k = \mathcal{S}_k/\mathcal{E}_k$, where $\mathcal{S}_k$ is the signal power of the kth subchannel. Equation (21) shows that the exact coefficient matrix can minimize the error power to enhance the subchannel SNR.

5 System Evaluation

In order to demonstrate the joint I/Q mismatch calibration, compensation and channel equalization approach, a *C* model, including channel impairment [6], is established to evaluate the STBC 2×2 MIMO-OFDM transceiver [1]. The amplitude- and phase-mismatch are assumed to be -5%–5% and $-5°$–$5°$ for examining the loop-back calibration. For $\pm 5\%$ of amplitude (gain) mismatch, the residual amplitude error is less than 0.086 dB as shown in left-side of Fig. 5. Similarly, for $\pm 5°$ of phase mismatch, the residual phase error is less than 0.1°

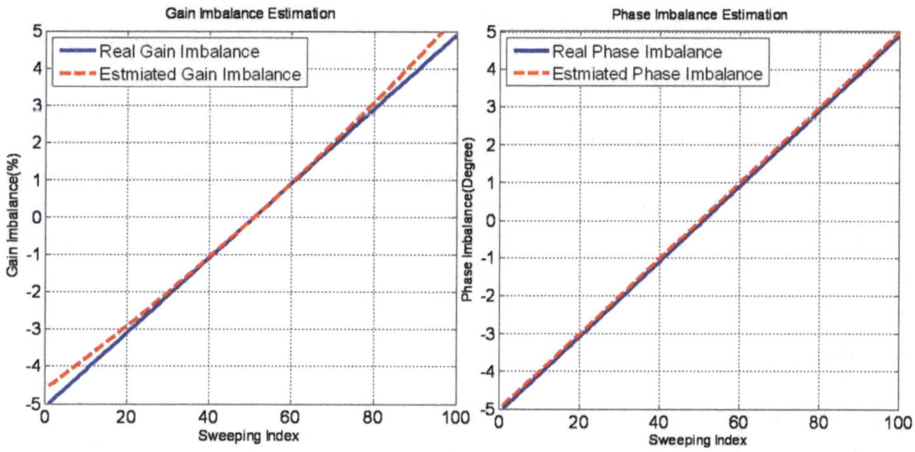

Fig. 5. Amplitude (left) and phase (right) imbalance sweeping result

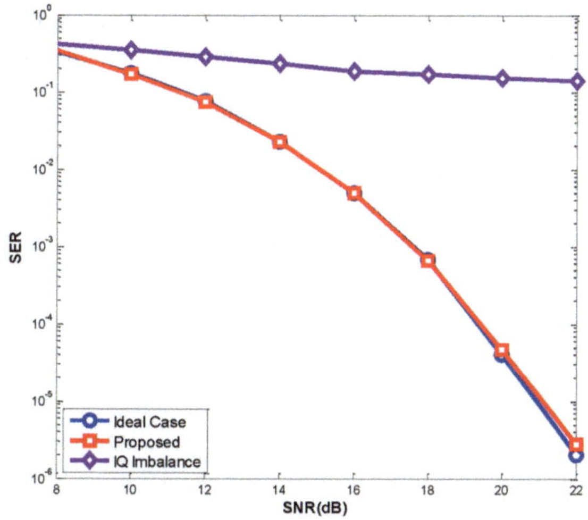

Fig. 6. SER comparisons for 2×2 MIMO-OFDM transceivers

as shown in right-side of Fig. 5. In summary, according to the calibrated results, the I/Q mismatches of Tx and Rx can be entirely estimated in the look-back calibration. Besides, assuming the remote Tx I/Q mismatch suffered from 3% amplitude and 3° phase mismatches, the symbol-error-rate (SER) is shown in Fig. 6, where the carrier frequency offset is considered as 232.2 KHz.

6 Conclusions

In this paper, the look-back calibration of I/Q mismatch is presented to estimate both amplitude- and phase-mismatch of Tx and Rx in the system power-on period. After look-back calibration, the estimated amplitude- and phase-mismatch are employed to construct a reparation block and then to pre-compensate I/Q mismatches of Tx and Rx mixers either end-user- or remote-side. Based on the evaluation results, the presented look-back calibration can really estimated the Tx and Rx I/Q mismatch since the residual amplitude- and phase-mismatch are 0.086 dB and 0.1°, respectively. Furthermore, the residual I/Q mismatch of Tx and Rx can be viewed as a portion of channel impairment. The residual I/Q mismatch and the quasi-static channel variation are jointly combated by the tracking process to minimize the subchannel MSE.

References

1. IEEE Std 802.11n-2009, 29 October 2009
2. Schenk, T.C.W., Smulders, P.F.M., Fledderus, E.R.: Estimation and compensation of TX and RX IQ imbalance in OFDM-based MIMO systems. In: IEEE Radio and Wireless Symposium, pp. 215–218, October 2006
3. Tarighat, A., Sayed, A.H.: On the baseband compensation of IQ imbalances in OFDM systems. In: Proceedings of the IEEE International Conference on Acoustics, Speech, and Signal Processing, vol. 4, pp. 1021–1024, May 2004
4. Schuchert, A., Hashozner, R., Buchholz, M.: Frequency domain equalization of IQ imbalance compensation in OFDM receivers. In: International Conference on Consumer Electronics (ICCE), pp. 28–29, June 2001
5. Lin, Y.H., Wu, C.F., Shiue, M.T., Wang, C.K.: I/Q imbalance compensation and channel equalization algorithm for MISO-OFDM systems. In: International Conference on ITS Telecom (ITST-2012), pp. 1–5, November 2012
6. TGn Channel Model: IEEE 802.11-03-0940-04-000n

A Constructive Problem-Based Course Design for Internet of Things

Xiaohu Ma[1], Yeh-Jong Pan[2], Fang Chen[1], Xinyi Ding[1], and Shih-Pang Tseng[2](✉)

[1] Software school, NingXia Polytechnic, Yinchuan, China
[2] Department of Computer Science and Entertainment Technology, Tajen University, No. 20, Weixin Rd., Yanpu Township 90741, Pingtung County, Taiwan
tsp@tajen.edu.tw

Abstract. Learning is the major human activity to acquire new, or reinforce existing knowledge and skills which can make people to accommodate the environment or the society better. The Internet Of Things (IOT) which integrate the sensing, networking and computing, will change the future living of humans totally, and provides more comfort and convenience. It has been an important issue to train enough qualified IOT programmers efficiently. This paper proposes a work which trains the undergraduate students to develop the mobile IOT via a constructive problem-based approach. It is a long term course project cooperated by NingXia Polytechnic and Tajen University, the second semester course is still running now. The study of the first semester basically shows the effectiveness of the proposed approach and the expectation of the students.

Keywords: Constructivism · Engineering education · Problem-based learning

1 Introduction

Learning is the major human activity to acquire new, or reinforce existing knowledge and skills which can make people to accommodate the environment or the society better. People can modify their behaviors, change the environment, or live more happily via learning. On the other hand, learning can improve the competitiveness and mobility of the society. Consequently, how to improve the effectiveness and efficiency of human learning is one important research issue.

Since the 1980s, the development of Internet technologies have dramatically changed the modern living of human beings in various areas. Originally, Internet is designed to connect the people. In the last 10 years, the Internet Of Things is proposed and developed to connect different things directly for more smart living environment, such as smart house [1], smart grid [2], and smart city [3].

As the IOT has been more important, the demand of IOT programmers has also been more strong in the second decade of the 21st century. Now, it has been

J.-S. Pan et al. (eds.), *Advances in Smart Vehicular Technology, Transportation, Communication and Applications*, Smart Innovation, Systems and Technologies 86,
https://doi.org/10.1007/978-3-319-70730-3_47

an important issue to train enough qualified IOT programmers efficiently. This paper proposes a work which trains the undergraduate students to develop the IOT project via a constructive problem-based approach. It is a long-term project cooperated by Tajen University, Taiwan, and NingXia Polytechnic, China. At this time, the second semester course is still running. Essentially, this paper only contains the results of the first semester. In the one semester course, the students learn the basics of IOT programming, finding the physical problem, system analysis, and completing the prototype.

The remainder of the paper is organized as follows. The related works are in Sect. 2. Section 3 provides a brief introduction about our courseware. Section 4 demonstrates' the implementation of the proposed courseware. Conclusion is given in Sect. 5.

2 Related Works

Constructivism [4,5] is considered as one theory about how people learn. In essence, learning is the knowledge re-creating (constructing) process in each learner brain according to the individual experience. Figure 1 shows the view of constructivism about learning. Experience from a specific context is the best way to learn the knowledge of the specific domain. The observation, experiment, navigation, exploration and practice are more important than the teaching the abstract knowledge by the blackboard. In the knowledge constructing process, the learner make the experienced thing meaning. The constructing process is influenced by the learner's background, culture, worldview, and previous learned knowledge. The learned knowledge can help the learner to explore more experiences in the similar context. Consequently, in the view of constructivism, learning should begin on the most accessible context around the learner, shown as Fig. 2. Step by step, the learner can explore the meanings of the context which is more difficult to access. For instance, the operation of gas engine can be easily observed by the learner. Relatively, the entropy (classical thermodynamics) is very difficult to measure and observe. Even though the second law of thermodynamics is the basis to design a gas engine, the operation of gas engine should be introduced before the second law of thermodynamics.

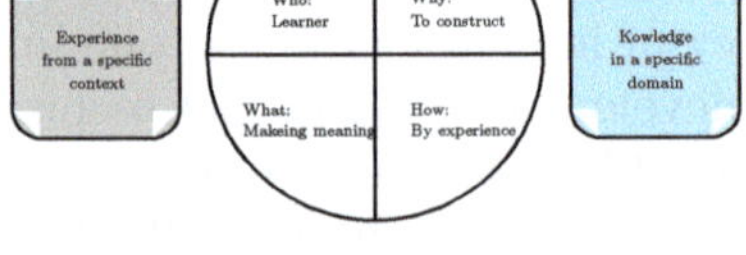

Fig. 1. Learning view of constructivism

Problem-based learning (PBL) [6,7] is a learner-centered instructional method in which students are given real-world situations or scenarios, denoted

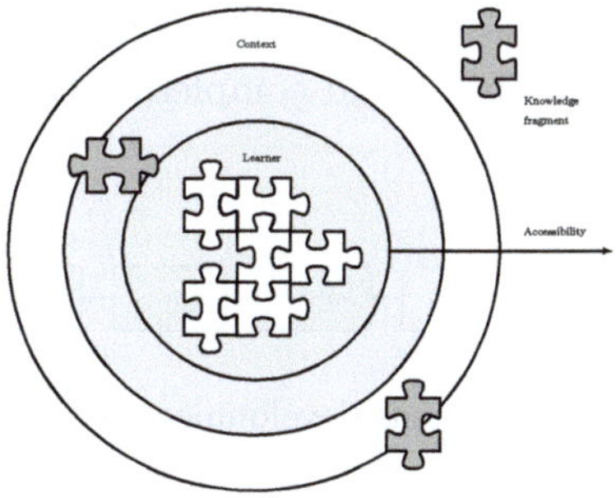

Fig. 2. Context accessibility in learning

as *problem*, and asked to offer insights regarding or solutions. Essentially, the teacher presents students the detail and importance of the problem to stimulate the students' interest and motivation. In general, Students identify what they need to learn in order to solve a problem by teamwork. The students are encouraged to perform self-directed learning. Finally, the students apply their new knowledge to the problem and evaluate the effectiveness of their solution. From the view of constructivism, the problem provides the context about the special domain, shown as Fig. 3. The students can explore more deep experiences via interaction and collaboration. The objectives of problem-based learning include helping students develop (1) intrinsic motivation, (2) flexible knowledge, (3) effective problem-solving skills, (4) effective self-directed learning skills, and (5) effective collaboration skills [7]. Problem-based learning began in medical schools and expanded to business, design and engineering education.

Fig. 3. The problem provides a context for the learner.

3 Design of the Curriculum

From the viewpoint of system engineering, the development process of IOT project is essentially the same as the systems development life cycle (SDLC) [8]. It is a two semesters project in Tajen University, Taiwan. At this time, the second semester course is still running. Essentially, this IOT training course is planned according to SDLC. The first semester course includes the *Ananlyze* and

Design stages. The students learn the basics of IOT programming, finding the physical problem, system analysis, and completing the prototype in the design stage (Fig. 4).

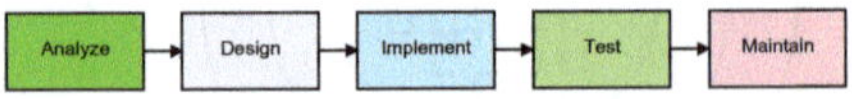

Fig. 4. Systems development life cycle

Before the first stage of SDLC, the student should decide what to do, or which problem his IOT project can solve. By the thinking of constructivism, the teacher does only provide the students a framework to support learning. This means that the student should find his problem by himself/herself. We think that the problem should be very close the student's experience, and then we decide the student's IOT project should enrich the living environment. The teacher illustrate how to discover the important things in the living environment. The student would be encouraged to think how to make the living environment more comfortable, secure, green, or economic. Relatively, the student has stronger motivation and self-regulation on learning how solve the problem he/she find. On the other hand, if the teacher directly give one imagined/simulated *case* to a student, the student often tries to search the similar and existent solution or model which were developed by the others previously. The student may use the *copy-and-paste* strategy to solve the case from the teacher. Consequently, the student must concretely learn the knowledge about the problem he/she find.

4 Performance Evaluation

This paper proposes a work which trains the undergraduate students to develop the IOT project via a constructive problem-based approach. This work is supported by the Smart Seed Union [9]. The Smart Seed Union includes nine universities on the south of Taiwan. The main work of Smart Seed Union is cultivating university students and teachers through interdisciplinary training. The objective of Smart Seed Union is to establish a new model of sustainable smart living space with consideration given to the humanities and technologies. The Smart Seed Union promotes a teaching model, denoted as D^3, which means *Discover*, *Develop*, and *Deliver*.

- *Discover*: the ability of discovering the practical problem of the living environment.
- *Develop*: the ability of developing the solution for the practical problem.
- *Deliver*: the ability of delivering the solution on the living environment.

The Smart Seed Union expects the students can focus on the practical problem of the living environment. The knowledge of various domains would be

Table 1. The study of D^3

	Very disagree	Disagree	Somewhat agree	Agree	Very agree
Discover	0	0	8.33%	66.67%	25.00%
Develop	0	8.33%	16.67%	58.33%	16.67%
Deliver	0	0	16.67%	66.67%	16.67%

needed to propose a total solution for this problem. Then the students are encouraged to learn the knowledge of different domains about this problem. Finally, the students should integrate the learned knowledge to propose a solution and try to apply this solution on the living environment.

After the first semester course, we do a study to evaluate the performance of the proposed work. We want to realize the enhancement of D^3 abilities. We use 5-point Likert Scale to measure the attitude of agreement. Each response obtains a corresponding value to measure the degree of agreement of the response. In this study, the response of **Agree** or **Very Agree** are classified as is consented to the content of the corresponding question. Table 1 shows the statistical result on the D^3 abilities. There are total 27 items, divided by 3 groups, in the questionnaire.

This study investigated the undergraduate students from the Software school, NingXia Polytechnic. And these student learn the IOT courses in the Department of Computer Science and Entertainment Technology, Tajen University. The valid response are 12 questionnaires. At the aspect of *Discover*, there are about 92% of students who consent to the proposed work can enhance their *Discover* ability effectively. It is reasonable because the main content of the first semester is about how to find a practical and valuable problem. As to the *Develop* ability, there are only 75% of students who consent to the proposed work is effective. This is significantly lower than the aspect of *Discover*. We think that the content of this course may be less on the technology than the traditional programming course, because of the problem finding in the living environment. Another possible explanation is that the students do not accommodate the new teaching model. In the past, the students usually get the course of traditional teaching model, in which they only passively listening and receiving. At the aspect of *Deliver*, there are only 83% of students who consent to our constructive problem-based teaching approach. Although this is only the result of the first semester, the result, at the aspect of *Deliver*, can reflect the expectation of the students.

5 Conclusion

The smart-phone which integrate the computing and the mobility, changes the modern living of humans totally, and provides more comfort and convenience. This paper proposes a work which trains the undergraduate students to develop the IOT prototype via a constructive problem-based approach. It is a two semesters course project, the second semester course is still running now.

The study of the first semester basically shows the effectiveness of the proposed approach and the expectation of the students.

Essentially, this paper only contains the results of the first semester. After the second semester, we will do another study about D^3 abilities and compare with the result of the first semester. In addition, we tries to use technology acceptance model to analyze the performance of the proposed approach.

Acknowledgment. This work was supported by Ministry of Science and Technology, Taiwan, under contact number MOST 105-2622-E-127-001-CC2.

References

1. Tseng, S.P., Li, B.R., Pan, J.L., Lin, C.J.: An application of internet of things with motion sensing on smart house. In: 2014 International Conference on Orange Technologies, pp. 65–68, September 2014
2. Luntovskyy, A., Spillner, J.: Smart grid, Internet of Things and fog computing, pp. 135–210. Springer, Wiesbaden (2017)
3. Cesana, M., Redondi, A.E.C.: IoT communication technologies for smart cities, pp. 139–162. Springer, Cham (2017)
4. Riegler, A.: Constructivism. In: L'Abate, L. (ed.) Paradigms in Theory Construction, pp. 235–255. Springer, New York (2012)
5. Tobias, S., Duffy, T.: Constructivist instruction: success or failure? Educational Psychology, Learning Sciences, Curriculum and Instruction. Routledge (2009)
6. Savery, J.R.: Overview of problem-based learning: definitions and distinctions. Interdiscip. J. Probl. Based Learn. **1**(1), 3 (2006)
7. Hmelo-Silver, C., Eberbach, C.: Learning theories and problem-based learning. In: Bridges, S., McGrath, C., Whitehill, T.L., (eds.) Problem-Based Learning in Clinical Education. Innovation and Change in Professional Education, vol. 8, pp. 3–17. Springer, Netherlands (2012)
8. Pressman, R.: Software Engineering: A Practitioner's Approach, 7th edn. McGraw-Hill Inc., New York (2010)
9. SmartSeedUnion (2017). http://ssu.ide.ncku.edu.tw/

Author Index

J.-S. Pan et al. (eds.), *Advances in Smart Vehicular Technology, Transportation, Communication and Applications*, Smart Innovation, Systems and Technologies 86,
https://doi.org/10.1007/978-3-319-70730-3

MIX
Papier aus verantwortungsvollen Quellen
Paper from responsible sources
FSC® C105338

If you have any concerns about our products,
you can contact us on
ProductSafety@springernature.com

In case Publisher is established outside the EU,
the EU authorized representative is:
Springer Nature Customer Service Center GmbH
Europaplatz 3, 69115 Heidelberg, Germany

Printed by Libri Plureos GmbH
in Hamburg, Germany